Proceedings of the 3rd Conference on Physical Modeling for Virtual Manufacturing Systems and Processes

Jan C. Aurich · Christoph Garth ·
Barbara S. Linke

Editors

Proceedings of the 3rd Conference on Physical Modeling for Virtual Manufacturing Systems and Processes

Editors
Jan C. Aurich
FBK - Lehrstuhl für Fertigungstechnik und
Betriebsorganisation
RPTU Kaiserslautern-Landau
Kaiserslautern, Germany

Christoph Garth
Scientific Visualization Lab
RPTU Kaiserslautern-Landau
Kaiserslautern, Germany

Barbara S. Linke
Mechanical and Aerospace Engineering
University of California Davis
Davis, CA, USA

ISBN 978-3-031-35781-7   ISBN 978-3-031-35779-4 (eBook)
https://doi.org/10.1007/978-3-031-35779-4

This Springer imprint is published by the registered company Springer Nature Switzerland AG
The registered company address is: Gewerbestrasse 11, 6330 Cham, Switzerland

Preface

From July 2014 to July 2023, the International Research Training Group (IRTG) 2057 "Physical Modeling for Virtual Manufacturing Systems and Processes" was carried out by the three partner universities TU Kaiserslautern, University of California Davis, and University of California. It received funding by the German Research Foundation (DFG).

The IRTG focused on a topic that has received a lot of attention in recent years: the implementation of physical properties in computer models for the planning of production processes. Within the IRTG, physical interactions on and between the three logical levels of a manufacturing system, factory, machine, and process were investigated. Toward this goal, the research agenda was driven by fundamental problems in both engineering and computer science and by the tight integration of both. As an international program, the IRTG 2057 brought together professors, senior researchers, and doctoral students from the three partner universities.

This volume contains the proceedings of the 3rd International Conference of the IRTG 2057, conducted on June 19–23, 2023, at the Asilomar Conference Grounds, Pacific Grove, California. The topics presented at the conference follow the program's research focus. The 16 contributions contained within this book underwent a two-stage, rigorous review by the international program committee. Submitted papers were accepted for presentation at the conference and subsequently reviewed again before final acceptance for publication.

We would like to express our immense gratitude to all authors that submitted a paper and the members of the program committee for their diligent work. Finally, we are indebted to the German Research Foundation (DFG) for the continued funding and support under contract number 252408385.

Acknowledgement

The research presented in all papers was funded by the Deutsche Forschungsgemeinschaft (DFG, German Research Foundation)—252408385—IRTG 2057.

Contents

List of Contributors

N. Altherr Institute for Manufacturing Technology and Production Systems, RPTU Kaiserslautern-Landau, Kaiserslautern, Germany

J. C. Aurich Institute for Manufacturing Technology and Production Systems (FBK), RPTU Kaiserslautern-Landau, Kaiserslautern, Germany;
Chair of Institute for Manufacturing Technology and Production Systems, RPTU Kaiserslautern-Landau, Kaiserslautern, Germany

C. R. D'Elia Department of Mechanical and Aerospace Engineering, University of California, Davis, CA, USA

K. M. de Payrebrune Institute for Computational Physics in Engineering, RPTU Kaiserslautern-Landau, Kaiserslautern, Germany

A. Ebert Human Computer Interaction Lab, Department of Computer Science, RPTU Kaiserslautern-Landau, Kaiserslautern, Germany

S. Ehmsen Institute for Manufacturing Technology and Production System, RPTU Kaiserslautern-Landau, Kaiserslautern, Germany

C. Garth Chair of Scientific Visualization Lab, RPTU Kaiserslautern-Landau, Kaiserslautern, Germany

M. Glatt Institute for Manufacturing Technology and Production System, RPTU Kaiserslautern-Landau, Kaiserslautern, Germany

D. Gond Laboratory of Engineering Thermodynamics (LTD), RPTU Kaiserslautern-Landau, Kaiserslautern, Germany

H. Hasse Laboratory of Engineering Thermodynamics (LTD), RPTU Kaiserslautern-Landau, Kaiserslautern, Germany

M. R. Hill Department of Mechanical and Aerospace Engineering, University of California, Davis, CA, USA

A. Jawaid Institute for Measurement and Sensor Technology, University of Kaiserslautern-Landau, Kaiserslautern, Germany

F. Jirasek Laboratory of Engineering Thermodynamics (LTD), RPTU Kaiserslautern-Landau, Kaiserslautern, Germany

E. Kinner Scientific Visualization Lab, RPTU Kaiserslautern-Landau, Kaiserslautern, Germany

B. Kirsch Institute for Manufacturing Technology and Production Systems (FBK), RPTU Kaiserslautern-Landau, Kaiserslautern, Germany

M. Klar Institute for Manufacturing Technology and Production Systems (FBK), RPTU Kaiserslautern-Landau, Kaiserslautern, Germany

F. Kästner Institute for Computational Physics in Engineering, RPTU Kaiserslautern-Landau, Kaiserslautern, Germany

H. Leitte Visual Information Analysis Group, RPTU Kaiserslautern-Landau, Kaiserslautern, Germany

B. S. Linke Department for Mechanical and Aerospace Engineering, University of California Davis, Davis, CA, USA

V. M. Memmesheimer Human Computer Interaction Lab, Department of Computer Science, RPTU Kaiserslautern-Landau, Kaiserslautern, Germany

J. Mertes Institute for Manufacturing Technology and Production Systems (FBK), RPTU Kaiserslautern-Landau, Kaiserslautern, Germany

R. Müller Institute for Mechanics, Technical University of Darmstadt, Darmstadt, Germany

B. Ravani Department of Mechanical and Aerospace Engineering, University of California Davis, Davis, CA, USA

S. Schmitt Laboratory of Engineering Thermodynamics (LTD), RPTU Kaiserslautern-Landau, Kaiserslautern, Germany

J. Seewig Institute for Measurement and Sensor Technology, University of Kaiserslautern-Landau, Kaiserslautern, Germany

J.-T. Sohns Visual Information Analysis Group, RPTU Kaiserslautern-Landau, Kaiserslautern, Germany

J. Staubach Laboratory of Engineering Thermodynamics (LTD), RPTU Kaiserslautern-Landau, Kaiserslautern, Germany

S. Stephan Laboratory of Engineering Thermodynamics (LTD), RPTU Kaiserslautern-Landau, Kaiserslautern, Germany

H. M. Urbassek Physics Department and Research Center OPTIMAS, RPTU Kaiserslautern-Landau, Kaiserslautern, Germany

V. H. Vardanyan Physics Department and Research Center OPTIMAS, RPTU Kaiserslautern-Landau, Kaiserslautern, Germany

D. Weber Institute for Manufacturing Technology and Production Systems, RPTU Kaiserslautern-Landau, Kaiserslautern, Germany

G. H. Weber Computational Research Division, Lawrence Berkeley National Laboratory, Berkeley, CA, USA

X. Wu Institute for Manufacturing Technology and Production Systems, RPTU Kaiserslautern-Landau, Kaiserslautern, Germany

S. Yan Institute of Applied Mechanics, University of Kaiserslautern-Landau, Kaiserslautern, Germany

L. Yi Institute for Manufacturing Technology and Production Systems, RPTU Kaiserslautern-Landau, Kaiserslautern, Germany

Discrete Filter and Non-Gaussian Noise for Fast Roughness Simulations with Gaussian Processes

A. Jawaid$^{(\boxtimes)}$ and J. Seewig

Institute for Measurement and Sensor Technology, University of Kaiserslautern-Landau,
Kaiserslautern, Germany
arsalan.jawaid@mv.rptu.de

Abstract. Rough surface simulations result in tight feedback loops in research procedures, such that they speed up studies for example about roughness' impact on tribology or fluid dynamics. To model and simulate a broad spectrum of rough surfaces, Gaussian processes (GP) have been suggested recently. However, these models are limited on surfaces with small sizes since computational time-costs and memory-costs of simulations with standard procedures scale cubically and quadratically, respectively. In this paper, we apply the discrete filter approach which is a special case of GPs. We use the discrete filter with the fast Fourier transform (FFT) algorithm to efficiently sample from a high-dimensional Gaussian distribution and we compare its computational costs with the contour integral quadrature algorithm. Our experiments show that GPs benefit from FFT and allow stationary rough surfaces with sizes as large as $30,000 \times 30,000$ to be efficiently sampled. Since this approach is complementary to the GP and noise model approach, we also show simulations of rough surfaces with underlying non-Gaussian noise models that can reduce computational complexity.

1 Introduction

Surface texture, hereinafter referred to as roughness or rough surface, is the scale-limited surface where small-scale components were removed (S-filter) followed by a form (F-operation) and low-scale components (L-filter) removal [1]. Historically, roughness has been used to control manufacturing processes [2, 3]. For example, if the roughness deviates from a reference, unwanted changes may have occurred on the machining tool. Another advantage of roughness, which is growing rapidly nowadays, is its direct contributions to various physical properties, e.g., fatigue [4] or heat transfer [5]. To study these functional contributions of roughness, a model-based approach for rough surfaces can be used. These roughness models should also provide an efficient simulation procedure so that they can be used to accelerate functional studies.

Commonly, stochastic process modeling is considered the golden standard in roughness modeling. Recently, there had a novel stochastic process approach emerged to model a broad spectrum of rough surfaces [6]. This approach utilized Gaussian processes (GPs) to model rough surfaces. Even though it can model and simulate more varieties of rough surfaces automatically or with predefined information than traditional approaches, its

J. C. Aurich et al. (Eds.): IRTG 2023, *Proceedings of the 3rd Conference on Physical Modeling for Virtual Manufacturing Systems and Processes*, pp. 1–15, 2023.
https://doi.org/10.1007/978-3-031-35779-4_1

computational bottleneck is its simulation procedure. Whereas other approaches have efficient implementations for rough surface simulations due to their nature, which allows the application of FFT [7, 8].

A simulation of a rough surface with the GP approach corresponds to a sample drawn from a multivariate Gaussian distribution. Using traditional methods for sampling multivariate Gaussian distributions is computationally expensive [6]. Particularly using Cholesky decomposition, sampling N data points have a computational cost of $\mathcal{O}(N^3)$ and a memory cost of $\mathcal{O}(N^2)$. On the one hand, several recent works in machine learning literature attends to fill this gap [9, 10] since efficient sampling is for example also beneficial in Bayesian optimization [11, 12]. On the other hand, we find that other approaches of traditional roughness modeling are complementary to the GP approach due to its linear process view.

In this paper, we address the computational expensive roughness simulation with a discrete filter approach with FFT for efficient GP sampling. We show that this approach approximates a stationary GP. And we compare the approach with two matrix factorization methods, the Cholesky decomposition and contour quadrature integrals (CIQ) [10]. Moreover, we show that a suitable non-Gaussian noise model can be used to reduce the number of GP samples for a honed surface, further reducing the computational cost. These additions are related to GP and noise model approach in [6]. Our contributions will result in a model which applies traditional methods to model rough surfaces and aims to get a more applicable GP approach.

2 Background

2.1 Roughness Model with Gaussian Processes

We follow the treatment given by [6, 13] and summarize their suggested roughness model. Their model-based approach follows [14] and characterizes a surface with two quantities, the autocovariance function (ACVF) and the probability density function (PDF). An ACVF describes the covariance or similarity between two locations of the surface (1) and, therefore, expresses a space-dependent influence in the model.

$$r(\mathbf{x}_i, \mathbf{x}_j) = \mathbb{E}\big[(Z(\mathbf{x}_i) - \mathbb{E}[Z(\mathbf{x}_i)]) \cdot (Z(\mathbf{x}_j) - \mathbb{E}[Z(\mathbf{x}_j)])\big], \qquad (1)$$

where $\mathbb{E}[\cdot]$ is the expectation operator and $Z(\mathbf{x}) \in \mathcal{Z} \subseteq \mathbb{R}$ is the surface height with position $\mathbf{x} \in \mathcal{X} \subseteq \mathbb{R}^D$. For rough surfaces, the traditional assumption is that the mean is zero which modifies the ACVF by $\mathbb{E}[Z(\mathbf{x})] = 0$ [14]. In contrast, a PDF does not take the locations of the surface into account and considers rather the surface height directly.

In the proposed model [6], these two quantities are assigned to two components of the roughness model. The first component is a GP, which is described by a zero-mean ACVF. The second component is a noise model specified by a PDF. The diagram in Fig. 1 illustrates this roughness model regarding a roughness simulation.

A specified model with given ACVF and PDF can simulate a rough surface by sampling from the GP and processing it with noise. The noise is sampled from the noise model given a sampled GP. The model and both sampling procedures are described as follows

$$G(\mathbf{x}) \sim \mathcal{GP}\big(0, r(\mathbf{x}_i, \mathbf{x}_j)\big), \qquad Z(\mathbf{x}_n)|g(\mathbf{x}_n) \sim p(z(\mathbf{x}_n)|g(\mathbf{x}_n)), \qquad (2)$$

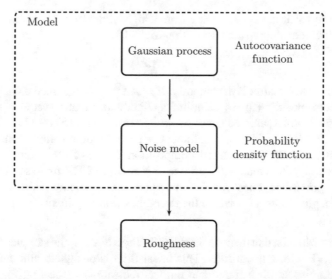

Fig. 1. Simulation procedure for a roughness simulation with model [6].

where $G(x)$ denotes the latent output and $Z(x)$ denotes the roughness.

If the noise model is Gaussian white noise, e.g., [6], the critical part regarding computational resources is the sampling from the GP. Thus, computational improvements in the GP sampling will lead to efficient simulations of rough surfaces with this model.

2.2 Simulation of Rough Surfaces

A GP is an infinite set of random variables $\{G(x)|x \in \mathcal{X}\}$ any finite set of which is multivariate Gaussian distributed [13]. Hence, a GP on a finite set $\mathcal{X}_N = \{x_n\}_{n=1}^{N}$ is a multivariate Gaussian distribution, e.g., (3). Usually, this Gaussian distribution is high-dimensional and has a non-diagonal covariance matrix. In the roughness model, the latent output is sampled from the following Gaussian distribution

$$G \sim \mathcal{N}(G; 0, R), \tag{3}$$

$$[R]_{ij} = r(x_i, x_j), \qquad (x_i, x_j \in \mathcal{X}_N), \tag{4}$$

where $G = \{G(x)|x \in \mathcal{X}_N\}$ is the finite set of a GP and R is the covariance matrix.

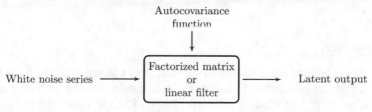

Fig. 2. Sampling procedure for latent outputs with a given ACVF by a matrix factorization or a linear filter.

Matrix Decomposition. The sampling from a high-dimensional Gaussian distribution traditionally includes the reparameterization trick [15]

$$G = A \cdot \varepsilon, \qquad \varepsilon \sim \mathcal{N}(\varepsilon; 0, I), \tag{5}$$

where the covariance matrix is decomposed $R = AA^T$.. We assumed a Gaussian distribution with zero-mean in (5). So, sampling a GP equals a matrix-vector multiplication with the matrix A and a sampled Gaussian white noise vector (5) (see Fig. 2). Since the covariance matrix R is positive-semidefinite, Cholesky decomposition is often applied to obtain the matrix A [13]. This matrix decomposition requires $\mathcal{O}(N^3)$ computational time and $\mathcal{O}(N^2)$ memory for a matrix R with size of $N \times N$ [16]. The finite set $\mathcal{X}_N$ is often an evenly spaced grid in roughness simulations, hence, the covariance matrix of a stationary ACVF is a Toeplitz matrix for which the decomposition requires a computational cost of $\mathcal{O}(N^2)$ [17].

Linear Filter. Models, that do not consider GPs directly, compute samples from a given stationary ACVF with a linear filter. This linear filter is designed with the ACVF and takes a white noise series as input and outputs the samples (see Fig. 2), e.g., [7]. In roughness literature, mostly a finite impulse response (FIR) filter is used and has the discretized form as follows

$$G_{p,q} = \sum_{k=0}^{V-1} \sum_{l=0}^{U-1} h_{k,l} \varepsilon_{p-k,q-l}, \quad p \in \mathcal{I}_{V-1}, q \in \mathcal{I}_{U-1}, \tag{6}$$

$$r_{vu} = \sum_{k=0}^{K-v-1} \sum_{l=0}^{L-w-1} h_{k,l} h_{v-k,u-l}, \quad v \in \mathcal{I}_{K-1}, u \in \mathcal{I}_{L-1}. \tag{7}$$

where we denote $\mathcal{I}_c \subset \mathbb{N}$ as an index set enumerated from 0 to c and where $G\left(x_p^{(0)}, x_q^{(1)}\right) = G_{p,q}$ is the latent output series, $r\left(\tau_v^{(0)}, \tau_u^{(1)}\right) = r_{v,u}$ are the ACVF coefficients, $h_{k,l}$ are the filter coefficients and $\varepsilon_{p,k}$ is a white noise series with unit variance. This filter shall mimic the ACVF (7) so that filtering a white noise series will result in an output series that has the imitated ACVF. To compute the filter coefficients with the ACVF, the nonlinear equations can be solved with the Newton method, or with the Fourier transform. More precisely, applying discrete Fourier transform (DFT) with the FFT algorithm supports efficiently computing the filter coefficients and the linear filter [8].

We found that a roughness simulation with this discrete filter can be applied within the GP roughness model. Thereby, the FFT algorithm is numerically fast and there are in-place implementations of the algorithm. This speed-up is significantly compared to traditional methods.

2.3   Related Work

One example to compute kernel-based methods and draw samples efficiently is the random Fourier features approach [18], e.g., in Bayesian optimization literature. It approximates a stationary ACVF $r(x_i, x_j) \approx \zeta(x_i)^\top \zeta(x_j)$ with a low-dimensional map with

Fourier features such that a simulation is efficiently computed by a scalar product between the low-dimensional feature map and a white noise vector. To reduce errors due to approximations other methods emerged recently in machine learning literature. For example, matrix root decomposition has been efficiently computed with Lanczos variance estimates [9] or CIQ [10], which are Krylov approaches. These methods are exact if the number of iterations is N, nevertheless, fewer iterations deal with small errors.

FFT has already been applied to simulate stationary GPs in literature. One commonly known method is circular embedding [19]. This approach extends the covariance matrix to a $2M \times 2M$ circulant matrix $M \geq N$, on which a FFT can be applied and lead to efficient matrix root decomposition. This method is exact and help to simulate efficiently. However, the positive-semidefiniteness of the circular matrix is not guaranteed for a minimal embedding which is a requirement. Similar to our method, [7, 20, 21] suggested simulating GPs with a filter by a Fourier transform approach. However, they simulated roughness only with GPs whereas we consider also a noise model. Furthermore, only [20] stated that their approach simulates a stationary GP.

3 Gaussian Process Filter

The zero-mean GP $G(x) \in \mathbb{R}$ has the following linear process view [6]

$$G(x) = \int_{\mathbb{R}^D} h(s)\varepsilon(x - s)\mathrm{d}s, \tag{8}$$

$$r(x_i, x_j) = \int_{\mathbb{R}^D} h(x_i - s)h(x_j - s)\mathrm{d}s, \tag{9}$$

where $\{\varepsilon(x), x \in \mathcal{X}\}$ is a continuous Gaussian white noise process, and $h : \mathbb{R}^D \mapsto \mathbb{R}$ is a map that characterizes the ACVF. The Gaussian white noise process has a GP representation [22] denoted as

$$\varepsilon(x) \sim GP\big(0, \delta(x_i - x_j)\big), \tag{10}$$

with the Delta function $\delta(\cdot)$ as its ACVF.

Considering a stationary GP, the ACVF is shift-invariant, and (9) has the following form

$$r(\tau) = \int_{\mathbb{R}^D} h(s)h(s + \tau)\mathrm{d}s, \tag{11}$$

where $\tau = x_i - x_j$ is the distance between two positions. Applying the Fourier transform on the above equation transforms the ACVF to the PSD $\tilde{r}(f)$ [23, 24] and transforms the integral into a multiplication

$$\tilde{r}(f) = \tilde{h}(f) \cdot \overline{\tilde{h}(f)} = \left|\tilde{h}(f)\right|^2, \tag{12}$$

where $\overline{(\cdot)}$ denotes the complex conjugate and $\tilde{h}(f)$ is the Fourier transform of $h(s)$. Thus, an approximate filter map $\hat{h}(s) \approx h(s)$ can be derived by the inverse Fourier transform

$$\hat{h}(s) = \int_{\mathbb{R}^D} (\tilde{r}(f))^{\frac{1}{2}} \cdot e^{i2\pi f^{\mathrm{T}} s} df. \tag{13}$$

If $|\tilde{h}(f)| = \tilde{h}(f)$ holds, then the approximated filter map is the true filter map. So, a stationary GP with zero-mean can theoretically be simulated by the continuous linear process

$$\hat{G}(x) = \int_{\mathbb{R}^D} \hat{h}(s)\varepsilon(x - s)ds. \tag{14}$$

To obtain the PSD $S(f)$ directly or the approximated filter map with given PSD requires solves of Fourier transforms or inverse transforms, respectively. Furthermore, the linear filtering process in (14) is only possible if the approximated filter map is square-integrable. Even if the feature map is exactly known, the above presentation is not practical due to the continuous white noise process. To avoid this presentation of the white noise process, the next section discusses the (discretized) linear process. With this discretization, a discrete white noise process is considered, and discrete-time Fourier transform (DTFT) can be used to compute the filter map.

3.1 Discrete Filter

We initially assume one-dimensionality ($D = 1$) for simplicity and extend this method to multidimensionality later. Sampling $G(x)$ on discrete points $G(n\Delta) = G_n$ with step size $\Delta \in \mathbb{R}_{<0}$, the series $\{G_n, n \in \mathbb{Z}\}$ has the ACVF

$$r(v\Delta) = \mathbb{E}[G_n G_{n+v}], \qquad v \in \mathbb{Z}, \tag{15}$$

which is sampled from the ACVF $r(\tau)$. The DTFT of the ACVF series is a $\frac{1}{\Delta}$ periodic continuation of $\tilde{r}(f)$

$$\tilde{r}_\Delta(f) = \sum_{k=-\infty}^{\infty} \tilde{r}(f + \frac{k}{\Delta}) \tag{16}$$

If the sampling theorem is fulfilled, then the Fourier transform $\tilde{r}(f)$ can be reconstructed. Otherwise, the DTFT is affected by errors due to aliasing.

Analogous to the continuous linear process (8), a discrete linear process presentation can be formulated with the ACVF series $\{r_v, v \in \mathbb{Z}\}$, which we refer to as a discrete filter

$$G_n = \sum_{k=-\infty}^{\infty} h_k \cdot \varepsilon_{n-k}, \tag{17}$$

$$r_v = \sum_{k=-\infty}^{\infty} h_k h_{k+v}, \tag{18}$$

where $\{\varepsilon_k, k \in \mathbb{Z}\}$ is a white noise series with the Kronecker delta as ACVF and the filter coefficients $\{h_k, k \in \mathbb{Z}\}$ that shall mimic the ACVF series. This discrete filter makes the linear process more applicable by avoiding the cumbersome continuous white noise process. If the filter coefficients are given so that (18) holds, a G_n can be computed by a convolution (17). To compute the filter coefficients the DTFT $\tilde{r}_\Delta(f)$ can be utilized. The inverse DTFT can be used to estimate these filter coefficients [7]

$$\hat{h}_k = \int_0^l (\tilde{r}_\Delta(f))^{\frac{1}{2}} \cdot e^{i2\pi f \frac{k}{l}} dk, \tag{19}$$

with $l = \frac{1}{\Delta}$.

The filter coefficients $\{h_k, k \in \mathbb{Z}\}$ are approximated by (19) since we assume that the DTFT of the filter coefficients $\tilde{h}_\Delta(f)$ equals $(\tilde{r}_\Delta(f))^{1/2}$. Furthermore, due to the periodic feature of $\tilde{r}_\Delta(f)$ aliasing errors are propagated to the filter coefficients. Nevertheless, the filter coefficients and the simulation of a latent surface can be computed efficiently with FFT in this approach.

3.2 Discrete Filter with FFT

It is not possible to compute the DTFT because a finite set $\{r_v, v \in \mathcal{I}_{V-1}\}$ instead of an infinite set is given. Therefore, DFT is applied to estimate the DTFT on a discrete set. The DFT of a finite sampled ACVF is

$$\tilde{r}_{\Delta,k} = \frac{1}{N} \sum_{v=0}^{N-1} r_v \cdot e^{-i\frac{2\pi}{N}kv}, \qquad k \in \mathcal{I}_{N-1}, \tag{20}$$

and by inverse DFT the filter coefficients can be estimated by

$$\hat{h}_k = \frac{1}{N} \sum_{v=0}^{N-1} (\tilde{r}_{\Delta,v})^{\frac{1}{2}} \cdot e^{i2\pi \frac{kv}{N}}, \qquad k \in \mathcal{I}_{N-1}. \tag{21}$$

With the estimated filter coefficients, a rough surface is simulated by

$$G_n = \sum_{k=0}^{N-1} \hat{h}_k \varepsilon_{n+k}, \qquad n \in \mathcal{I}_{N-1}, \tag{22}$$

where $\hat{h}_k$ is assumed to be a N periodic series due to the inverse DFT.

This will result in the following relation between the finite sampled ACVF and the estimated filter coefficients

$$r_v = \sum_{k=0}^{N-1} \hat{h}_k \hat{h}_{k+v}, \qquad v \in \mathcal{I}_{N-1}, \tag{23}$$

with $\hat{h}_{k+N} = \hat{h}_k$.

The extension to two-dimensionality is straightforward

$$\tilde{r}_{\Delta,k,l} = \frac{1}{VU} \sum_{v=0}^{V-1} \sum_{u=0}^{U-1} r_{v,u} e^{-i2\pi\left(\frac{kv}{V} + \frac{lu}{U}\right)}, \qquad k \in \mathcal{I}_{V-1}, \quad l \in \mathcal{I}_{U-1} \tag{24}$$

$$\hat{h}_{k,l} = \frac{1}{VU} \sum_{v=0}^{V-1} \sum_{u=0}^{U-1} \left(\tilde{r}_{\Delta,v,u}\right)^{\frac{1}{2}} e^{i2\pi\left(\frac{kv}{V} + \frac{lu}{U}\right)}, \qquad k \in \mathcal{I}_{V-1}, \quad l \in \mathcal{I}_{U-1} \tag{25}$$

$$G_{p,q} = \sum_{k=0}^{V-1} \sum_{l=0}^{U-1} \hat{h}_{k,l} \varepsilon_{p+k,q+l}, \qquad p \in \mathcal{I}_{V-1}, \quad q \in \mathcal{I}_{U-1} \tag{26}$$

where $G_{p,q}$ is the latent series and has size $V \times U$.

The DFTs and inverse DTFs can be efficiently computed with the FFT algorithm. The Eq. (26) is a convolution operation and, thus, can also be computed by the FFT algorithm[1]. Therefore, this simulation procedure samples a latent GP in a computational time of $\mathcal{O}((3N - 2)\log(3N - 2))$ with $N = V \cdot U$.

4   Experiments

In this section, we conduct benchmarks to compare the discrete filter's efficiency. Initial experiments are performed to study the timings of the proposed method with different FFT implementations. Afterward, the discrete filter is compared against matrix factorization methods Cholesky and CIQ. In all cases, we focus on ground surfaces with their inherent ACVF

$$r(\boldsymbol{\tau}; \boldsymbol{\phi}) = \sigma_k^2 \exp\left(-\left(\boldsymbol{\tau}'^{\top} \boldsymbol{\Lambda}^{-2} \boldsymbol{\tau}'\right)^{\frac{1}{2}}\right), \qquad \boldsymbol{\tau}' = \boldsymbol{T}_\phi \boldsymbol{\tau}^{\top}, \tag{27}$$

where $\boldsymbol{\Lambda} \in \mathbb{R}_{>0}^{2\times2}$ is a diagonal matrix and has diagonal elements λ_a and λ_b, ϕ is the angle of the grinding grooves, $\boldsymbol{T}_\phi$ is the inverse rotation matrix and σ_k is a scaling hyperparameter. Furthermore, we conducted the experiments on an equidistant mesh

$$\mathcal{X}_N = \{(v\Delta, u\Delta), \Delta \in \mathbb{R}_{>0}, v, u \in \mathcal{I}_{B-1}\}, \tag{28}$$

where Δ is the sample step size in x-direction and y-direction, and the surfaces are quadratic $N = B^2$. All timings have been performed either on the CPU hardware Intel Xeon Gold 6126 Processor or the NVIDIA Tesla V100 GPU.

---

[1] The explicit computing of the filter coefficients is therefore redundant and not necessary.

4.1 Timings of the Discrete Filter with SciPy and CuFFT

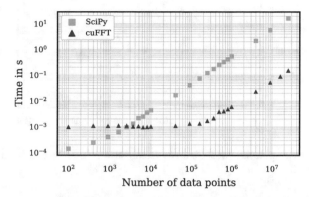

Fig. 3. Timings for computing discrete filter and surface with SciPy and cuFFT.

The parameters were chosen with $\sigma_k = 1$ μm, $\lambda_a = 500$ μm, $\lambda_b = 5$ μm, $\phi = 0$, $\Delta = 0.5$ μm. We utilized the implemented FFT in the SciPy library (version 1.9.3) for CPU and the CUDA FFT library cuFFT (version 11.8) for GPU. For the comparison, we measured the performance of computing the filter coefficients and the simulation process for different surface sizes. This procedure has been conducted 10, 000 times for each surface size. We use this comparison, given the aforementioned hardware, to discuss the SciPy and cuFFT implementations for the discrete filter.

Figure 3 shows the timings of the discrete filter of both FFT implementations. To note is that no uncertainties are assigned to the data because the uncertainties are relatively small. These experiments show that cuFFT and SciPy's FFT scale equally since both are based on the FFT algorithm. However, cuFFT leads to a significant acceleration against SciPy's FFT implementation for large surfaces. It should also be noted that because of the memory advantage over GPUs, surfaces with $9 \cdot 10^8$ points could be generated with the CPU, while the GPU reached its limits.

4.2 Benchmarking Discrete Filter

Since the discrete filter approach is more efficient through the GPU, we benchmark it with the cuFFT implementation. And the Krylov approach CIQ is also performant on GPUs due to its inherent matrix-vector multiplications [10].

Except for the grinding groove angle $\phi = \frac{\pi}{6}$, we have chosen the hyperparameters of the ACVF identical to Sect. 4.1. Then we computed the speed-up of the discrete filter over Cholesky factorization (see Fig. 4). Since the Cholesky decomposition is memory inefficient, we performed Cholesky on the CPU, and the surface size was at most 70×70. The Cholesky method has been compared with the cuFFT implementation in Fig. 4, although both were run on different hardware the cuFFT was slower than the SciPy implementation on a CPU for small surfaces (see Fig. 3). The visualization shows that sampling from the high-dimensional Gaussian distribution is much faster with the discrete filter method (more than 900 times faster than Cholesky).

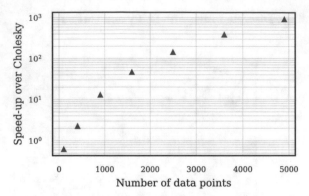

Fig. 4. Speed-up of discrete filter over Cholesky factorization. The achieved speed-up is computed by sampling 10, 000 times.

For the CIQ approach, we chose its experimental setting according to its publication [10]. For the Krylov method, we selected the number of iterations with $J = 100$. We also selected the number of quadrature points with $Q = 8$. To reduce the memory complexity, they leveraged symbolic tensors [25] for computing [10]. Figure 5 shows the timings and the error of the discrete filter and the CIQ method. The error is computed by drawing 1000 surfaces and computing the mean squared error between the sample covariance matrix and the true covariance matrix element-wise.

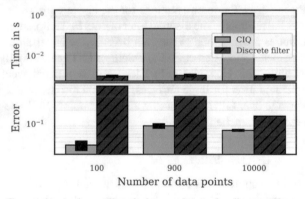

Fig. 5. Computing and sampling timings and error by discrete filter and CIQ.

Similar to [10], CIQ produces accurate samples of the latent surface. The error of the discrete filter is greater than that of the CIQ approach because the discrete filter method implies approximations of the GP. Firstly, the filter-map is approximated by assuming its Fourier transform equals the squared root of the PSD (19). Secondly, aliasing might occur due to the discrete Gaussian stochastic process. Even though the error can be reduced by upsampling, resulting in a tradeoff in computational speed, the discrete filter with FFT is much faster than CIQ. We noted that a larger latent surface simulation with constant sampling step size reduces the errors, which is because the given ACVF converges to 0 for large τ.

5 Applications

We apply the discrete filters to simulate rough surfaces with the standard additive Gaussian noise model $p(z(\boldsymbol{x}_n)|g(\boldsymbol{x}_n)) = \mathcal{N}\big(z(\boldsymbol{x}_n); g(\boldsymbol{x}_n), \sigma^2\big)$ and we show an application for a non-Gaussian noise model.

The usual approach to simulate honed surfaces is by simulating multiple ground surfaces and superposing them [6, 21, 26]. Alternatively, we use a non-Gaussian noise model with a generalized ACVF to simulate a honed surface in only one simulation procedure. This approach can reduce the computational time since a one-step honed surface needs only one rather than two simulations. The generalized ACVF is

$$ r(\tau) = \frac{\sigma_k^2}{2}\left(\exp\left(-\left(\boldsymbol{\tau}'^{\mathsf{T}}\boldsymbol{\Lambda}^{-2}\boldsymbol{\tau}'\right)^{\frac{1}{2}}\right) + \exp\left(-\left(\boldsymbol{\tau}^{*\mathsf{T}}\boldsymbol{\Lambda}^{-2}\boldsymbol{\tau}^{*}\right)^{\frac{1}{2}}\right)\right), \qquad (29) $$

with $\boldsymbol{\tau}' = \boldsymbol{T}_\phi\boldsymbol{\tau}^{\mathsf{T}}$, $\boldsymbol{\tau}^* = \boldsymbol{T}_{-\phi}\boldsymbol{\tau}^{\mathsf{T}}$.

In the following, we simulated a one-step honed surface with the proposed approach with an additive non-Gaussian noise model defined as follows

$$ Z(\boldsymbol{x}) = g(\boldsymbol{x}) + \varepsilon, \qquad \boldsymbol{x} \in \mathcal{X}_N \qquad (30) $$

where ε is an i.i.d. Pearson type III distributed random variable. We use the Pearson type III distribution that defines a skewness parameter because a honed surface often comes with skewed distributions.

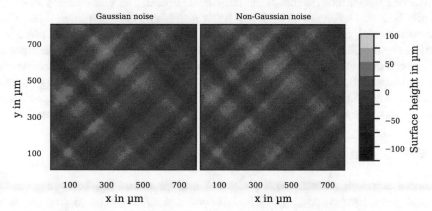

Fig. 6. Simulated honed surfaces with Gaussian and non-Gaussian noise models but the same underlying latent surface.

To observe only the influence of the noise models, the honed surfaces have the same latent surface once simulated with the generalized ACVF (29). So, the same sampled latent surface $g = \{g(x), x \in \mathcal{X}_N\}$ was passed into the standard Gaussian white noise model and the non-Gaussian noise model to generate a roughness sample $z = \{z(x), x \in \mathcal{X}_N\}$ in each case. The Fig. 6 clearly shows that both surfaces have the same latent surface whereas only different additive noise models have been applied. However, the surfaces do not have continuous grooves which is a characteristic of honed surfaces. This error is due to the structure of the generalized ACVF that has a distinct peak in the center.

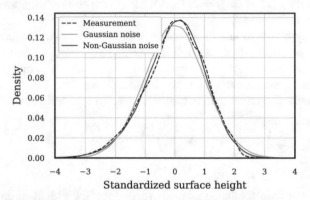

Fig. 7. Estimated distributions of a real honed surface and simulated two-step honed surface with an additive Gaussian and an additive non-Gaussian noise model.

To emphasize the differences in both simulations, we compare the sampled distributions of the simulated honed surfaces with a real honed surface in Fig. 7. The real surface data were measured with a confocal measurement tool and processed with a form operator after a noise filter. We obtained the sampling distributions in each case by considering the surface heights of the individual surface. The visualization shows that the non-Gaussian noise model leads to a better match with the real surface than a Gaussian noise model, especially at the tails. This is because the Pearson type III noise leads also to a skewed distribution of roughness. In fact, skewness estimations of the measured surface and the roughness with the non-Gaussian noise model have a comparable magnitude, whereas the Gaussian noise model has a.s. no skewness. Explaining skewness by additive noise models alone, however, will most likely lead to incorrect surfaces. Nevertheless, this approach is helpful for an adjustment in case of small skewness.

6   Conclusion

A model of rough surfaces with a GP and a noise model has been suggested by [6] that generalizes current approaches. However, simulations are limited due to the computational complexity associated with sampling from the latent GP.

We addressed this problem for stationary rough surfaces in this paper. Similar to [7, 20, 21], we applied the discrete filter together with the FFT algorithm since it can be motivated from the GP linear process view. Compared to Cholesky and CIQ, this approach leads to a significant speed-up. However, the discrete filter comes with approximation errors and is only limited to stationary surfaces. The errors can be reduced by sampling a larger latent surface. Additionally, we applied an additive non-Gaussian noise model for honed surfaces that have distributions with small skewness. This approach can further reduce computational complexity.

Even though we benchmarked the discrete filter with Cholesky and CIQ in this paper, an extensive comparison with other state-of-the-art methods is required for further classification of this approach. For example, a comparison should be made with circulant embedding [19], random Fourier features [18], Lanczos variance estimates [9], and also CIQ [10] with inducing point methods [27]. Moreover, future research should study the approximation errors of this approach if the PSD is known or even the continuous filter map is known from the ACVF. Another focus could be the application of non-additive noise models for roughness simulations because we discussed only additive noise models in this work.

Acknowledgments. This work was funded by Deutsche Forschungsgemeinschaft (DFG, German Research Foundation) – 252408385 – IRTG 2057.

References

1. ISO 25178–2. Geometrical product specifications (GPS) - Surface texture: Areal - Part 2: Terms, definitions and surface texture parameters. International Organization of Standardization (2021)
2. Jiang, X., Scott, P.J., Whitehouse, D.J., Blunt, L.: Paradigm shifts in surface metrology. part I. historical philosophy. Proc. R. Soc. Math. Phys. Eng. Sci. **463**(2085), 2049–2070 (2007). https://doi.org/10.1098/rspa.2007.1874
3. Jiang, X., Scott, P.J., Whitehouse, D.J., Blunt, L.: Paradigm shifts in surface metrology. part II. the current shift. Proc. R. Soc. Math. Phys. Eng. Sci. **463**(2085), 2071–2099 (2007). https://doi.org/10.1098/rspa.2007.1873
4. Taylor, D., Clancy, O.M.: The fatigue performance of machined surfaces. Fatigue Fract. Eng. Mater. Struct. **14**(2–3), 329–336 (1991). https://doi.org/10.1111/j.1460-2695.1991.tb00662.x
5. Li, X., Meng, J., Li, Z.: Roughness enhanced mechanism for turbulent convective heat transfer. Int. J. Heat Mass Transf. **54**(9–10), 1775–1781 (2011). https://doi.org/10.1016/j.ijheatmasstransfer.2010.12.039
6. Jawaid, A., Seewig, J.: Model of rough surfaces with Gaussian processes. Surf. Topogr. Metrol. Prop. **11**(1), 015013 (2023). https://doi.org/10.1088/2051-672X/acbe55
7. Hu, Y.Z., Tonder, K.: Simulation of 3-D random rough surface by 2-D digital filter and fourier analysis. Int. J. Mach. Tools Manuf. **32**(1–2), 83–90 (1992). https://doi.org/10.1016/0890-6955(92)90064-N
8. Wu, J.-J.: Simulation of rough surfaces with FFT. Tribol. Int. **33**(1), 47–58 (2000). https://doi.org/10.1016/S0301-679X(00)00016-5
9. Pleiss, G., Gardner, J.R., Weinberger, K.Q., Wilson, A.G.: Constant-time predictive distributions for Gaussian processes. In: Proceedings of the 35th international conference on machine learning, **80**, pp. 4114–4123 (2018). https://proceedings.mlr.press/v80/pleiss18a.html
10. Pleiss, G., Jankowiak, M., Eriksson, D., Damle, A., Gardner, J.R.: Fast matrix square roots with applications to gaussian processes and bayesian optimization. In: Advances in neural information processing systems, **33**, pp. 22268–22281 (2020). https://proceedings.neurips.cc/paper/2020/file/fcf55a303b71b84d326fb1d06e332a26-Paper.pdf

11. Thompson, W.R.: On the likelihood that one unknown probability exceeds another in view of the evidence of two samples. Biometrika **25**(3/4), 285 (1933). https://doi.org/10.2307/233 2286

12. Hernández-Lobato, J.M., Hoffman, M.W., Ghahramani, Z.: Predictive entropy search for efficient global optimization of black-box functions. In: Advances in neural information processing systems, **27** (2014). https://proceedings.neurips.cc/paper/2014/file/069d3bb002acd8d7dd 095917f9efe4cb-Paper.pdf

13. Rasmussen, C.E., Williams, C.K.I.: Gaussian Processes for Machine Learning. MIT Press, Cambridge (2006)

14. Whitehouse, D.J., Archard, J.F.: The properties of random surfaces of significance in their contact. Proc. R. Soc. Lond. Math. Phys. Sci. **316**(1524), 97–121 (1970). https://doi.org/10. 1098/rspa.1970.0068

15. Kingma, D.P., Welling, M.: Auto-encoding variational bayes. In: International Conference on Learning Representations (2014)

16. Gardner, J.R., Pleiss, G., Bindel, D., Weinberger, K.Q., Wilson, A.G.: GPyTorch: Blackbox matrix-matrix gaussian process inference with GPU acceleration. In: Advances in neural information processing systems, **31** (2018). https://proceedings.neurips.cc/paper/2018/file/ 27e8e17134dd7083b050476733207ea1-Paper.pdf

17. Kailath, T.:: A theorem of i. schur and its impact on modern signal processing. In: I. Schur Methods in Operator Theory and Signal Processing, 18, I. Gohberg, Ed. Basel: Birkhäuser Basel, pp. 9–30 (1986). https://doi.org/10.1007/978-3-0348-5483-2_2

18. Rahimi, A., Recht, B.: Random features for large-scale kernel machines. In: Advances in Neural Information Processing Systems, 20 (2007). https://proceedings.neurips.cc/paper/2007/ file/013a006f03dbc5392effeb8f18fda755-Paper.pdf

19. Dietrich, C.R., Newsam, G.N.: Fast and exact simulation of stationary gaussian processes through circulant embedding of the covariance matrix. SIAM J. Sci. Comput. **18**(4), 1088–1107 (1997). https://doi.org/10.1137/S1064827592240555

20. Shinozuka, M., Jan, C.-M.: Digital simulation of random processes and its applications. J. Sound Vib. **25**(1), 111–128 (1972). https://doi.org/10.1016/0022-460X(72)90600-1

21. Rief, S., Ströer, F., Kieß, S., Eifler, M., Seewig, J.: An approach for the simulation of ground and honed technical surfaces for training classifiers. Technologies **5**(4), 66 (2017). https:// doi.org/10.3390/technologies5040066

22. Tobar, F., Bui, T.D., Turner, R.E.: Learning stationary time series using gaussian processes with nonparametric kernels. In: Advances in Neural Information Processing Systems, **28** (2015). https://proceedings.neurips.cc/paper/2015/file/95e6834d0a3d99e9ea881185 5ae9229d-Paper.pdf

23. Bochner, S., Tenenbaum, M., Pollard, H., Bochner, S.: Lectures on Fourier integrals. Princeton University Press, Princeton (1959)

24. Chatfield, C.: The Analysis of Time Series: An Introduction, 6th ed. Chapman and Hall/CRC, Boca Raton (2003). https://doi.org/10.4324/9780203491683

25. Charlier, B., Feydy, J., Glaunès, J.A., Collin, F.-D., Durif, G.: Kernel operations on the GPU, with autodiff, without memory overflows. J. Mach. Learn. Res. **22**(74), 1–6 (2021)

26. Pawlus, P.: Simulation of stratified surface topographies. Wear **264**(5–6), 457–463 (2008). https://doi.org/10.1016/j.wear.2006.08.048

27. Wilson, A., Nickisch, H.: Kernel interpolation for scalable structured gaussian processes (KISS-GP). In: Proceedings of the 32nd international conference on machine learning, **37**, pp. 1775–1784 (2015). https://proceedings.mlr.press/v37/wilson15.html

Phase Field Simulations for Fatigue Failure Prediction in Manufacturing Processes

S. Yan[1]([✉]), R. Müller[2], and B. Ravani[3]

[1] Institute of Applied Mechanics, University of Kaiserslautern-Landau, Kaiserslautern,
Germany
`sikang.yan@rptu.de`
[2] Institute for Mechanics, Technical University of Darmstadt, Darmstadt, Germany
[3] Department of Mechanical and Aerospace Engineering, University of California Davis, Davis,
USA

Abstract. Fatigue failure is one of the most crucial issues in manufacturing and engineering processes. Stress cycles can cause cracks to form and grow over time, eventually leading to structural failure. To avoid these failures, it is important to predict fatigue crack evolution behavior in advance. In the past decade, the phase field method for crack evolution analysis has drawn a lot of attention for its application in fracture mechanics. The biggest advantage of the phase field model is its uniform description of all crack evolution behaviors by one evolution equation. The phase field method simultaneously models crack nucleation and crack propagation which will be particularly useful manufacturing problems. In this work, we show that the phase field method is capable to reproduce the most important fatigue features, e.g., Paris' law, mean stress effect, and load sequence effects. For efficient computing, a "cycle"- "time" transformation is introduced to convert individual cycle numbers into a continuous time domain. In order to exploit the symmetry property of the demonstrated examples, a phase field model in cylindrical coordinates is presented. Finally, the fatigue modeling approach presented is applied to study a cold forging process in manufacturing.

1 Introduction

The phase field model was initially used to solve the interfacial problem, like ferromagnetism, ferroelectrics, and solidification dynamics [1]. Moreover, the phase field model can also be applied in fracture mechanics [2–6]. The method has the advantage that it takes a monolithic approach to simulate crack initiation, branching, bifurcation, and unification. It also overcomes stress singularity, displacement jumps, or interface tracking during the fracture simulation. Differing from other methods, neither remeshing nor finite elements with special shape functions are needed in the phase field model; the simulation is performed on a fixed mesh. The core idea of a phase field fracture model is to introduce an additional field variable to represent cracks. This scalar field variable interpolates smoothly between the values of 0 and 1, representing cracked and undamaged material, respectively. The relevant equations are derived from the total energy of

J. C. Aurich et al. (Eds.): IRTG 2023, *Proceedings of the 3rd Conference on Physical Modeling for Virtual Manufacturing Systems and Processes*, pp. 16–31, 2023.
https://doi.org/10.1007/978-3-031-35779-4_2

the system by a variational principle (one equation models the equilibrium of the stress field, and a second models the evolution of the crack field). Consequently, contour plots of a scalar field variable used allow for the visualization of the progression of fracture and reproduce the crack situation. A phase field model has been successfully applied for quasi-static [7–9] and dynamic cases [6, 10–14]. Further recent model extensions also allow the consideration of ductile fracture [15–18], anisotropic fracture properties [19–21], and the evolution of fracture in various multi-physics scenarios [22–25].

In manufacturing processes like cutting, the dynamic loads typically do not cause an immediate failure of a tool; instead, tool failure can occur due to fatigue fracture development over numerous loading cycles. Thus, a phase field model which can handle the fatigue scenarios is required. In this paper the application of the model presented is focused on the fatigue failure of manufacturing tools. Since the driving mechanisms of fatigue failure significantly differ from those of classical linear elastic fracture mechanics, it was necessary to make appropriate adjustments to the evolution equation of the fracture field to model fatigue crack growth in manufacturing. Time-resolved simulations are impractical since fatigue failure only happens after a significant number of cycles; hence the evolution equation must be written in the context of cycles. The numerical implementation must be able to consolidate multiple cycles into a pseudo time to achieve the efficiency needed for the use of the model in actual production processes. In this work, we present a phase field model for cyclic fatigue. Since fatigue cracks won't appear until several loading cycles have been completed, fatigue simulations generally consume high computing time. We introduce an adaptive cycle increment algorithm, which provides a moderate computing time without losing accuracy compared to the classical computing strategies.

This paper proceeds as follows: in Sect. 2, a phase field model for cyclic fatigue is presented. In addition, a "cycle"- "time" transfer is proposed to bundle several cycles to a pseudo time domain for efficient computing. An adaptive cycle increment algorithm is then developed to reduce the computational cost without losing accuracy. In Sect. 3, an example of a manufacturing problem is modeled by the phase field fatigue model. In Sect. 4, the conclusions are stated.

2 A Phase Field Model for Cyclic Fatigue

The phase field fracture model introduces an additional field variable to represent cracks [7, 26]. The crack field s is 1 if the material is undamaged and if it is 0 where cracks occur. Furthermore, it is postulated that the displacement field $\mathbf{u}$ and crack field s locally minimize the total energy of a loaded body Ω. This yields the equilibrium of the stress field and the evolution of the crack field for fatigue fracturing. The extended total energy $\mathcal{E}$ with t as the external traction and f as the volume forces on the body is given by

$$\mathcal{E} = \int_{\Omega} \psi \, dV - \int_{\partial\Omega} t \, dA - \int_{\Omega} f \, dV \tag{1}$$

where ψ denotes the total energy density of the body

$$\psi = (g(s) + \eta)\psi^e(\boldsymbol{\varepsilon}) + \psi^s(s, \nabla s) + h(s)\psi^{\mathrm{ad}}(D), \tag{2}$$

which consists of three parts: elastic part, fracture surface part and additional fatigue part.

The strain energy density

$$\psi^e(\varepsilon) = \frac{1}{2}\varepsilon : \mathbb{C}(\varepsilon) \tag{3}$$

is the elastic energy stored inside of a body with $g(s)$ as a degradation function, which models the loss of stiffness of the broken material. The tensor ε is the infinitesimal strain, defined by

$$\varepsilon = \begin{bmatrix} \varepsilon_{xx} & \varepsilon_{xy} & \varepsilon_{xz} \\ \varepsilon_{yx} & \varepsilon_{yy} & \varepsilon_{yz} \\ \varepsilon_{zx} & \varepsilon_{zy} & \varepsilon_{zz} \end{bmatrix} = \begin{bmatrix} \frac{\partial u_x}{\partial x} & \frac{1}{2}\left(\frac{\partial u_y}{\partial x} + \frac{\partial u_x}{\partial y}\right) & \frac{1}{2}\left(\frac{\partial u_x}{\partial x} + \frac{\partial u_z}{\partial z}\right) \\ \frac{1}{2}\left(\frac{\partial u_y}{\partial x} + \frac{\partial u_x}{\partial y}\right) & \frac{\partial u_y}{\partial y} & \frac{1}{2}\left(\frac{\partial u_y}{\partial z} + \frac{\partial u_z}{\partial y}\right) \\ \frac{1}{2}\left(\frac{\partial u_z}{\partial x} + \frac{\partial u_x}{\partial z}\right) & \frac{1}{2}\left(\frac{\partial u_z}{\partial y} + \frac{\partial u_y}{\partial z}\right) & \frac{\partial u_z}{\partial z} \end{bmatrix} \tag{4}$$

The crack surface density

$$\psi^s(s, \nabla s) = \mathcal{G}_c\left(\frac{(1-s)^2}{4\epsilon} + \epsilon|\nabla s|^2\right) \tag{5}$$

is the energy required to separate the material to generate a crack, which is assumed to be proportional to the crack surface. The parameter $\mathcal{G}_c$ denotes fracture resistance and can be related to fracture toughness. The numerical parameter ϵ – not to be confused with strain tensor - models the width of the smooth transition zone between the broken and unbroken material.

The fatigue energy density

$$\psi^{ad}(D) = q < D - D_c >^b \quad \text{with} \quad D = D_0 + dD \tag{6}$$

is introduced to account for the accumulated fatigue driving forces, which is associated with a fatigue damage parameter D. This parameter D models the damage related to fatigue, inspired by Miner rule [28], which is accumulated during the simulation. The parameter D_0 is the previous damage and

$$dD = \frac{dN}{n_D}\left(\frac{\hat{\sigma}}{A_D}\right)^k \tag{7}$$

is the damage increment, which is associated with the cycle increment dN, where the parameters n_D, A_D and k are extracted from the Wöhler curve of experiments [29]. This formulation allows the phase field fatigue model to incorporate all the influences from the environment into the fatigue propagation behavior [30]. In the phase field model, the first principal stress $\hat{\sigma}_1$ from the undegraded stress field

$$\hat{\sigma}_1 = [\mathbb{C}\varepsilon]_1 \tag{8}$$

is used as the fatigue driving force for high cycle fatigue. It is noted that it is not claimed that this choice of the driving force is suitable for all materials. Other effective stress quantities, e.g. the von-Mises stress, might be more suitable for ductile material and low cycle fatigue [31, 32]. Moreover, a mean stress corrector can be applied to include the mean stress effect on the fatigue crack propagation [27, 33]. The parameter D_c is a damage threshold, which models the crack nucleation process. With the Macauley brackets ($< \bullet >$), the additional fatigue energy ψ^{ad} will not contribute when the damage D is below this threshold. After the crack nucleation stage, the parameters q and b are parameters controlling how intense the additional fatigue energy drives the crack. A discussion of different choices of the parameters q and b can be found in [34]. The degradation function $h(s)$ - similarly as $g(s)$ - models the loss of the stiffness of broken material due to cyclic fatigue. A discussion of different choices of the degradation functions can be found in [34, 36].

With the variational principle of Eq. (1), four coupled equations are derived

$$\text{div}\frac{\partial \psi}{\partial \nabla \mathbf{u}} + f = 0 \tag{9}$$

$$\frac{\partial \psi}{\partial s} - \text{div}\frac{\partial \psi}{\partial \nabla s} = 0 \tag{10}$$

$$\frac{\partial \psi}{\partial \nabla s} \cdot \mathbf{n} = 0 \quad \text{on } \partial\Omega_{\nabla s} \tag{11}$$

$$\left(\frac{\partial \psi}{\partial \nabla \mathbf{u}}\right)\mathbf{n} = t \quad \text{on } \partial\Omega_t \tag{12}$$

Equation (9) describes the equilibrium condition of the stress field; Eq. (10) described the evolution behavior of the crack field; Eq. (11) and Eq. (12) are the Neumann boundary conditions for the crack field and displacement field. Those equations define the fatigue fracture problem.

The phase field fatigue model can reproduce the most important fatigue properties. In the following evaluation, the material parameters are taken from [26] with a CT specimen [37] as a numerical example. The crack growth rate is depicted in Fig. 1 for various maximum stress amplitude values. It is to observe that even though different stress amplitudes for the simulation are applied, the rate of crack growth can be described with the same Paris' law. The result matches Paris' law with $m = 5.54$ very well. Radhakrishnan [38] shows that in some materials the constant C and the slope m depend on the stress ratio R. The stress ratio R is defined as the ratio between the minimum stress and the maximum stress. At high positive mean stress, a decrease in fatigue life is associated with multiple crack initiation sites at the specimen surface. Fatigue limit is highly affected by the tensile mean stress and stress ratio since the maximum stress approaches near yield stress and it causes cyclic ratcheting [39]. Figure 2 displays the effect of mean stress on the crack growth rate, which reflects the fact that higher mean stress increases the rate of crack growth [40]. Figure 3 reports the effect of the loading sequence on the crack growth rate. Results show that a high-low loading sequence results in short fatigue life. This phenomenon is called the loading sequence effect [41, 42]. It has been shown that the material with a low-high load sequxsence results in a longer fatigue life because the low load level is mostly involved in the crack nucleation and the high load level is contributed to the crack propagation [43]. This effect can be explained by the residual stresses and crack closure near the crack tip [44]. Although Miner's rule does not include the loading sequence effect, the damage quantity D with a low load level increases slowly, such that it reaches the critical damage state D_c later than a high load level.

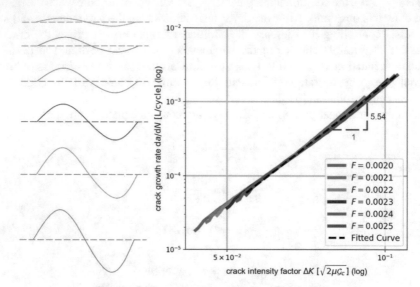

Fig. 1. Different maximum load amplitude [35].

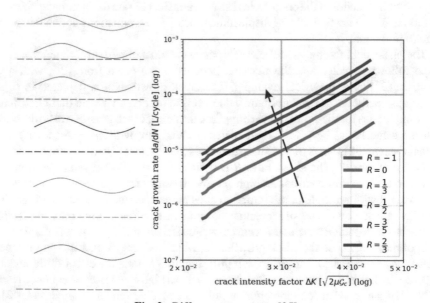

Fig. 2. Different mean stress [35].

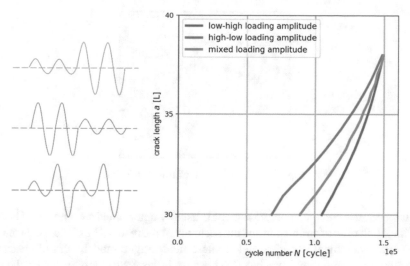

Fig. 3. Different loading sequences.

2.1 A Time-Cycle Transformation in the Phase Field Fatigue Model

As the discussion in the previous section, the phase field fatigue model is more suitable for high cycle fatigue. Speaking of high cycle fatigue, the number of cycles to failure is usually around tens of millions or even more. Thus, it is not feasible to simulate the accumulated cycles one after another.

The first step of an efficient integration concept is proposed by Chaboche [45] with a non-linear cumulative damage model, where cycles with similar loading are bundled into blocks. The "time"- "cycle" transfer of the phase field model is similar to this idea. It is to assume a constant block size of cycle number per time $\frac{dN}{dt}$ representing a certain evolution of fatigue damage [26]. Thus, the individual single loading cycle is not used in the proposed phase field fatigue model; rather, the cycle is converted into continuous pseudo "time" as illustrated in Fig. 4. The red line in Fig. 4 represents the envelope loading, which approximates the actual discrete cyclic loading. In addition, several load cycles are combined into one block in order to reduce the overall number of load cycles: in one simulation step, the incremental change in pseudo "time" is connected to a specific number of load cycles.

In addition, for irregular loading sequences, the rain flow algorithm is used to convert a loading sequence of varying stress into an equivalent set of constant amplitude stress [19, 46].

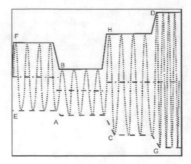

Fig. 4. A "cycle" - "time" transformation

However, the cycle number increment is usually determined by a trade-off between the computing time of simulation and the accuracy of the result. The choice of the number of the cycle increment is critical in the phase field fatigue model, not only because it determines the simulation time, but also because it has a strong influence on the crack topology [34].

The damage parameter D is introduced in the phase field model to model material damage caused by fatigue. Additionally, the "cycle"- "time" transform captures the loading with similar fatigue damage influence together. To reduce the computational effect, the adaptive cycle number adjustment algorithm (*ACNAA*) works by associating the cycle number increment with the damage increment. The simulation of fatigue fracture is divided into three stages based on the damage state (see Fig. 5):

1. $D < D_c$: The fatigue energy term disappears at this point, so it can be viewed as a pure static mechanical state. The cycle increment should be as large as possible in order to reach the critical fatigue state as quickly as possible.
2. $D \approx D_c$: The material is about to break at this point, and the cycle number increment dN should be chosen so that the damage increment dD is small enough to simulate the transient process.
3. $D > D_c$: The fatigue crack begins to propagate. The damage increment dD is regulated at this stage to achieve a moderate growth rate of the fatigue energy.

Our method has been shown to reduce computing time to nearly 3% when compared to constant cycle number increments with $dN = 5$ [34]. The reason is that the huge computing time involved in the crack nucleation is dramatically reduced. Additionally, the adaptive cycle number adjustment method is also suitable for parallel computing [35]. With parallel computing (e.g. MPI), an additional significant decrease in computing time can be obtained, which keeps a 3D simulation within a reasonable time limit [35].

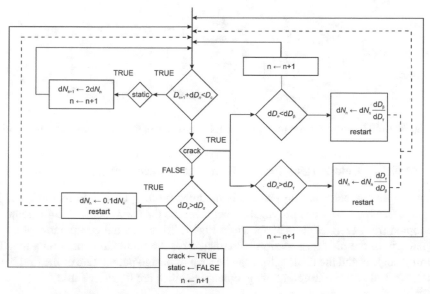

Fig. 5. A flowchart illustrates the idea of ACNAA [35]. dD_α, dD_β, dD_γ are suitable numerical parameters.

3   Phase Field Model in the Context of Manufacturing Process

3.1   Application in the Cold Forging Process

In the past decades, cold forging has gained a lot of attention and has become a economic production method for complex geometries with net-shaped or near-net-shaped surfaces. The cold forge is characterized by the circumstance that the forming of the workpiece begins at room temperature and without external heating. The major advantages of cold forging are close dimensional tolerances, good surface finish quality, and interchangeability as well as reproducibility due to its simple process [47, 48]. During the cold forging process, the material of a metal billet is put into a container (called a die). The material, compressed by a ram, flows through the container and is formed into the desired shape. In general, the cold forging process involves 5 steps (see Fig. 6):

a. lubrication: the workpiece is lubricated to avoid sticking to the die and to maintain a low temperature.
b. insertion: the workpiece is inserted onto a die with the shape of the final part.
c. stroke: a great force is stroked onto the workpiece to create the desired form.
d. flash: the excess metal around the dice is trimmed.
e. removing: the workpiece is removed from the die.

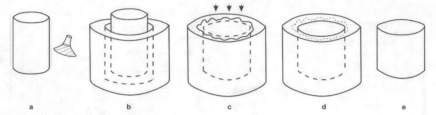

Fig. 6. Cold forging process: (a: lubrication; b: insertion; c: stroke; d: flash; e: removing).

3.2 Modeling Cold Forging Process Using Phase Field Method

In this paper, the cold forging process is modeled by the phase field method, and the fatigue life, where the crack propagation behavior are the main focus. The die geometry is adopted from Lang et al. [49] shown in Fig. 7. To reduce the computational cost, a 2D slice from the die cross-section is extracted for the finite element simulation. The opening angle α and the die length L can be seen as design parameters of the die. In this paper, we evaluate two different die geometries, which are listed in Table 1.

Table 1. Die geometry

α	L
45°	23 mm
60°	25 mm

This design of the die enables high stresses at the fillet radius to generate the fatigue crack initiation and crack growth after a short number of production cycles [49]. The material of the die is AISI 2D [50, 51]. The simulation loading settings are motivated by the experiments of Dalbosco et al. [52]. One contribution of his work for this application is the different assumptions regarding the interference between the workpiece and the die.

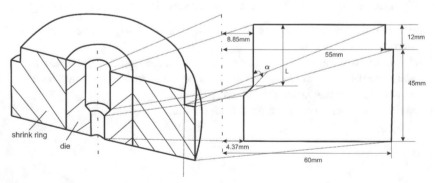

Fig. 7. The cold forging tool geometry presented in [49] and a 2D slice for finite element simulation.

In our first simulation setting (see Fig. 8a) the entire inner face of the die is assumed compressed with a constant distributed load, and the bottom of the cold forging tool is fixed by Dirichlet boundary conditions. In a different design of the die (see Fig. 8b), there is no interference from the point of transition radius until the bottom of the die in the second example. This is caused by a lack of material apposition, resulting in stress vacancy along this area of the die. For the sake of simplicity, we assume a constant load only applying it to the inner face of the die. Moreover, as an alternative design, it is also considered that the inner wall can shrunk less due to the lower shrink-fit of the die material on this part as shown in Fig. 8c. As shown in this last Figure, only the fillet of the die is under the tension loading.

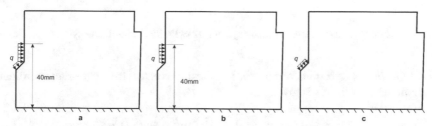

Fig. 8. **a**: both fillet and inner wall are loaded; **b**: only inner wall is loaded; **c**: only fillet is loaded.

3.3   Phase Field Fatigue Model in Cylindrical Coordinate System

In the *cartesian* coordinate system, the positions of points are determined with respect to three mutually perpendicular planes, giving the length-, width- and height coordinates. For a suitable computational cost, a 2D slice from the cross-section of the die is chosen for the finite element calculation Fig. 9a. This simplification in a sense of a cartesian coordinate system is to assume that the width of the body is infinite and all the derivatives regarding z-direction are zero. However, the cold forging die does not have an endless width, rather say, it is symmetric around its axis. Thus, a proper way to simulate the cold forging process with less computational resources is to bring this 2D slice cross-section of the die into a cylindrical coordinate system to exploit its rotational symmetry. A cylindrical coordinate system is specified by a radial position, an angular position, and a height position as shown in Fig. 9b.

The total energy of the body reads

$$\mathcal{E} = \int \left[(g(s) + \eta) \psi^e \left(\boldsymbol{\varepsilon}^{\mathrm{cyl}} \right) + \psi^s \left(s, \nabla^{\mathrm{cyl}} s \right) + h \left(s, \nabla^{\mathrm{cyl}} s \right) \psi^{\mathrm{ad}}(\mathrm{D}) \right] \mathrm{d} V^{\mathrm{cyl}}, \quad (13)$$

where $\boldsymbol{\varepsilon}^{\mathrm{cyl}}$ is the strain tensor in the cylindrical coordinates and $\mathrm{d} V^{\mathrm{cyl}}$ is the infinite cylinder volume element.

Let r be the radius, θ be the circumferential angle and z be the height, the transformation between the cartesian coordinates (x, y, z) and cylindrical coordinates (r, θ, z) can be given as

$$x = r \cos \theta \quad y = r \sin \theta \quad z = z, \quad (14)$$

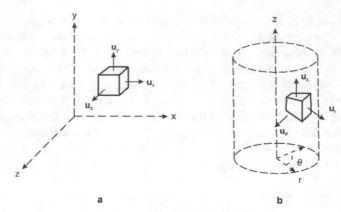

Fig. 9. a: cartesian coordinate system; **b**: cylindrical coordinate system.

and the *Jacobian* matrix transforming the infinitesimal vectors from cartesian coordinates to cylindrical coordinates is given as

$$J = \begin{bmatrix} \frac{\partial x}{\partial r} & \frac{\partial y}{\partial r} & \frac{\partial z}{\partial r} \\ \frac{\partial x}{\partial \theta} & \frac{\partial y}{\partial \theta} & \frac{\partial z}{\partial \theta} \\ \frac{\partial x}{\partial z} & \frac{\partial y}{\partial z} & \frac{\partial z}{\partial z} \end{bmatrix} = \begin{bmatrix} \cos\theta & \sin\theta & 0 \\ -r\sin\theta & r\cos\theta & 0 \\ 0 & 0 & 1 \end{bmatrix}. \tag{15}$$

The displacement vector in the cylindrical coordinate system with rotational symmetry properties is given as

$$\mathbf{u}^{\text{cyl}} = [u_r, u_\theta, u_z]^T \underset{rot.\ sym.}{\Longrightarrow} [u_r, 0, u_z]^T, \tag{16}$$

where u_r and u_z are the width and height components of the displacement vector.

For rotational sysmmetry, the derivative in angular direction vanishes, thus, the strain tensor is given by.

$$\varepsilon^{\text{cyl}} = \begin{bmatrix} \frac{\partial u_r}{\partial r} & 0 & \frac{1}{2}\left(\frac{\partial u_r}{\partial z} + \frac{\partial u_z}{\partial r}\right) \\ 0 & \frac{u_r}{r} & 0 \\ \frac{1}{2}\left(\frac{\partial u_z}{\partial r} + \frac{\partial u_r}{\partial z}\right) & 0 & \frac{\partial u_z}{\partial z} \end{bmatrix}. \tag{17}$$

It is noted that the entry in the middle $\frac{u_r}{r}$ provides an additional contribution into the energy density, which is omitted in the cartesian coordinates system for 2D. The fatigue driving force $\hat{\sigma}^{\text{cyl}}$ can be given with the constitutive law and taking as the first principal stress

$$\hat{\sigma}^{\text{cyl}} = \left[\mathbb{C}\varepsilon^{\text{cyl}}\right]_1 \tag{18}$$

where the stiffness tensor $\mathbb{C}$ remains the same as it is in the cartesian coordinate system because of its isotropic character. The crack field s itself does not need to be modified into a cylindrical coordinate system since it is a scalar variable to indicate the broken state of the material. The gradient of the crack field $\nabla^{cyl}s$ in the cylindrical coordinate system is given as

$$\nabla^{cyl}s = [\frac{\partial s}{\partial r} 0 \frac{\partial s}{\partial z}]^T .$$

3.4 Phase Field Simulation of Cold Forging Process

In the our first analysis, it is assumed that the fillet and the inner wall of the die are completely loaded, two different geometries of the die are investigated. The angle of crack propagation is nearly 30° in Fig. 10a and nearly 40° in Fig. 10b. Those angles of the crack propagation directions can be explained by the mixed energy fracture criterion [53], since the tools are under a mixed mode I/II load situation. The bigger angle of fracture initiation in Fig. 10b can be explained by the dominant influence of shear stress from mode II in comparison to the tension stress from mode I. Furthermore, the first initialized crack can be found after around 3,000 production cycles at the forging tool with an opening angle of 45° and the fatigue life of the second tool ($\alpha = 60°$) is only around 500 cycles of production. This analysis reveals the fact that the dominated shear stress on the inner wall of the die shortens the fatigue life of the tool.

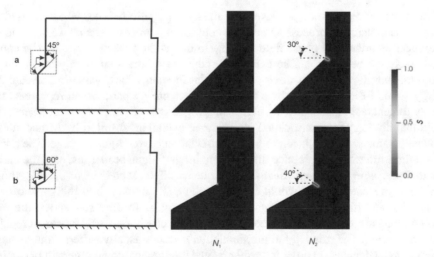

Fig. 10. The simulation of the cold forging process at first crack N_1 and final stage N_2 (**a**: opening angle $\alpha = 45°$; **b**: opening angle $\alpha = 60°$).

For further investigations, the cold forging tools are simulated with different loading assumptions (Fig. 8) as shown in Fig. 11. Results show that loading acts merely on the inner wall of the die and can dramatically increase the fatigue life of the die. In the analysis that was performed itt yielded the highest fatigue life at around 55,000 production cycles for the opening angle of 45° (Fig. 11b). Different loading assumptions lead to different patterns of crack propagation. In Fig. 11a and Fig. 11b, the crack propagates first sloping downward, which is influenced by a mixed mode loading situation. After these stages, the crack curves moves in a nearly horizontal direction because of the mainly vertical tensile stress. In contrast, loads acting only on the inner wall yield almost the same crack propagation patterns, where the angle of crack propagation is around 70°. This can be explained by a pure shear mode II loading situation. These crack propagation behaviors from the phase field simulations have been found similarly in reported experiments [52].

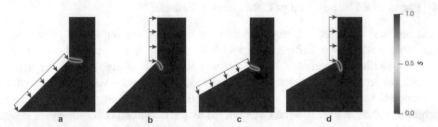

Fig. 11. The simulation of the cold forging process by different loading conditions.

4   Conclusion

In this paper, we presented a phase field model for cyclic fatigue, which is used to analyze manufacturing process namely the cold forging process. The phase field model introduces an additional phase field variable to model the broken material. The entire crack evolution behavior can be derived by considering the total energy of the body. The total energy consists of three parts: an elastic energy part, which represents the energy stored inside of the body; a fracture surface energy part, which represents the energy to generate cracks; and an additional fatigue energy part, which represents the additional driving forces associated with fatigue evolution. Inspired by Miner rule, a damage parameter is introduced to model the accumulative fatigue damage. The phase field fatigue model can reproduce the most important fatigue properties, e.g., the Paris' law, the mean stress effect, and the loading sequence effect. Moreover, a "cycle"- "time" transfer is presented which would transform the cycle domain into the pseudo time domain for an efficient fatigue simulation. For irregular loading sequences, the rain flow counting algorithm is used to convert the load cycles into several blocks of regular uniform loading. The existing fatigue simulation methods usually suffers from its huge computational demand. In order to further reduce the computational time without losing accuracy, different numerical strategies are proposed. The core idea of the *ACNAA* is to associate the damage increment with the cycle increment. Additional computing time reduction can be obtained by applying parallel computing.

The main contribution of this work is that we apply the phase field model to the manufacturing problem to predict fatigue life and crack patterns. We used the cold forging process as the demonstrated example since it is an important manufacturing methods for producing parts with complex geometries. To exploit the rotational symmetry property of the problem, a phase field fatigue model for cylindrical coordinates is introduced. Different cold forging die geometries and load conditions in the processing are presented to analyze the fatigue life and crack patterns. Results show that the phase field model can be effectively applied to cold forging process. This enables a physics-based prediction of the lifetime of manufacturing tools and the identification of process parameters relevant to detect the onset of damage.

Acknowledgment. Funded by the Deutsche Forschungsgemeinschaft (DFG, German Research Foundation) – 252408385 – IRTG 2057.

References

1. Steinbach, I.: Phase-field models in materials science. Modell. Simul. Mater. Sci. Eng. **17**(7), 073001 (2009)
2. Griffith, A.A.: VI. The phenomena of rupture and flow in solids. Philosophical Transactions of the Royal Society of London. Series A, Containing Papers of a Mathematical or Physical Character **221**(582–593), 163–198 (1921)
3. Francfort, G.A., Marigo, J.J.: Revisiting brittle fracture as an energy minimization problem. J. Mech. Phys. Solids **46**(8), 1319–1342 (1998)
4. Bourdin, B., Francfort, G.A., Marigo, J.J.: Numerical experiments in revisited brittle fracture. J. Mech. Phys. Solids **48**(4), 797–826 (2000)
5. Bourdin, B.: Numerical implementation of the variational formulation for quasi-static brittle fracture. Interfaces and Free Boundaries **9**(3), 411–430 (2007)
6. Bourdin, B., Larsen, C.J., Richardson, C.L.: A time-discrete model for dynamic fracture based on crack regularization. Int. J. Fract. **168**(2), 133–143 (2011)
7. Kuhn, C., Müller, R.: A continuum phase field model for fracture. Eng. Fract. Mech. **77**(18), 3625–3634 (2010)
8. Amor, H., Marigo, J.J., Maurini, C.: Regularized formulation of the variational brittle fracture with unilateral contact: numerical experiments. J. Mech. Phys. Solids **57**(8), 1209–1229 (2009)
9. Miehe, C., Welschinger, F., Hofacker, M.: Thermodynamically consistent phase-field models of fracture: variational principles and multi-field FE implementations. Int. J. Numer. Meth. Eng. **83**(10), 1273–1311 (2010)
10. Larsen, C.J., Ortner, C., Süli, E.: Existence of solutions to a regularized model of dynamic fracture. Math. Models Methods Appl. Sci. **20**(07), 1021–1048 (2010)
11. Borden, M.J., Verhoosel, C.V., Scott, M.A., Hughes, T.J., Landis, C.M.: A phase-field description of dynamic brittle fracture. Comput. Methods Appl. Mech. Eng. **217**, 77–95 (2012)
12. Hofacker, M., Miehe, C.: Continuum phase field modeling of dynamic fracture: variational principles and staggered FE implementation. Int. J. Fract. **178**(1), 113–129 (2012)
13. Hofacker, M., Miehe, C.: A phase field model of dynamic fracture: robust field updates for the analysis of complex crack patterns. Int. J. Numer. Meth. Eng. **93**(3), 276–301 (2013)
14. Schlüter, A., Willenbücher, A., Kuhn, C., Müller, R.: Phase field approximation of dynamic brittle fracture. Comput. Mech. **54**(5), 1141–1161 (2014). https://doi.org/10.1007/s00466-014-1045-x
15. Miehe, C., Aldakheel, F., Raina, A.: Phase field modeling of ductile fracture at finite strains: a variational gradient-extended plasticity-damage theory. Int. J. Plast **84**, 1–32 (2016)

16. Ambati, M., De Lorenzis, L.: Phase-field modeling of brittle and ductile fracture in shells with isogeometric NURBS-based solid-shell elements. Comput. Methods Appl. Mech. Eng. **312**, 351–373 (2016)
17. Borden, M.J., Hughes, T.J., Landis, C.M., Anvari, A., Lee, I.J.: A phase-field formulation for fracture in ductile materials: finite deformation balance law derivation, plastic degradation, and stress triaxiality effects. Comput. Methods Appl. Mech. Eng. **312**, 130–166 (2016)
18. Kuhn, C., Noll, T., Müller, R.: On phase field modeling of ductile fracture. GAMM-Mitteilungen **39**(1), 35–54 (2016)
19. Li, B., Peco, C., Millán, D., Arias, I., Arroyo, M.: Phase-field modeling and simulation of fracture in brittle materials with strongly anisotropic surface energy. Int. J. Numer. Meth. Eng. **102**(3–4), 711–727 (2015)
20. Nguyen, T.T., Réthoré, J., Baietto, M.C.: Phase field modelling of anisotropic crack propagation. Eur. J. Mech.-A/Solids **65**, 279–288 (2017)
21. Schreiber, C., Ettrich, T., Kuhn, C., Müller, R.: A phase field modeling approach of crack growth in materials with anisotropic fracture toughness. In: 2nd International Conference of the DFG International Research Training Group 2057–Physical Modeling for Virtual Manufacturing (iPMVM 2020). Schloss Dagstuhl-Leibniz-Zentrum für Informatik (2021)
22. Xu, B.X., Schrade, D., Gross, D., Mueller, R.: Phase field simulation of domain structures in cracked ferroelectrics. Int. J. Fract. **165**(2), 163–173 (2010)
23. Miehe, C., Schaenzel, L.M., Ulmer, H.: Phase field modeling of fracture in multi-physics problems. Part I. Balance of crack surface and failure criteria for brittle crack propagation in thermo-elastic solids. Comput. Methods Appl. Mech. Eng. **294**, 449–485 (2015)
24. Miehe, C., Hofacker, M., Schänzel, L.M., Aldakheel, F.: Phase field modeling of fracture in multi-physics problems. Part II. Coupled brittle-to-ductile failure criteria and crack propagation in thermo-elastic–plastic solids. Comput. Methods Appl. Mech. Eng. **294**, 486–522 (2015)
25. Miehe, C., Mauthe, S.: Phase field modeling of fracture in multi-physics problems. Part III. Crack driving forces in hydro-poro-elasticity and hydraulic fracturing of fluid-saturated porous media. Comput. Methods Appl. Mech. Eng. **304**, 619–655 (2016)
26. Schreiber, C., Kuhn, C., Müller, R., Zohdi, T.: A phase field modeling approach of cyclic fatigue crack growth. Int. J. Fract. **225**(1), 89–100 (2020). https://doi.org/10.1007/s10704-020-00468-w
27. Schreiber, C., Müller, R., Kuhn, C.: Phase field simulation of fatigue crack propagation under complex load situations. Arch. Appl. Mech. **91**(2), 563–577 (2020). https://doi.org/10.1007/s00419-020-01821-0
28. Miner, M.A.: Cumulative damage in fatigue (1945)
29. ASTM E739 10 - Standard Practice for Statistical Analysis of Linear or Linearized Stress-Life (S-N) and Strain-Life (ε-N) Fatigue Data
30. Yan, S., Müller, R., Ravani, B.: Simulating Fatigue Crack Growth including Thermal Effects Using the Phase Field Method (2022)
31. Mises, R.V.: Mechanik der festen Körper im plastisch-deformablen Zustand. Nachrichten von der Gesellschaft der Wissenschaften zu Göttingen, Mathematisch-Physikalische Klasse **1913**, 582–592 (1913)
32. Bathias, C., Pineau, A.: Fatigue of Materials and Structures: Fundamentals. Hoboken, (2010)
33. Walker, K.: The effect of stress ratio during crack propagation and fatigue for 2024-T3 and 7075-T6 aluminum (1970)
34. Yan, S., Schreiber, C., Müller, R.: An efficient implementation of a phase field model for fatigue crack growth. Int. J. Fract., 1–14 (2022). https://doi.org/10.1007/s10704-022-00628-0
35. Yan, S., Müller, R.: An efficient phase field model for fatigue fracture. In: 15th World Congress on Computational Mechanics (WCCM-XV) and 8th Asian Pacific Congress on Computational Mechanics (APCOM-VIII) (2022)

36. Kuhn, C.: Numerical and analytical investigation of a phase field model for fracture. Technische Universität Kaiserslautern (2013)
37. ASTM: ASTM E399-09, Standard test method for linear-elastic plane-strain fracture toughness k ic of metallic materials. http://www.astm.org (2009)
38. Radhakrishnan, V.M.: Parameter representation of fatigue crack growth. Eng. Fract. Mech. **11**(2), 359–372 (1979)
39. Pradhan, D., et al.: Effect of stress ratio and mean stress on high cycle fatigue behavior of the superalloy IN718 at elevated temperatures. Mater Res Express **6**(9), 0965a6 (2019)
40. Kamaya, M., Kawakubo, M.: Mean stress effect on fatigue strength of stainless steel. Int. J. Fatigue **74**, 20–29 (2015)
41. Paepegem, W.V., Degrieck, J.: Effects of load sequence and block loading on the fatigue response of fiber-reinforced composites. Mech. Adv. Mater. Struct. **9**(1), 19–35 (2002)
42. Kamaya, M., Kawakubo, M.: Loading sequence effect on fatigue life of type 316 stainless steel. Int. J. Fatigue **81**, 10–20 (2015)
43. Stephens, R.I., Fatemi, A., Stephens, R.R., Fuchs, H.O.: Metal Fatigue in Engineering. John Wiley & Sons (2000)
44. Stephens, R.I.: Fatigue crack growth under spectrum loads. ASTM International (2011)
45. Chaboche, J.L., Lesne, P.M.: A non-linear continuous fatigue damage model. Fatigue Fract. Eng. Mater. Struct. **11**(1), 1–17 (1988)
46. Yan, S., Müller, R., Ravani, B.: A phase field fatigue model for complex loading situations. In: 15th International Conference on Fracture (2023)
47. Fritz, A.H., Schulze, G. (eds.): Fertigungstechnik. Springer Berlin Heidelberg, Berlin, Heidelberg (2004)
48. Fritz, A.H.: Umformen. In: Fritz, A.H. (ed.) Fertigungstechnik. S, pp. 133–223. Springer, Heidelberg (2018). https://doi.org/10.1007/978-3-662-56535-3_3
49. Lange, K., Hettig, A., Knoerr, M.: Increasing tool life in cold forging through advanced design and tool manufacturing techniques. J. Mater. Process. Technol. **35**(3–4), 495–513 (1992)
50. Bringas, J.E.: Handbooks of comparative world steel standards (2004)
51. Pyun, Y.S., et al.: Development of D2 tool Steel trimming knives with nanoscale microstructure. In: AISTECH-Conference Proceedings, vol. 2, p. 465. Association for Iron Steel Technology
52. Dalbosco, M., da Silva Lopes, G., Schmitt, P.D., Pinotti, L., Boing, D.: Improving fatigue life of cold forging dies by finite element analysis: a case study. J. Manuf. Process. **64**, 349–355 (2021)
53. Kfouri, A.P., Brown, M.W.: A fracture criterion for cracks under mixed-mode loading. Fatigue Fract. Eng. Mater. Struct. **18**(9), 959–969 (1995)

Embedding-Space Explanations of Learned Mixture Behavior

J.-T. Sohns[1]([✉]), D. Gond[2], F. Jirasek[2], H. Hasse[2], G. H. Weber[3], and H. Leitte[1]

[1] Visual Information Analysis Group, RPTU Kaiserslautern-Landau, Kaiserslautern, Germany
j_sohns12@cs.uni-kl.de
[2] Laboratory of Engineering Thermodynamics (LTD), RPTU Kaiserslautern-Landau, Kaiserslautern, Germany
[3] Computational Research Division, Lawrence Berkeley National Laboratory, Berkeley, CA, USA

Abstract. Data-driven machine learning (ML) models are attracting increasing interest in chemical engineering and already partly outperform traditional physical simulations. Previous work in this field has mainly focused on improving the models' statistical performance while the thereby imparted knowledge has been taken for granted. However, also the structures learned by the model during the training are fascinating yet non-trivial to assess as they are usually high-dimensional. As such, the interpretable communication of the relationship between the learned model and domain knowledge is vital for its evaluation by applying engineers. Specifically, visual analytics enables the interactive exploration of data sets and can thus reveal structures in otherwise too large-scale or too complex data.

This chapter focuses on the thermodynamic modeling of mixtures of substances using the so-called activity coefficients as exemplary measures. We present and apply two visualization techniques that enable analyzing high-dimensional learned substance descriptors compared to chemical domain knowledge. We found explanations regarding chemical classes for most of the learned descriptor structures and striking correlations with physicochemical properties.

1   Introduction

Machine learning is rapidly entering the field of engineering. The data-driven prediction using such methods is already outperforming traditional engineering algorithms for multiple properties [1–3]. With the transition from a computer science gimmick to appliance in real-world scenarios, the stakes rise significantly. Whether human lives are on the line or the planning of an expensive production step, the confidence in the algorithm needs to be exceptional. An emerging solution is to provide human-understandable explanations for the decisions of machine learning, which can spark trust and suspicion where necessary [4, 5].

Recent research has introduced the concept of matrix completion methods (MCM) to predict the thermodynamic properties of mixtures, or, in other words, the mixture behavior, from a sparse data set of experimental values [1–3, 6, 7]. Among others, these methods allow predicting activity coefficients, which are a measure for the non-ideality of

© The Author(s) 2023
J. C. Aurich et al. (Eds.): IRTG 2023, *Proceedings of the 3rd Conference on Physical Modeling for Virtual Manufacturing Systems and Processes*, pp. 32–50, 2023.
https://doi.org/10.1007/978-3-031-35779-4_3

a mixture. In the present work, models for the prediction of activity coefficients of solutes at infinite dilution in solvents at a constant temperature of 298.15 K [1, 7] are taken as a prototype to create an algorithmic pipeline that is transferable to a broader series of use cases. To give an instance in the context of process level production planning, an accurate and trusted machine learning algorithm empowers precise, fast and, most importantly, cheap simulations, thereby avoiding costly and time-consuming experiments.

We order the data set in matrix form with solutes as one axis, solvents as the other axis and mixture behavior, i.e., the activity coefficients, as cell entries. The assumption is that the resulting matrix is of low rank, i.e., that it can be described by a few factors. The MCM algorithm learns a predefined number of latent features (factors) per row and column that are optimized to reproduce the existing entries through vector products of the factors. Here, four latent features have proven to yield excellent results [1, 7]. We name them $u1$ to $u4$, though the numbers do not induce an order. Different starting conditions of the algorithm could result in a switch in the numbering. The latent features are called *latent*, because they are intermediate features in the mixture prediction workflow and are typically not shown in practice. However, we consider them the point of interest of the algorithm, since they contain all information within the learning algorithm for each individual substance. Subsequent processing is a trivial vector multiplication.

An explanation of the latent features could describe the learned compressed model of each substance's mixture behavior and thereby increase trust in the current model, where justified, possibly superseding the empirical model [8, 9] that is currently used in practice. Ideally, explanations also open up future models to be substance-data-driven instead of mixture-data-driven. This would alleviate a current drawback of MCM in that it, in its pure form, cannot extrapolate to substances outside the training set.

We base our explanations of the substances on a comparison with chemical knowledge captured in two additional data sets. First, a chemist has annotated each substance with its most defining chemical class. Second, we gathered a set of readily available physicochemical descriptors, e.g. molar mass, on each substance. The questions we are trying to answer throughout this chapter are:

- Is there structure in the learned latent space that is sensible to a human, i.e. does it coincide with domain knowledge?
- Are there correlations with physicochemical descriptors and properties that explain certain latent features, ideally allowing bidirectional reasoning?

Since the latent space is spanned by four dimensions, communicating its information is hard, since a direct visualization is impossible. Therefore, we rely on two interactive visual analytics tools [10, 11] that employ dimension reduction techniques to create two-dimensional and thereby viewable embeddings.

Throughout this chapter, we provide the following contributions:

- We provide an analysis of the feature space learned by MCM with two visual analytics tools regarding their relationship to two types of physicochemical knowledge in Sects. 2.3, 3.2 and 3.3.2.
- We propose an extension of a decision boundary visualization tool towards regression models in Sect. 3.3.1.

2   Rangesets

We will first introduce the challenges and possibilities of interpreting high-dimensional embeddings, present a solution with rangesets proposed in [10], and then provide a rangeset analysis of latent features in matrix completion with regard to domain knowledge.

2.1   Motivation

Reading attribute information out of high-dimensional embeddings is difficult as the reduction of dimensions aggregates the original data on typically just two viewing axes. The interpretation of these axes depends on the type of projection. Linear projections like principal component analysis (PCA) [12] as presented in Sect. 3 can still be meaningfully annotated with axes. However, the linearity in projection can also be a constraint when the original dimensionality is too high impeding cluster analysis and outlier detection. In these cases, non-linear techniques, which try to untangle the complex coherence of data points, often work better to uncover structures in high-dimensional space. Even though corresponding methods like multidimensional scaling (MDS) [13], t-distributed stochastic neighbor embedding (t-SNE) [14] and uniform manifold approximation (UMAP) [15] are commonly used in computer science and engineering fields, these techniques share that they inhibit the direct annotation of original dimension axes in projection space.

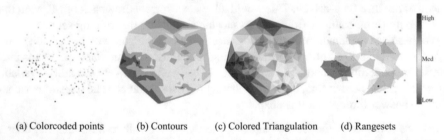

(a) Colorcoded points   (b) Contours   (c) Colored Triangulation   (d) Rangesets

Fig. 1. [10] Comparison of different augmentation strategies of an original attribute in a non-linear embedding on an exemplary dataset [16, 17]. (a) Colorcoded points require mental grouping for outlier detection. (b + c) Field-based approaches fail to capture regions with diverse values. (d) Rangesets alleviate both problems.

However, the visual retrieval of original data attributes is vital for the interpretation of these otherwise abstract plots. An augmentation of the embedding with color can provide this information. Nonato and Aupetit [18] classify augmentation strategies of non-linear dimension reductions into three main categories: Direct enrichment, spatially structured enrichment and cluster-based enrichment. In direct enrichment the layout is enriched per point [19–22]. The most common technique, color-coding each point can be seen in Fig. 1 (a). While simple to implement and understand, these techniques suffer from occlusion and overplotting, making it hard to identify clusters and respective outliers [23]. Spatially structured enrichments encode the embedding space based on a geometrical abstraction. These provide an immediate sense of attribute value distribution, but resort to averaging, as in the iso-contours of Fig. 1 (b), or fine-grained tessellation as in the triangulation of Fig. 1 (c). Cluster- or set-based approaches group points based on their visual or data-space proximity and plot abstractions of these groups [24–26]. The technique used in this chapter belongs to this third option, while integrating parts of the previous two to increase readability.

2.2   Rangeset Construction

Rangesets [10], shown in Fig. 1 (d), first bin data points with similar attribute values and then draw geometric contours based on visual proximity for a set-based visualization that captures both visual and data-space proximity. Clusters of points with similar attribute values are conveyed through non-convex α-hulls, while outliers are kept as points. Users are enabled to quickly observe structure and detect outliers.

As this approach first groups in data attribute space and then in embedding space, we outline the algorithm illustrated in Fig. 2 in the following. It is designed to show the distribution of a specific data attribute in an arbitrary (non-linear) embedding. This data attribute does not necessarily need to be considered for the creation of the embedding beforehand.

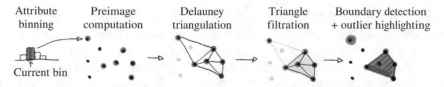

Fig. 2. [10] Key steps of the rangeset algorithm to compute contours and outliers.

As a set-based visualization, the attribute values to be displayed need to be in categories. Categorical data can be used directly, numerical data needs to be binned. For each bin, the corresponding data points are extracted and a Delauney triangulation of the filtered points is computed. From this Delauney triangulation all triangles that contain an edge longer than a defined threshold ε are removed. The remaining connected triangles form α-hulls that describe connected regions, while the unconnected points are highlighted as dots of increased size.

Both α-hulls and outliers are colored based on their respective bin. Visualizing non-linear attribute distribution as rangesets instead of as a continuous field (ref. Figure 1 (b) (c)) polygons can overlap, which is accounted for by semi-transparent rendering.

The choice of parameter ε strongly influences the visual appearance of rangesets. The effects of various ε values are shown in Fig. 3. For $\varepsilon = 0$ all points are outliers and drawn as dots, Fig. 3 (b). For small values of ε, tight contours are created with many points considered outliers, Fig. 3 (c). Larger values of ε lead to larger polygons up to the convex hull of the considered set of points. A default value is proposed in [7] based on Wilkinson [27]:

$$\varepsilon = q_{75} + 1.5 \cdot (q_{75} - q_{25})$$

With q_{25} and q_{75} being the 25th and 75th percentile of the edge lengths in the minimal spanning tree.

While the mathematical formulation of rangesets is well defined, the best parameter choice for interpretation varies based on the individual use case. The shape of rangesets depends both on the choice of ε and the choice of bins for numerical data. While default values have been stated in the previous paragraphs, users can refine bin ranges and shown attributes in an interactive browser tool called NoLiES [10]. NoLiES further provides comparison of attribute distributions via small multiples [28] and colored histograms. The tool is built in Jupyter Notebook with common plotting libraries [29–31]. A demo is available at *bndr.it/96wza* and the code at *github.com/Jan-To/nolies*.

2.3 Application to Process-Level

With the technique introduced above, we are able to collate the latent feature space of solutes learned by MCM with available chemical knowledge. We first check whether the learned structure is sensible at all through a comparison with chemical classes, then we analyze the structure of the learned latent space itself and lastly look for correlations with physicochemical substance descriptors.

2.3.1 Chemical Class as Descriptor of Learned Solute Features

The chemical sensibleness of the learned latent space spanned by *u1-u4* can be initially reviewed by the distribution of chemical classes. We know from empirically designed models that structural groups are often well-suited for characterizing the mixture behavior [8, 9]. Hence, chemical classes that are defined by these structural groups should be a good high-level descriptor to check whether the learned latent distribution correlates with expectations.

The MDS projection on latent features *u1-u4* in Fig. 3 is optimized to preserve high-dimensional distances between points in the 2D environment. Substances with similar latent feature values are generally projected closer to each other than substances with dissimilar values. Consequently, substances with the same chemical class should be close to each other as well and form visible groups. To encode this visually, chemical classes are chosen as the attribute for rangesets.

Chemical class is already a categorical variable and needs no further discretization to define the rangesets, but the filtration parameter ε is still indeterminate. Varying the values of ε confirms that first, coloring per point, Fig. 3 (a), is inferior at communicating distribution, clustering and outliers. Second, too high values of ε, Fig. 3 (c), integrate outliers into clusters, leading to inexpressive polygons. Lastly, the default ε value 0.54, Fig. 3 (b), and values slightly above it, Fig. 3 (d), give a good balance between connected components and outliers in this dataset. Further analysis is performed in this configuration.

Figure 3 (d) shows a striking coherence of chemical classes and the similarity in latent features, which constitute the positioning in the embedding. The sparsely overlapping rangeset polygons for most colors (blues, oranges, browns, light green, light purple) indicate that chemical class can be a distinct descriptor of the solute's learned latent features. The polygons for aromatics, alkanes and alkenes span a wider space and have

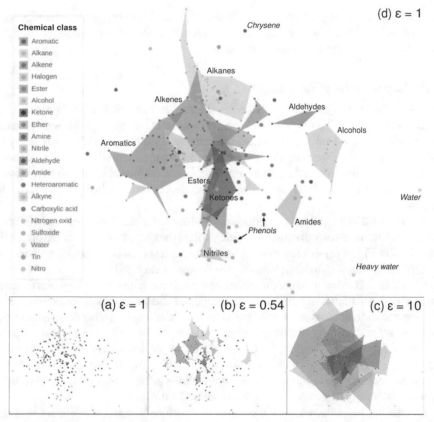

Fig. 3. Visual effect of various values for contour parameter ε. MDS-embedding of 240 solutes based on latent features *u1-u4* with rangesets colored by chemical class.

minor overlap with other classes. These classes have a common, but not unique latent feature profile. Each of the rangesets for nitriles, alcohols and aldehydes is clustered yet separated from the rest, hence indicating that for these solutes a distinct characteristic latent feature combination is learned. The polygons for ester and ketones are overlapping, indicating similar learned solute properties. All three observations fit with chemical knowledge.

Analyzing the coherence within each chemical class, we look at the outliers, highlighted by bigger dots, with respect to the same colored polygon(s). We observe that alcohols, nitriles, amides and alkanes have one or less outliers, indicating uniform latent features and hence learned solute behavior. On the other hand, aromatics generally share latent features, but aromatics like chrysene or phenol differ significantly, in line with their unique chemical structure. Water and heavy water are isolated as well, again due to their unique chemical structure.

From the analysis above, we can conclude that chemical classes generally coincide with the learned distribution on latent features. The cases where position and therefore latent feature values are ambiguous with chemical classes can mostly be explained with the chemical knowledge of an expert. For the considered set of solutes, chemical class therefore is a suitable descriptor of MCM features, even though the features were purely derived from the respective mixture behavior.

2.3.2 Latent Feature Distribution

The MDS projection used as the base for the analysis in this section is a non-linear projection technique. The tradeoff of such projections is that the high-dimensional axes are not readable anymore as there is no direct mapping. We lose the ability to quickly find high/low values, the direction along which the values are increasing and the occurring value combinations. While the point-based non-linear definition of the MDS projection forbids a perfect reconstruction of the axis, rangesets provide insight into these lost attributes.

Since the MDS projection is conducted to reduce latent features $u1$-$u4$ to two dimensions, the distribution of individual u's could explain the spatial structure of the dimension reduction. In Fig. 4 (a)–(d) the rangeset attributes are set the individual latent features discretized into five equidistant bins from very low to very high.

For $u2$ and $u3$ there are clear directions of increasing values, hinted by black arrows, which give these directions a simple meaning. For $u1$ and $u4$ the trends are non-linear and not monotonically increasing. For $u1$ the values in the orange *high* bin form a connected patch but are enclosed by and overlapping with the yellow medium bin. The deduction of $u1$ value from the embedding position is therefore ambiguous for this area. The same phenomenon occurs for $u4$ with the blue *very low* bin. We further observe a plethora of rangeset outliers in $u1$ and $u4$ that are not following the overall trend, which further hampers the ability to guess u values from the MDS projection.

Comparing the rangesets of the original dimensions in a small multiples setting in Fig. 4 also reveals common occurring feature combinations. The matching trends of increasing values from top left to bottom right in the rangesets of $u3$ and $u4$ implies a positive correlation between the dimensions. On the flipside, the trends of $u2$ and $u3$ are perpendicular and therefore uncorrelated. Comparing the patches of $u1$ and $u3$, we recognize that substances with both very high and very low $u1$ values have high or very high values in $u3$.

In essence, the analysis of the projection dimensions with rangesets unveil the lost 'axes' of the projection and their interaction, even though both can be too complex to grasp.

2.3.3 Physicochemical Descriptors of Learned Solute Features

The analysis in Sect. 2.3.1 showed that while chemical classes work as general descriptors of learned solute features, they are too coarse-grained to describe the feature combinations precisely. However, any precise correlation between readily available information and MCM features would be essential to enhance the MCM to a data-driven virtual approach. As physicochemical descriptors are available for most substances, we apply rangesets to analyze possible correlations.

Figure 4 (e) and (f) show two simple descriptors, molar mass and polarity, where correlations can be seen. As before, the properties are discretized in five equidistant bins

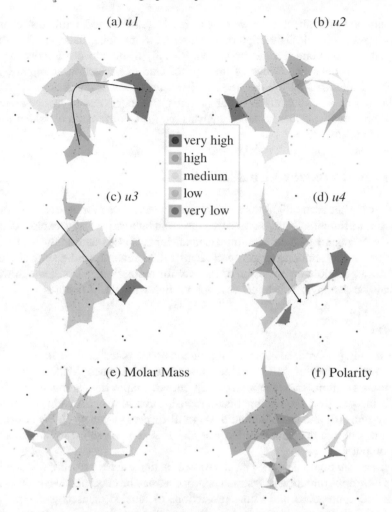

Fig. 4. 240 solutes in an MDS-embedding based on latent features *u1-u4*. Rangesets are chosen to interpret the distribution of individual latent features (a)–(d) or physicochemical descriptors of each solute (e)–(f) in the dimension reduction. Black arrows are added manually to indicate major value trends where applicable.

and $\varepsilon = 1$. Considering molar mass, the red and blue distribution of outlier points at the opposite sides with overlapping regions in the center, hints that extreme molar mass values are characteristic for solute features, but medium values are not. The findings for polarity are even more clear. The big blue polygon in Fig. 4 (f) indicates that non-polar substances share common solute features. From this region, polarity is gradually increasing with the change in similarity, analogous to *u3*, suggesting that polarity is rather continuously captured in solute features.

Some of the descriptors may be good for describing individual MCM features or at least be captured in combinations thereof. However, rangesets capture only the trends of the continuous relationship between changes in features and attributes, chemical classes or physicochemical descriptors. Rangesets are grouping data points based on their neighborhood in one specific projection. Visually filling the space between points suggests that we have knowledge of this space. Due to projection ambiguity, these neighborhood relationships are not necessarily monotonous or continuous, as seen in the overlap in Fig. 4. To get further insight which parameters need to change exactly to achieve a certain value, a more detailed analysis requires a different tool, which we will present in the next section.

3   Decision Boundary Visualization

The relationship between a high-dimensional space and a related variable can be modeled as a multivariate function. In this section we present an interactive tool to explore decision functions with regard to their high-dimensional input spaces and apply it to deepen our analysis on the relationship between latent MCM features and chemical classes. Afterwards, we propose an extension of the tool for regression analysis, which we can then expand on the physicochemical descriptors' link with MCM features.

3.1   CoFFi

Machine learning approaches span a high-dimensional space in their input or, in the application in this chapter, MCM, in the latent features. As such, MCM is considered a black box algorithm, since the relationship between input data and generated latent features is inaccessible. Explaining this relationship can improve trust in properly performing systems [6] and can point out flaws in ill-formed systems [7]. We abstract the black box model to a decision function $y = f(x)$, that can be probed for any input x to generate output y.

Visual explanations of black-box decision functions are a pressing research field that leads in various directions. Common approaches can be classified in two categories. Sample-based approaches find fitting projections of labeled datasets and explain the changes based on the contrastive juxtaposition of discrete data points. Rangesets fall into this category, but more specialized approaches exist that focus on individual regions [32–34]. The other direction is to compute visual maps by probing the input space densely in a fixed two-dimensional embedding [35]. Literature can be united under the conclusion that the interesting parts of the decision function lie where the output value changes significantly [32, 35]. Since humans internally reason by comparisons [36], counterfactual reasoning, reasoning over what would need to change to achieve a different result, is another preferable approach for explaining decision functions [37]. The visual analytics tool Counterfactual Finder (CoFFi) [11] unites sample-based analysis with visual maps and counterfactual reasoning and we therefore use it to further analyze the data at hand. While CoFFi was previously only intended for classification problems, we introduce modifications for regression analysis after a tool overview.

The interface features five components marked in Fig. 5 that are linked and interactively explorable. A data table (A) displays the data set in an accustomed fashion. The topology view (B) provides multiple algorithms for non-linear dimension reduction to assess overall class distribution and data set separability. This view is identical to rangesets with $\varepsilon = 0$ or plain glyphs colored by class affiliation. The partial dependence view (C) shows the expected outcome for changes to individual input parameters of a currently selected reference point, while holding the other parameters fixed. This uni-variate behavior analysis, which is typically called partial dependence analysis [38], is displayed as horizon charts [39], where the vertical baseline is the decision boundary. A higher prediction per class is indicated by vertical areas, each indicating 25% increase in prediction. Areas are colored according to the most probable class with more confident predictions in richer colors.

The embedding view (D) advances the partial dependence analysis to multivariate space. A PCA projection based on a local neighborhood of data points is regularly sam-pled to produce a visual map of a slice of the output space. Due to the unique properties of PCA, the original high-dimensional axis can be overlaid as a gray biplot [40] capturing feature variance and correlation. Positively correlated axes point in similar directions, negatively correlated axes in opposite directions. The necessary feature changes to reach the white decision boundaries can be read with regard to the axes. Since PCA is linear, feature values increase linearly in the direction of each axis, but do not change orthogo-nal to the axis. Decision boundaries orthogonal to axes are relying on the respective axis feature value to cross a threshold, which is shown on mouse-over. Finally, axes with little importance or of little interest can be fixed and thereby excluded from the multivariate analysis in the feature selection (E). For more details on functionality and theoretical background, we refer the reader to the original publication [8]. A demo is available at *bndr.it/cqk5w* and the open source code on Github at *github.com/Jan-To/COFFI*.

3.2 Chemical Classes in Latent Feature Space

An analysis of the class distribution in CoFFi may provide more hints on how each class is encoded in the MCM's latent feature values. The latent features *u1-u4* are the input dimensions that define the feature space and the chemical class is the output dimension that defines the color. We hope to read out which changes are necessary to flip between classes.

However, two preprocessing steps are necessary. We filter the data set to the eight most occurring classes, since saturation was previously used to expand the colormap and saturation is overloaded here as a probability indicator already. Further, MCM is by design only defined on the training samples and can therefore not be probed at intermediate samples. As this is a disadvantage of MCM we would like to overcome in the future, we train a surrogate model instead to predict the chemical class from latent features. While a surrogate model creates a continuous decision function, the exact values are only interpolated from the model and need to be taken carefully. Nonetheless, surrogate models are an established explanation approach [41] and the general decision areas are expressive enough to deduce explanations. Our experiments showed that a three-layer fully-connected neural network is able to capture all decision boundaries while keeping them simple.

We first gain an overview by centering the PCA at the data set mean in Fig. 5. Comparing the MDS embedding in (B) with the one in Fig. 3 (d) reveals that the classes alcohols (pink), alkanes (cyan), alkene (purple) and aromatics (orange) are even more

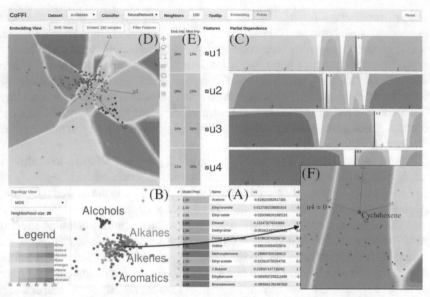

Fig. 5. CoFFi interface with chemical classes for solutes relative to latent feature space of MCM. (A) Data table (B) Non-linear projection (C) Partial dependence (D) Sampled linear projection (E) Importance distribution (F) Manually added linear projection of decision boundaries in the neighborhood of Cyclohexene. Alkanes (cyan) differ from alkenes (blue) in $u4 = 0$.

distinct than without the filtration to eight classes. The model importance (E) is highest for $u3$ and $u4$ and lowest for $u1$, which therefore seems to be less discriminative, but still relevant regarding chemical class. In Fig. 5 we find that $u2 > 1$ is characteristic for alcohols. We deduce that alcoholic mixture behavior is encoded in high $u2$ values by MCM. In both the linear (D) and non-linear projection (B), alkanes and alkenes are still neighboring as expected from their similar molecular structure. The deciding feature between the two seems to be a threshold of $u4$, since it is orthogonal to the axis. We confirm this assumption by updating the embedding view to a representative of alkenes, cyclohexene, and its neighbors in Fig. 5 (F) finding a threshold of $u4 \approx 0$.

We pick an outlier in the MDS plot to check whether the exceptional latent feature values align with the chemical expectations. 2,2,2-Trifluoroethanol is a halogen that is highly reactive and has an isolated position in MDS. In the focused plot of Fig. 6 (A) the neighboring classes are aromatics (orange) and alcohols (pink). The closeness with alcohols is sensible, since 2,2,2-Trifluoroethanol, as the name suggests, contains a hydroxyl group and therefore can be considered an alcohol as well. By probing the decision boundary with alcohols (gray cross), we learn that alcohols differ by slightly higher values in all features, which is something that was not visible in the univariate and non-linear analysis. Another interesting finding is that $u1$ and $u3$ as well as $u2$ and $u4$ are highly correlated when restricting to this local neighborhood, which is contrary to without the exclusion of alkanes and alkenes in Fig. 5 (D).

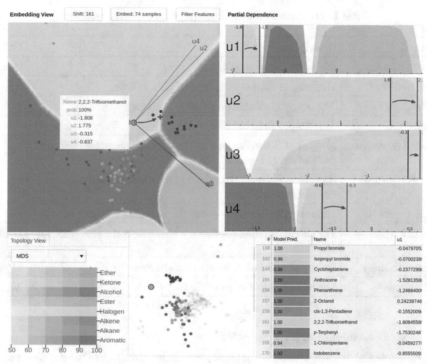

Fig. 6. Chemical classes in latent feature space focused on 2,2,2-Trifluoroethanol surrounded by similar substances. A gray cross marks the counterfactual probe on the decision boundary to alcohols. In the usual univariate analysis with partial dependence the same change, marked by dark gray lines, is not visible as a counterfactual.

3.3 Latent Features in Physicochemical Descriptor Space

A reproducible link between latent MCM features and physicochemical descriptors would significantly improve our capability to build machine learning algorithms – possibly up to the point of extrapolation. The previous rangeset analysis fell short in uncovering usable relationships. As a next step, we relate individual latent features in CoFFi to a set of readily available physicochemical descriptors of the solutes. The descriptors – dipole moment, polarizability, anisotropy, normed anisotropy, H-bond acceptance, H-bond donation, HOMO-LUMO gap, ionization energy, electron affinity and molar mass – are described in Table 1.

We again learn a surrogate model to predict the latent features from the set of physicochemical descriptors. We specifically decided against direct u-to-property scatterplots, which would avoid the surrogate model, but possibly miss higher-dimensional effects. We use the same model as in the previous section, as it proved to provide a good balance between simplicity and accuracy. CoFFi has previously only been used for classification problems, but predicting a continuous latent feature value is a regression problem. In the following section, we propose an adaptation to the existing workflow to handle regression problems in CoFFi.

Table 1. Overview over considered physicochemical descriptors.

Name	Description	Unit
Dipole moment	Strength of permanent dipoles	Debye
Polarizability	Tendency to acquire an induced dipole moment	a_0^3
Anisotropy	Anisotropy of electric polarizability tensor	
Normed anisotropy	Anisotropy / Polarizability	
H-bond acceptance	Number of H-bond acceptor sides / molar weight for molecules that can accept h-bonds	
H-bond donation	Number of H-bond donor sides / molar weight for molecules that can donate h-bonds	
HomoLumoGap	The energy difference between the molecule's frontier orbitals	eV
Ionization energy	The energy difference between a molecule and its isostructural cation	kJ/mol
Electron affinity	Energy released when an electron attaches	kJ/mol
Molar mass	The mass of 1 mol of molecules	g/mol

3.3.1 CoFFi Adaptation for Regression

The fundamental basis of counterfactual reasoning is the definition of the decision boundary. With categorical outcome variables, this boundary is trivial as the change in the most probable class. For continuous outcome variables this is no longer given. As such, we segment the continuous space into ranges that can be described together. Thereby, the regression can be handled like a classification with decision boundaries at the segment crossings.

In practice, the transformation requires some precautions. Contrary to the five equidistant bins in rangesets, the segment edges should always be designed according to the modality of the distribution. Since the edges are even more pronounced here, they need to be chosen and communicated explicitly. A histogram in the bottom left of Fig. 7 displays both the distribution and the currently chosen segment edges. In this case, we chose to set the segment to low, median and high, since there is an unimodal distribution. The physicochemical descriptors then serve as the input for the surrogate model to predict whether one specific u is within a certain segment.

3.3.2 Latent Feature Analysis

In this section, we analyze the individual latent features $u1$-$u3$ by associating the physical properties in the regression variant of CoFFi and draw conclusions on the relevance for model explanation. The analysis for $u4$ is omitted due to space constraints.

$u1$. Figure 7 shows the CoFFi interface for $u1$ containing all solutes. The distribution of $u1$ values is unimodal with the peak at 0. Therefore, $u1$ has little to no influence on the mixture computation in MCM for most solutes. We deduce that $u1$ is not encoding a common mixture behavior, but is rather specialized to encode a few rather exotic solutes.

In the MDS and PCA projections we observe that the solutes with low $u1$ values (blue) are spread on the outside of the plots, hence holding unusual physicochemical descriptor combinations. Selecting the blue dots in the MDS updates the data table to reveal that these are aromatics and esters with high organic solubility. Selection of high $u1$ values (red) reveals that these characteristics are not united in physicochemical descriptors, but are rather taken by phenol, chloroform and their variants. The model importances are rather evenly distributed within 5% and 14%, which is why we cannot deduce any dominant relationship between a particular physicochemical descriptor and the $u1$ value. We conclude that the $u1$ value is encoding mixture behavior which is not captured in the current set of physicochemical descriptors.

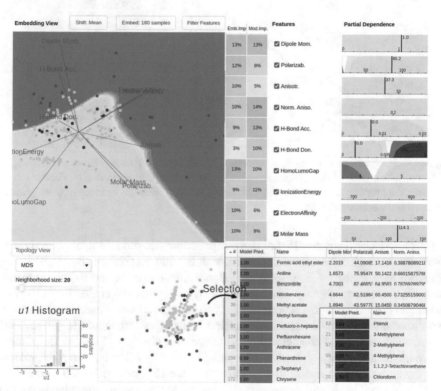

Fig. 7. Default CoFFi interface for physicochemical descriptors' influence on $u1$. $u1$ is unimodal distributed with most values close to zero. Solutes with low $u1$ values (blue) hold uncommon physicochemical descriptor values (peripheral distribution in embeddings), while phenols and chloroform hold high values (red) and blend in with the average points (yellow).

$u2$. The distribution of $u2$ is bimodal as shown in the bottom left of Fig. 8. We therefore segment into negative (blue), close-to-zero (green), medium (orange) and high (red) values. The UMAP projection shows clusters of distinct physicochemical descriptor combinations in line with the segmentation. High $u2$ values are reached by alcohols while medium values encode ketones. We select an evenly class-distributed subgroup of substances with the lasso tool in UMAP to contrast the high and medium solutes

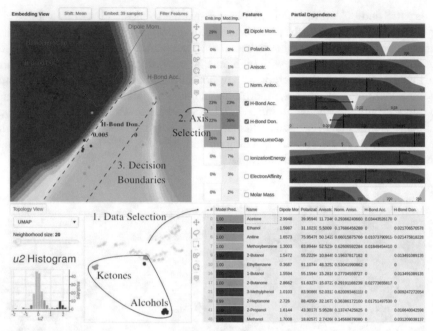

Fig. 8. CoFFi interface with physical properties related to *u2*. After selection of a balanced group of green, yellow and red data points, the other components update accordingly. Further filtration to influential property axes updates the embedding to cover only the variance in these properties centered at the selection mean.

to the close-to-zero solutes. Partial dependence view and model importance show that H-bond characteristics are most important, with H-bond acceptance above 0.018 changing from high *u2* to medium *u2* and H-Bond donation of below 0.005 changing to the main close-to-zero group (horizontal arrows). We filter the embedding view to the influential physicochemical descriptors only to extend our uni-dimensional analysis to two-dimensional dependencies. The decision boundaries orthogonal to the respective axis reveal that H-bond characteristics dominate over changes of similar magnitude in dipole moment. As we wonder about the curved boundary in the bottom right, we hover over the embedding to realize that it starts when the H-bond donation is already zero, but HomoLumoGap is still decreasing. We conclude that HomoLumoGap is therefore only influential on *u2* for solutes that are not H-bond donors, e.g. Ketones.

u3. The distribution of *u3* shown in Fig. 9 A is different from the previous ones in that all values are negative. Most values lie between −1.5 and 0 (red), with a segment of medium (yellow) and significantly negative (blue) values. The MDS and the data table in Fig. 9 B + C reveal a string cluster of medium-size alkanes (yellow) transitioning into long-chained alkanes (blue). We concentrate our analysis via selection on just this cluster to retrieve the responsible physicochemical descriptor. Polarizability (39%) and molar mass (21%) are the most influential descriptors in our surrogate model (Fig. 9 D). This explanation aligns with chemical knowledge, since polarizability and molar mass increase with chain-size in alkanes. We notice the strong but sensible correlation

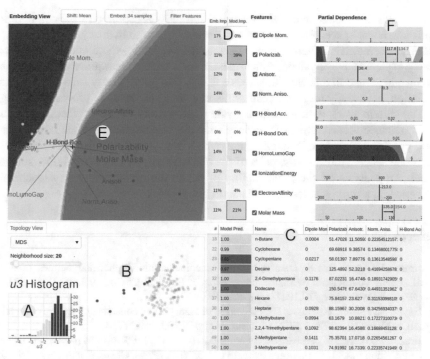

Fig. 9. CoFFi interface relating physical attributes and bins of *u3* (A). Selection of a string cluster (B) reveals that solutes with small u3 are different length alkanes (C). The model importance analysis (D) shows polarizability and/or molar mass to be the explaining descriptors. Their axes (E) coincide signaling high positive correlation, to which probing (gray cross in E)) reveals the split point between bins (gray lines in F).

in the embedding – the polarizability and molar mass axes point in exactly the same direction in Fig. 9 E – and probe the embedding (gray cross) to read in Fig. 9 F that above 135 a_0^3 polarizability and 154 g/mol molar mass (arrows) the solutes have *u3* < -2.6 (blue). However, we need to note that this is merely an explanation and not a physical dependency. Other hidden properties may be the actual influential factor, but the surrogate model identified polarizability and molar mass as specific descriptors to distinguish alkanes from each other and the other substances, which is sensible and can be used for the development of data-driven MCM algorithms.

4 Conclusion and Future Work

Matrix completion methods have proven to be more accurate than current state-of-the-art solutions for prediction of thermodynamic properties of mixtures. In this chapter, we analyzed the latent feature space of such a matrix completion model with regard to chemical knowledge. Within two interactive visual analytics tools, we were able to provide explanations for the learned solute features. We found that chemical classes coincide with the structure of the learned feature space wherever the chemical class is defining for a substance's solute behavior and that chemical similarity is captured by the neighborhood relation in latent space. Alcohols and other substances with hydroxyl groups were particularly exceptional in their learned characteristics. Finally, some latent features were clearly explained by physicochemical descriptors, while others only revealed trends. The insight gained in this chapter serves a first step towards a fully data-driven mixture prediction, potentially reducing costs while increasing accuracy in process planning.

The current work is limited in analyzing one attribute or latent feature at a time. The simultaneous correlation of physicochemical descriptors to multiple latent features could provide a global understanding into which descriptors are reflected in which part of learned space. Additionally, parts of the current analysis rely on a simple surrogate model. An analysis on the predictability of solute descriptors from physicochemical descriptors in a sophisticated model is up to future work.

References

1. Jirasek, F., Alves, R.A.S., Damay, J., et al.: Machine learning in thermodynamics: prediction of activity coefficients by matrix completion. J. Physical Chemistry Lett. **11**(3), 981–985 (2020). https://doi.org/10.1021/acs.jpclett.9b03657
2. Hayer, N., Jirasek, F., Hasse, H.: Prediction of Henry's law constants by matrix completion. AIChE J. **68**(9), e17753 (2022). https://doi.org/10.1002/aic.17753
3. Jirasek, F., Hayer, N., Abbas, R., Schmid, B., Hasse, H.: Prediction of parameters of group contribution models of mixtures by matrix completion. Phys. Chem. Chem. Phys. **25**, 1054–1062 (2023). https://doi.org/10.1039/D2CP04478A
4. Bussone, A., Stumpf, S., and O'Sullivan, D.: The role of explanations on trust and reliance in clinical decision support systems. In: Proceedings of 2015 International Conference on Healthcare Informatics. Dallas, USA, pp. 160–169 (2015). https://doi.org/10.1109/ICHI.2015.26
5. Schramowski, P., Stammer, W., Teso, S., et al.: Making deep neural networks right for the right scientific reasons by interacting with their explanations. Nat Mach Intell **2**, 476–486 (2020). https://doi.org/10.1038/s42256-020-0212-3
6. Jirasek, F., Hasse, H.: Perspective: machine learning of thermophysical properties. Fluid Phase Equilib. **549**, 113206 (2021). https://doi.org/10.1016/j.fluid.2021.113206
7. Jirasek, F., Bamler, R., Mandt, S.: Hybridizing physical and data-driven prediction methods for physicochemical properties. Chem. Commun. **56**, 12407–12410 (2020). https://doi.org/10.1039/D0CC05258B
8. Fredenslund, A., Jones, R.L., Prausnitz, J.M.: Group contribution estimation of activity coefficients in nonideal liquid mixtures. AIChE J. **21**, 1086–1099 (1975)
9. Fredenslund, A., Gmehling, J., Rasmussen, P.: Vapor-Liquid Equilibria Using UNIFAC, A Group-Contribution Method. Elsevier, Amsterdam, Netherlands (1977)
10. Sohns, J.-T., Schmitt, M., Jirasek, F., et al.: Attribute-based explanation of non-linear embeddings of high-dimensional data. IEEE TVCG **28**(1), 540–550 (2022). https://doi.org/10.1109/TVCG.2021.3114870
11. Sohns, J.-T., Garth, C., Leitte, H.: Decision boundary visualization for counterfactual reasoning. Comp. Graphics Forum (2022). https://doi.org/10.1111/cgf.14650
12. Hotelling, H.: Relations between two sets of variates. Biometrika **28**(3/4), 321–377 (1936). https://doi.org/10.2307/2333955
13. Borg, I., Groenen, P.: Modern multidimensional scaling: theory and applications. Springer Series in Statistics (2005). https://doi.org/10.1007/978-1-4757-2711-1

14. van der Maaten, L., Hinton, G.: Visualizing data using t-SNE. J. Mach. Learn. Res. **9**(86), 2579–2605 (2008)
15. McInnes, L., Healy, J.: UMAP: Uniform manifold approximation and projection for dimension reduction, ArXiv e-prints 1802.03426 (2018)
16. Aeberhard, S., Coomans, D., De Vel, O.: Comparative analysis of statistical pattern recognition methods in high dimensional settings. Pattern Recogn. **27**(8), 1065–1077 (1994)
17. Dua, D., Graff, C.: UCI Machine Learning Repository (2017)
18. Nonato, L., Aupetit, M.: Multidimensional projection for visual analytics: linking techniques with distortions, tasks, and layout enrichment. IEEE TVCG **25**, 2650–2673 (2019). https://doi.org/10.1109/TVCG.2018.2846735
19. Dowling, M., Wenskovitch, J., Fry, J., et al.: Sirius: dual, symmetric, interactive dimension reductions. IEEE TVCG **25**(1), 172–182 (2018). https://doi.org/10.1109/TVCG.2018.286 5047
20. Lee, H., Kihm, J., Choo, J., et al.: iVisClustering: an interactive visual document clustering via topic modeling. Comp. Graphics Forum **31**(3pt3), 1155–1164 (2012). https://doi.org/10. 1111/j.1467-8659.2012.03108.x
21. Lehmann, D.J., Theisel, H.: General projective maps for multidimensional data projection. Comp. Graphics Forum **35**, 443–453 (2016). https://doi.org/10.1111/cgf.12845
22. Stahnke, J., Dork, M., Müller, B., Thom, A.: Probing projections: Interaction techniques for interpreting arrangements and errors of dimensionality reductions. IEEE TVCG **22**, 629–638 (2016). https://doi.org/10.1109/TVCG.2015.2467717
23. Mayorga, A., Gleicher, M.: Splatterplots: overcoming overdraw in scatter plots. IEEE TVCG **19**, 1526–1538 (2013). https://doi.org/10.1109/TVCG.2013.65
24. Collins, C., Penn, G., Carpendale, S.: Bubble sets: revealing set relations with isocontours over existing visualizations. IEEE TVCG **15**(6), 1009–1016 (2009). https://doi.org/10.1109/ TVCG.2009.122
25. Schreck, T., Schußler, M., Zeilfelder, F., Worm, K.: Butterfly Plots for Visual Analysis of Large Point Cloud Data (2008)
26. Joia, P., Petronetto, F., Nonato, L.G.: Uncovering representative groups in multidimensional projections. Comp. Graphics Forum **34**, 281–290 (2015). https://doi.org/10.1111/cgf.12640
27. Wilkinson, L., Anand, A., Grossman, R.: Graph-theoretic scagnostics. IEEE INFOVIS 2005, 157–164 (2005)
28. Tufte, E.: Envisioning Information. Graphics Press 67 (1990)
29. Bokeh Development Team: Bokch: Python Library for Interactive Visualization (2020)
30. Gillies, S., et al.: Shapely: Manipulation and Analysis of Geometric Objects (2007)
31. P. D. Team. Panel: A High-level App and Dashboarding Solution for Python
32. Ma, Y., Maciejewski, R.: Visual analysis of class separations with locally linear segments. IEEE TVCG **27**(1), 241–253 (2021). https://doi.org/10.1109/TVCG.2020.3011155
33. Tatu, A., Maass, F., Färber, I., et al.: Subspace search and visualization to make sense of alternative clusterings in high-dimensional data. IEEE VAST, pp. 63–72 (2012). https://doi. org/10.1109/VAST.2012.6400488
34. Jeong, D.H., Ziemkiewicz, C., Fisher, B., et al.: IPCA: an interactive system for PCA-based visual analytics. Comp. Graphics Forum **28**, 767–774 (2009). https://doi.org/10.1111/j.1467-8659.2009.01475.x
35. Espadoto, M., Rodrigues, F., Telea, A.: Visual analytics of multidimensional projections for constructing classifier decision boundary maps. Int. .erence Inf. Visualization Theory Appl. **10**, 28–38 (2019). https://doi.org/10.5220/0007260800280038
36. Lipton, P.: Contrastive explanation. R. Inst. Philos. Suppl. **27**, 247–266 (1990). https://doi. org/10.1017/S1358246100005130

37. Wachter, S., Mittelstadt B.D., Russell, C.: Counterfactual explanations without opening the black box: automated decisions and the GDPR. ArXiv e-prints (2017). http://arxiv.org/abs/1711.00399

38. Friedman, J.H.: Greedy function approximation: a gradient boosting machine. The Annals of Statistics **29**(5), 1189–1232 (2001). ISSN: 00905364

39. Few, S.: Time on the Horizon (2008). http://www.perceptualedge.com/articles/visual_business_intelligence/time_on_the_horizon.pdf

40. Gabriel, K.R.: The biplot graphic display of matrices with application to principal component analysis. Biometrika **58**(3), 453–467 (1971). ISSN: 00063444

41. Ribeiro, M.T., Singh, S., Guestrin, C.: "Why should i trust you?": explaining the predictions of any classifier. ACM KDD **16**, 1135–1144 (2016). https://doi.org/10.1145/2939672.2939778

Insight into Indentation Processes of Ni-Graphene Nanocomposites by Molecular Dynamics Simulation

V. H. Vardanyan$^{(\boxtimes)}$ and H. M. Urbassek

Physics Department and Research Center OPTIMAS, RPTU, Erwin-Schrödinger-Straße, 67663 Kaiserslautern, Germany
vardanya@rhrk.uni-kl.de

Abstract. Molecular dynamics simulations provide insight into the processes underlying material plasticity and hard-ness. We demonstrate its uses here for the special case of a metal-matrix nanocomposite, viz. Ni-graphene. A series of increasingly more complex simulation scenarios is established, starting from a single-crystalline matrix over bi-crystal samples to fully polycrystalline arrangements. We find that the nanocomposite is weaker than the single-crystalline metal, since the graphene flakes are opaque to dislocation transmission and thus constrain the size of the dislocation network produced by the indenter. However, the flakes increase the hardness of a polycrystalline metal matrix. This is caused by dislocation pile-up in front of the flakes as well as dislocation absorption (annihilation) by the graphene flakes.

1 Introduction

The topic of physical modeling in the framework of the graduate school *Physical Modeling for Virtual Manufacturing Systems and Processes* spans many spatial and temporal scales. On the finest scale, atomistic modeling techniques can be used, of which the method of molecular dynamics simulation is an important example. Atomistic modeling allows one to obtain insights into the basic processes underlying manufacturing. In this chapter, we present an example of the potential of atomistic simulation to identify basic proccsscs undcrlying machining of materials.

In materials-science simulation studies, surface machining processes are often sub-divided into the classes of indentation, scratching and cutting. Of these, indentation is the most basic process, as it only requires penetration of a tool in normal direction into the workpiece, but not any lateral movement. We therefore focus in this chapter on the indentation of a tool into the workpiece. In order to allow comparison with the majority of other studies, the tool will be assumed to be spherical; or, in other words, the part of the tool that actually penetrates the surface is considered to be curved with a fixed radius of curvature. Also, the tool will be considered to be rigid; this is a model of a hard diamond indenter and simplifies the discussion, since all aspects of tool wear are excluded.

© The Author(s) 2023
J. C. Aurich et al. (Eds.): IRTG 2023, *Proceedings of the 3rd Conference on Physical Modeling for Virtual Manufacturing Systems and Processes*, pp. 51–69, 2023.
https://doi.org/10.1007/978-3-031-35779-4_4

Indentation processes have been simulated for many materials using molecular dynamics, including metals [1–11] and ceramics [12–18]. Such studies provided information on the mechanisms underlying material plasticity during indentation, see Refs. [19, 20] for a review. However, the plasticity in composite materials differs from that in homogeneous materials and still poses open questions, such as to the pile-up, transmission and absorption of dislocations at interfaces. We therefore investigate here graphene-reinforced metal-matrix materials, focusing on Ni-graphene nanocomposites. Such graphene-metal composites have excellent mechanical properties [21–28] as they combine the superior mechanical properties of graphene [29, 30] with the ductility of the matrix metal.

The plasticity in these nanocomposites is based on dislocation nucleation and migration in the matrix metal and on how these dislocation-based processes are influenced by the graphene flakes [31–33]. These issues are well suited to investigation by molecular dynamics simulation [25, 34–41]. Previous studies focused on the improvement of the composite hardness in the metal matrix [34, 36, 39, 42, 43]. In recent work [37, 38, 44], we studied the effects of the graphene interfaces in absorbing dislocations and hindering their propagation; the present chapter reviews these studies.

2 Method

The Ni blocks used for the simulations have typical sizes of 40 nm × 40 nm in lateral directions and a depth of 30 nm, containing around 5×10^6 atoms. They are filled with graphene flakes of various sizes and morphologies; the exact configurations vary with each simulation and are described in the following sections. The indenter is modeled as a rigid, hard and frictionless sphere of radius $R = 5$ nm which penetrates in normal direction into the sample with a velocity of 20 m/s.

Ni atoms interact with each other by a many-body interaction potential developed by Mishin et al. [45]. The interaction among C atoms is modeled by the so-called adaptive intermolecular reactive empirical bond order (AIREBO) potential of Stuart *et al.* [46]. The interaction between C and Ni is described by a pairwise Lennard-Jones potential,

$$V(r) = 4\varepsilon\left[(\sigma/r)^{12} - (\sigma/r)^6\right], \tag{1}$$

with length parameter σ and energy parameter ε. These parameters are provided by Huang et al. [47] as $\varepsilon = 23.049$ meV and $\sigma = 2.852$ Å. Finally, the interaction of the indenter with the sample (Ni or C) atoms is modeled by a repulsive potential as proposed by Kelchner *et al.* [48],

$$V(r) = k(R - r)^3, \tag{2}$$

for all atoms with distance r to the indenter center smaller than the indenter radius R. The indenter stiffness has been set to $k = 10$ eV Å^{-3}, following Refs. [5, 48].

The simulations are performed with the open-source code LAMMPS [49] using a constant time step of 1 fs. The indentation is performed in the so-called displacement-controlled mode; i.e., the indenter position is advanced each time step according to its velocity; then the resulting forces on the sample atoms are calculated and the atom positions are updated. Further details on the molecular dynamics method used in nanoindentation are found in Refs. [19, 37, 38, 44].

Common-neighbor analysis [50] is used to identify the crystalline structure and the dislocation detection algorithm DXA [50–52] to determine the length of dislocations. The software OVITO [53] is used to visualize the simulation results.

3 Ni Single Crystal

In a first step of our molecular dynamics simulations, a Ni single crystal with a (111) surface is indented with an indenter of radius $R = 5$ nm. Figure 1a shows the corresponding force-depth curve. It is characterized by an elastic part extending up to around 0.7 nm indentation following the Hertzian $F \sim d^{3/2}$ behavior; the ensuing load drop is caused by the formation of dislocations which relieves the stress exerted on the substrate. With continuing indentation, the force increases; the fluctuations are caused by the stochastic nature of dislocation generation and emission. The defects generated in the Ni material after an indentation to 4.1 nm are displayed in Fig. 1b. The stacking faults visible in the environment of the indentation pit are the traces left over after the movement of dislocations away from the high-stress environment of the indenter.

We contrast these findings with the indentation of a Ni (111) single crystal into which a graphene flake is embedded at a depth of 3 nm. The flake extends on all sides 17 nm from the indentation point, such that its extension can be considered 'infinite' in lateral direction in the sense that dislocations created by the indentation process cannot interact with the flake boundaries. The force-depth curve, Fig. 1a, shows that almost at all indentation depths, less force is required to indent the graphene-loaded Ni matrix than the pure Ni crystal; in other words, the Ni-graphene nanocomposite is *weaker* than the pure Ni crystal. Figure 1b shows that the graphene layer prevents dislocations to be created in the sub-graphene region and even prevents the migration of dislocations into that region. We conclude that the blocking of dislocation transmission reduces the force necessary for indentation.

The role of the edges of the graphene flakes is investigated in a further series of simulations. As Fig. 2 shows, here 9 different systems are studied. All of them are based on a Ni single-crystalline matrix with a (100) surface. The depth below the surface at which the flake is situated is varied from 3 nm ('top') to 5 nm ('middle') and 10 nm ('bottom'). Also the lateral position of the flake is varied; if it ends below the center of the indenter, it is denoted as system '1'; if it extends beyond the center by 5.3 nm or 10.7 nm, as system '2' or '3'.

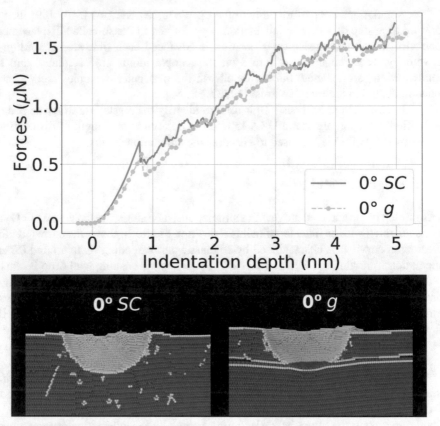

Fig. 1. Comparison of the indentation into single-crystalline (*SC*) Ni with Ni containing a graphene flake (*g*) at a depth of 3 nm. (a) Dependence of the indentation force on depth. (b) Microstructure formed at an indentation depth of $d = 4.1$ nm, immediately before dislocation nucleation in the lower Ni block. Atoms are colored according to common-neighbor analysis. Purple: fcc; red: stacking faults; cyan: other defects; brown: graphene. Data taken from Ref. [38] under CC BY 4.0.

Figure 3 exemplifies the evolution of the dislocation network for the systems '2', in which the flake extends roughly by the size of the indenter radius beyond the indenter and compares it with that for a pure Ni matrix. The reference case, pure Ni, follows that of previous studies of indentation into face-centered cubic (fcc) metals [5, 54–56]. The generated dislocation network is characterized by the ejection of loops in the <110> glide directions of the fcc matrix; in addition, a dense network adherent to the indent pit forms.

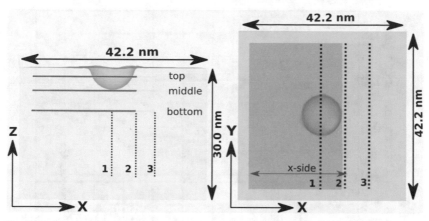

Fig. 2. Side view (left) and top view (right) of the setup of the simulation system. x and y are the cartesian directions parallel to the surface, and z normal to the surface. 9 different positions of the graphene flake (brown) are simulated, which differ in their depth below the surface ('top', 'middle', and 'bottom') as well as in their lateral extension in x direction, designated by '1', '2', and '3'. Taken with permission from Ref. [37].

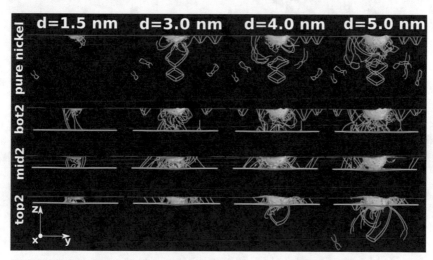

Fig. 3. Side views onto the y–z plane of the indented pure Ni sample, as well as the *bot2, mid2,* and *top2* Ni-graphene systems for indentation depths of $d = 1.5, 3, 4$, and 5 nm; the view direction is along the $-x$ axis, see Fig. 2. The volume is cut immediately under the indenter such that only half of the plastic zone is visible for clarity. The light blue color highlights the graphene flake as well as the indentation pit. Dislocations are colored according to their Burgers vector. Green: 1/2 <112>; dark blue: 1/2 <110>; pink: 1/6 <110>; yellow: 1/3 <001>; red: other. Taken with permission from Ref. [37].

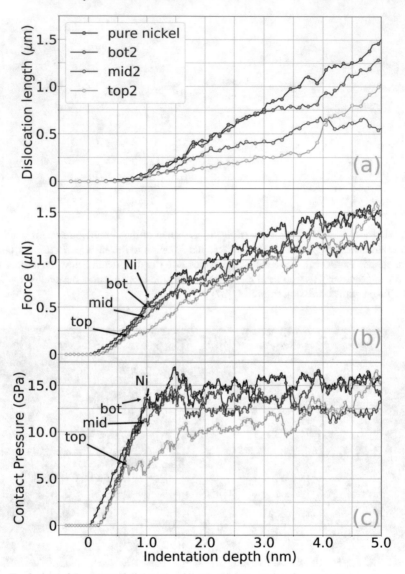

Fig. 4. Evolution of the (a) total dislocation length, (b) indentation force, and (c) contact pressure with indentation depth. The first load drops are indicated in (b) and (c) by arrows. Taken with permission from Ref. [37].

The insertion of a graphene flake stops the emission of dislocation loops and also constrains the evolution of the network adherent to the indent pit. A novel situation arises for the 'top' conformation of the flake, where it is situated only 3 nm below the surface. After the indenter is pushed deeper into the material – $d = 4$ and 5 nm in Fig. 3 – dislocations are generated also in the lower part of the Ni matrix. Note that the flake does not rupture; the indenter does never touch the Ni material below the flake, but the stresses it generates are sufficient to lead to dislocation formation and even loop emission into the lower Ni matrix. These simulations thus show that the main role of the graphene flake consists in constraining the dislocation networks building up in the Ni matrix. The dislocation network, i.e., the plastic zone generated by the indenter, adapts its geometry to the non-transmitting flake.

Figure 4a demonstrates that not only the geometry, but also the total length of dislocations in the plastic zone is affected by the presence of the flake. In all cases, the total length of the dislocations is reduced with respect to that in the pure Ni matrix; the amount of reduction is determined by the distance of the flake from the surface. For the 'bottom' geometry, the influence of the flake is only noticeable for indentation depths beyond 3 nm, while for the'top' geometry, already at $d = 1$ nm, a reduction in dislocation length is visible. These data quantify the qualitative information provided in Fig. 3 of the influence of the flake position on the generated dislocation network.

Figures 4b and c show how the normal force and the build-up of contact pressure during indentation are affected by the presence of the graphene flake. The contact pressure is determined as the ratio of the normal force to the contact area of the tip. The normal forces show a clear ordering with the position of the flake, at least for indentation depths up to 3 nm: the farther the flake is buried inside the matrix, the higher the load that the system can carry. These data thus confirm and extend the conclusions made with the discussion of Fig. 1b. Since we use the same indenter in all these studies, this result also carries through to the contact pressure, Fig. 4c, and as the average contact pressure can be identified with the hardness of the sample, we find that the hardness of the nanocomposite material is decreased if the flake is positioned closer to the surface.

4 Ni Bi-crystal

In real composites, graphene flakes will usually not be incorporated within a single-crystalline matrix, but favor sites at grain boundaries. We investigated such a scenario by studying Ni bi-crystals and inserting a graphene flake in the grain boundary. To be specific, we stayed with a Ni(111) crystal; a twist grain boundary is inserted at a depth of 3 nm, such that the Ni crystal in the bottom grain also has a (111) surface, which is, however, twisted with respect to the top grain by an angle θ. We investigated the cases of $\theta = 30°$ and $60°$ and also compared to the reference case of $\theta = 0°$ (no grain boundary). The $\theta = 60°$ tilt boundary was selected as it is the lowest-energy grain boundary in fcc Ni [57]; it constitutes actually a coherent $\Sigma 3$ twin boundary and has a specific energy of only 0.06 Jm^{-2}. On the other hand, the $\theta = 30°$ twist boundary has the highest energy among all (111) twist boundaries; its specific energy amounts to 0.49 Jm^{-2} [57]. In the following, we will also denote the $\theta = 30°$ grain boundary as the *high-energy* grain boundary, since its defect energy is high, and the $\theta = 60°$ grain boundary as the *low-energy* grain boundary since its defect energy is low and the material hence more closely resembles an ideal Ni crystal.

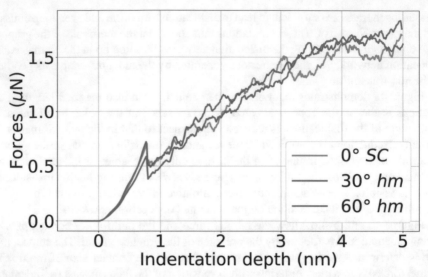

Fig. 5. Comparison of the force-depth curves for indentation into bi-crystals containing a 30° or a 60° twist grain boundary (*hm*) with that of a single crystal (*SC*). Data taken from Ref. [38] under CC BY 4.0.

Let us first focus on the effect of a graphene-free grain boundary on indentation into pure Ni; these pure Ni bi-crystals will be denoted as *hm* (homointerface) systems. Figure 5. Shows the force-depth curve for the three systems and Fig. 6 visualizes the dislocations generated. Clearly, the high-energy $\theta = 30°$ grain boundary prevents dislocations from crossing, while dislocation transmission is possible for the low-energy $\theta = 60°$ grain boundary. These findings are in agreement with earlier simulation work on fcc metals [58–61]. Interestingly, the system with the low-energy grain boundary, $\theta = 60°$, requires the highest load for indentation. This observed 'hardening' may be attributed to the fact that dislocation transmission through the twin boundary requires external strain [58, 59, 61, 62].

In contrast, the high-energy $\theta = 30°$ grain boundary requires less force for indentation, as observed in Fig. 5, at least until the indenter touches the grain boundary at d = 3 nm. Here, dislocation absorption at the grain boundary decreases the hardness of the Ni matrix. This behavior is in agreement with previous simulations [63, 64].

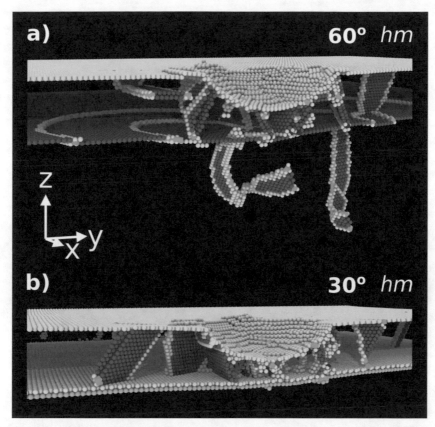

Fig. 6. Dislocation network building up after indentation to a depth of 2.1 nm into a Ni bi-crystal containing (a) a 60° and (b) a 30° twist grain boundary at 3 nm depth. Atoms are colored according to common-neighbor analysis. Red: stacking faults; yellow (cyan): defective atoms in the upper (lower) Ni grain. Fcc atoms have been removed for clarity. Green lines show Shockley partials. Data taken from Ref. [38] under CC BY 4.0.

The effect of filling the low-energy $\theta = 60°$ grain boundary with a graphene flake is shown in Fig. 7. Figure 7a demonstrates that the effect of graphene is to considerably decrease the force necessary for indentation; the material is thus weakened. As the graphene flake blocks the transmission of dislocations through the grain boundary, Fig. 7b, this effect is even stronger than what was observed for the Ni single-crystal, Fig. 1. The blocking of dislocation transmission by the graphene sheet is responsible for the reduced force necessary for indentation of the graphene-Ni composite.

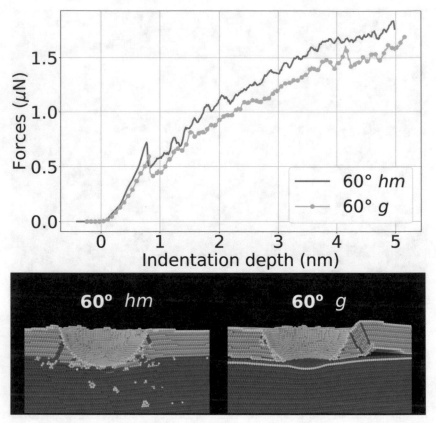

Fig. 7. Comparison of the indentation into a Ni bi-crystal with a 60° twist boundary without (*hm*) and with (*g*) graphene. (a) Evolution of force with indentation depth. (b) Snapshots showing the microstructure at an indentation depth of $d = 4.2$ nm. Atoms are colored according to common-neighbor analysis. Purple: fcc (lower Ni block); gold: fcc (upper Ni block); red: stacking faults; cyan: other defects; brown: graphene. Data taken from Ref. [38] under CC BY 4.0.

For the high-energy $\theta = 30°$ grain boundary, the effect of graphene coating the grain boundaries is negligible, see Fig. 8a. The reason hereto is that dislocations cannot cross this grain boundary even if graphene is absent, cf. Figure 8b; hence the effect of graphene on dislocation transmission is absent and consequently also the effect on the force-depth curve and the material hardness.

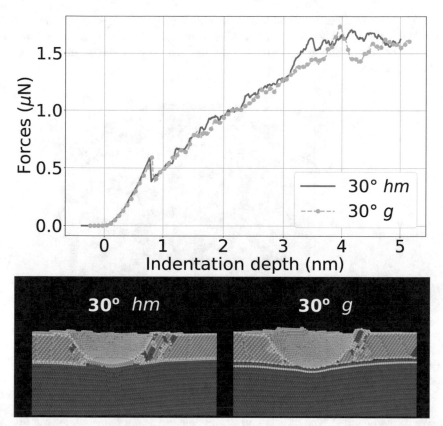

Fig. 8. Comparison of the indentation into a Ni bi-crystal with a 30° twist boundary without (*hm*) and with (*g*) graphene. (a) Evolution of force with indentation depth. (b) Snapshots showing the microstructure immediately before dislocation nucleation in the lower Ni block: at $d = 3.4$ nm (*hm*) and at $d = 4.0$ nm (*g*). Atoms are colored as in Fig. 7. Data taken from Ref. [38] under CC BY 4.0.

5 Ni Polycrystal

Finally, molecular dynamics simulation was used to study the effects of coating grain boundaries with graphene in a Ni polycrystal. The polycrystal was created using the method of rapid quenching from the melt [65, 66]; by choosing the quench rate of 1 K/ps, we obtain a Ni polycrystal with an average grain size of 4.1 nm. Figure 9 shows the final structure of the polycrystalline Ni matrix after quenching.

In order to create a Ni-graphene polycrystalline nanocomposite, 64 square graphene flakes with a side length 2.6 nm were added to the melt before quenching. We observed that the morphology of these flakes in the melt changed considerably during the 14 ns equilibration time at 2300 K. During this time, the flakes are rather mobile and change their curvature and their positions in the Ni melt. They tend to roll up and assume a wrinkled or folded structure as shown in Fig. 10 right; the C-C attraction of flake atoms stabilizes the tubular structures similar as in nanotubes. We denote this structure as the *wrinkled morphology*; it survives the quench process and is also seen in the nanocomposite.

Fig. 9. Polycrystalline structure of Ni after quenching from the melt. Colors represent different grains. Taken with permission from Ref. [44].

We created another type of nanocomposite denoted as *flat morphology*, see Fig. 10 left. This was possible by increasing the C-Ni attraction of the flake edge atoms by a factor of around 10; more precisely, we chose $\varepsilon = 200$ meV and $\sigma = 1.514$ Å in Eq. (1). This is justified by the fact that the twofold coordinated C atoms at the flake edges are free to form covalent bonds with surrounding Ni atoms; the Lennard-Jones parameters used here were provided by Tavazza *et al.* [67] in a density-functional-theory (DFT) study of the Ni-C interaction. As Fig. 10 left shows, this modification lets the flakes keep a more planar structure in the melt and the creation of tubular structures is avoided.

We determine the hardness in these nanocomposites using nanoindentation with an $R = 4$ nm indenter. In order to take the spatial inhomogeneity of the systems into account, 5 replicas of the nanocomposite systems were created containing free surfaces; these free surfaces were indented at a total of 50 points. These 50 indentations thus varied in the local stoichiometry of the C content, in the local grain size and orientation and in the vicinity and structure of the grain boundaries. For reference, we note that our polycrystalline Ni exhibits a hardness of 9.95 ± 0.13 GPa.

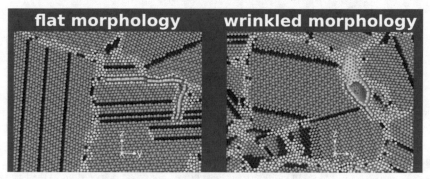

Fig. 10. Crystallographic structure of Ni-graphene composites with (left) flat morphology and (right) wrinkled morphology. Atoms are colored according to common-neighbor analysis. Green: fcc nickel; dark blue: stacking faults in nickel; yellow: grain boundaries; light blue: graphene; red: edge atoms of graphene flakes. Taken with permission from Ref. [44].

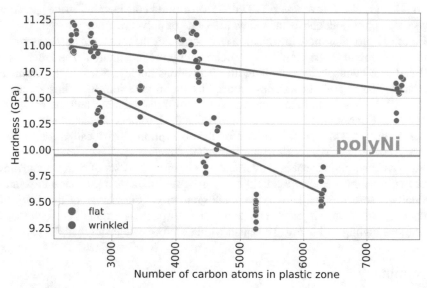

Fig. 11. Dependence of composite hardness on the number of carbon atoms in the plastic zone. Lines are to guide the eye. The gray line gives the average hardness of polycrystalline Ni. Taken with permission from Ref. [44].

Figure 11 shows the result of these hardness measurements. The main observation is that nanocomposite systems with flat graphene are systematically stronger than those with wrinkled graphene flakes; the difference is sizable and amounts to 10%. In other words, wrinkled graphene offers less resistance against deformation than flat graphene. This is plausible, since out-of-plane deformations, i.e., curving, folding and crumpling, can be effected with small forces while in-plane deformation of graphene is considerably harder. Note that the hardness of all flat nanocomposites (average of 10.86 GPa) lies above that of pure polycrystalline Ni with a comparable grain size, while the hardness

t = 165.0 ps   t = 165.5 ps   t = 166.0 ps

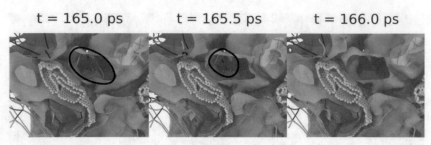

Fig. 12. Dislocation interaction with a graphene flake during indentation of the wrinkled composite. Light blue atoms represent folded graphene flakes. Atoms of Ni are removed for clarity. The gray shades represent grain boundaries. Dislocations are colored according to their Burgers vector. Green: $1/2 <112>$; dark blue: $1/2 <110>$; red: other. The black loop highlights a set of dislocations that are eventually absorbed at the graphene flake. Taken with permission from Ref. [44].

of wrinkled graphene (average of 10.04 GPa) exhibits considerably more spread and shows data points both above and below the value of polycrystalline pure Ni.

Figure 11 provides more information as it correlates the measured hardness with the amount of C found in the plastic zone generated by the indentation. The latter zone is identified as a hemisphere containing all dislocations nucleated inside the grains and shear activation in grain boundaries induced by the indentation. Its radius amounts to roughly 10–15 nm for the indenter radius used in the simulation and the composites studied here. For both morphologies, we find that the hardness decreases with the local graphene content. This effect is caused by the absorption of dislocations at the flakes discussed above.

Such an absorption event is shown in detail in Fig. 12. A set of dislocations moves under the inhomogeneous stress field of the indenter towards a graphene flake and is eventually absorbed there. Due to the small grain sizes, the change of the dislocation network – nucleation, migration and eventual absorption – is quite fast such that the dislocations highlighted in Fig. 12 vanish within a time scale faster than 0.5 ps.

6 Summary

In this chapter, we gave an example of how the method of molecular dynamics simulation allows to obtain detailed insights into the plastic processes underlying nanoindentation. Even for a complex material – in this example a Ni-graphene nanocomposite – the generation of dislocations in the metal matrix and their propagation could be studied in detail and the effects on the material hardness could be quantified.

For this example, a series of increasingly more complex simulation scenarios was established, starting from a single-crystalline matrix over bi-crystal samples to fully polycrystalline arrangements. This series of simulations allowed us to draw the following conclusions.

1. Single-crystalline Ni is harder than a Ni-graphene nanocomposite. The nanocomposite hardness decreases as the graphene flake is situated closer to the surface.

2. The graphene flake is opaque to dislocation transmission and thus constrains the size of the dislocation network produced by the indenter. This is the cause for the decreased hardness of the nanocomposite.
3. Grain boundaries in the metal matrix generally hinder dislocation transmission and therefore reduce the material hardness. An exception is given by low-energy grain boundaries – such as twin boundaries – which are partially transparent to dislocations.
4. If graphene coats a single grain boundary – such as in a bi-crystal matrix – the dislocation transmission is further reduced resulting in a decrease in hardness.
5. In a polycrystalline metal matrix, coating of the grain boundaries leads to an increase of the material hardness. This is caused by dislocation pile-up in front of the flakes [31, 32, 68]. For high C concentrations in the plastic zone, also dislocation absorption (annihilation) by the graphene flakes is observed.
6. The morphology of the graphene flakes has a strong effect on the nanocomposite hardness. While flat flakes generally increase the hardness, wrinkled flakes have less effect or even reduce the composite hardness.

Further studies are required in particular in order to relate the nanocomposite hardness to the graphene concentration. Experiments find that there is an optimum graphene concentration for improving the composite hardness [69, 70], while in simulations the hardness generally decreases with graphene content, cf. Figure 11. The verification and explanation of the optimum graphene concentration thus poses an interesting problem for atomistic simulation.

Acknowledgments. We acknowledge support by the Deutsche Forschungsgemeinschaft (DFG, German Research Foundation) – project number 252408385 – IRTG 2057. Simulations were performed at the High Performance Cluster Elwetritsch (RHRK, TU Kaiserslautern, Germany).

References

1. Van Vliet, K.J., Li, J., Zhu, T., Yip, S., Suresh, S.: Quantifying the early stages of plasticity through nanoscale experiments and simulations. Phys. Rev. B **67**, 104105 (2003)
2. Ma, X.-L., Yang, W.: Molecular dynamics simulation on burst and arrest of stacking faults in nanocrystalline Cu under nanoindentation. Nanotechnology **14**, 1208 (2003)
3. Ting Zhu, J., Li, K.J., Vliet, V., Ogata, S., Yip, S., Suresh, S.: Predictive modeling of nanoindentation-induced homogeneous dislocation nucleation in copper. J. Mech. Phys. Sol. **52**, 691 (2004)
4. Ziegenhain, G., Urbassek, H.M.: Effect of material stiffness on hardness: a computational study based on model potentials. Philos. Mag. **89**, 2225–2238 (2009)
5. Ziegenhain, G., Hartmaier, A., Urbassek, H.M.: Pair vs many-body potentials: influence on elastic and plastic behavior in nanoindentation of fcc metals. J. Mech. Phys. Sol. **57**, 1514–1526 (2009)
6. Ziegenhain, G., Urbassek, H.M., Hartmaier, A.: Influence of crystal anisotropy on elastic deformation and onset of plasticity in nanoindentation: a simulational study. J. Appl. Phys. **107**, 061807 (2010)

7. Hagelaar, J.H.A., Bitzek, E., Flipse, C.F.J., Gumbsch, P.: Atomistic simulations of the formation and destruction of nanoindentation contacts in tungsten. Phys. Rev. B **73**, 045425 (2006)
8. Biener, M.M., Biener, J., Hodge, A.M., Hamza, A.V.: Dislocation nucleation in bcc Ta single crystals studied by nanoindentation. Phys. Rev. B **76**, 165422 (2007)
9. Alcala, J., Dalmau, R., Franke, O., Biener, M., Biener, J., Hodge, A.: Planar defect nucleation and annihilation mechanisms in nanocontact plasticity of metal surfaces. Phys. Rev. Lett. **109**, 075502 (2012)
10. Naveen Kumar, N., Tewari, R., Durgaprasad, P.V., Dutta, B.K., Dey, G.K.: Active slip systems in bcc iron during nanoindentation: a molecular dynamics study. Comput. Mater. Sci. **77**, 260 (2013)
11. Ruestes, C.J., et al.: Atomistic simulation of tantalum nanoindentation: effects of indenter diameter, penetration velocity, and interatomic potentials on defect mechanisms and evolution. Mat. Sci. Eng. A **613**, 390–403 (2014)
12. Szlufarska, I., Kalia, R., Nakano, A., Vashishta, P.: Atomistic mechanisms of amorphization during nanoindentation of SiC: a molecular dynamics study. Phys. Rev. B **71**, 174113 (2005)
13. Chen, H.-P., Kalia, R.K., Nakano, A., Vashishta, P., Szlufarska, I.: Multimillion-atom nanoindentation simulation of crystalline silicon carbide: orientation dependence and anisotropic pileup. J. Appl. Phys. **102**, 063514 (2007)
14. Szlufarska, I., Kalia, R.K., Nakano, A., Vashishta, P.: A molecular dynamics study of nanoindentation of amorphous silicon carbide. J. Appl. Phys. **102**, 023509 (2007)
15. Mishra, M., Szlufarska, I.: Possibility of high-pressure transformation during nanoindentation of SiC. Acta Mater. **57**, 6156–6165 (2009)
16. Dong Earn Kim and Soo Ik Oh: Deformation pathway to high-pressure phases of silicon during nanoindentation. J. Appl. Phys. **104**, 013502 (2008)
17. Zhang, Z., Stukowski, A., Urbassek, H.M.: Interplay of dislocation-based plasticity and phase transformation during Si nanoindentation. Comput. Mater. Sci. **119**, 82–89 (2016)
18. Luo, X., et al.: Atomistic simulation of amorphization during AlN nanoindentation. Ceram. Int. **47**, 15968–15978 (2021)
19. Ruestes, C.J., Bringa, E.M., Gao, Y., Urbassek, H.M.: Molecular dynamics modeling of nanoindentation. In: Tiwari, A., Natarajan, S. (eds.) Applied Nanoindentation in Advanced Materials (Wiley, Chichester, UK, 2017), Chap. 14, pp. 313–345 (2017)
20. Ruestes, C., Alhafez, I., Urbassek, H.: Atomistic studies of nanoindentation—a review of recent advances. Crystals **7**(10), 293 (2017). https://doi.org/10.3390/cryst7100293
21. Ramanathan, T., et al.: Functionalized graphene sheets for polymer nanocomposites. Nat. Nanotechnol. **3**(6), 327–331 (2008). https://doi.org/10.1038/nnano.2008.96
22. Zhang, P., et al.: Fracture toughness of graphene. Nat. Commun. **5**, 3782 (2014)
23. Xiong, D.-B., et al.: Graphene-and-copper artificial nacre fabricated by a preform impregnation process: bioinspired strategy for strengthening-toughening of metal matrix composite. ACS Nano **9**, 6934–6943 (2015)
24. Yang, Z., Wang, D., Zixing, L., Wenjun, H.: Atomistic simulation on the plastic deformation and fracture of bio-inspired graphene/Ni nanocomposites. Appl. Phys. Lett. **109**, 191909 (2016)
25. Liu, X.Y., Wang, F.C., Wang, W.Q., Wu, H.A.: Interfacial strengthening and self-healing effect in graphene-copper nanolayered composites under shear deformation. Carbon **107**, 680–688 (2016)
26. Guo, Q., Kondoh, K., Han, S.M.: Nanocarbon-reinforced metal-matrix composites for structural applications. MRS Bull. **44**(1), 40–45 (2019). https://doi.org/10.1557/mrs.2018.321
27. Feng, Q., Song, X., Xie, H., Wang, H., Liu, X., Yin, F.: Deformation and plastic coordination in WC-Co composite – molecular dynamics simulation of nanoindentation. Mater. Des. **120**, 193–203 (2017)

28. Shuang, F., Aifantis, K.E.: Modelling dislocation-graphene interactions in a bcc Fe matrix by molecular dynamics simulations and gradient plasticity theory. Appl. Surf. Sci. **535**, 147602 (2021)
29. Lee, C., Wei, X., Kysar, J.W., Hone, J.: Measurement of the elastic properties and intrinsic strength of monolayer graphene. Science **321**, 385 (2008)
30. Cao, K., et al.: Elastic straining of free-standing monolayer graphene. Nat. Commun. **11**, 284 (2020)
31. Wang, J., Misra, A.: An overview of interface-dominated deformation mechanisms in metallic multilayers. Curr. Opin. Solid State Mater. Sci. **15**, 20–28 (2011)
32. Beyerlein, I.J., Demkowicz, M.J., Misra, A., Uberuaga, B.P.: Defect-interface interactions. Prog. Mater Sci. **74**, 125–210 (2015)
33. Weng, S., et al.: Strengthening effects of twin interface in Cu/Ni multilayer thin films – a molecular dynamics study. Mater. Des. **111**, 1–8 (2016)
34. Chang, S.-W., Nair, A.K., Buehler, M.J.: Nanoindentation study of size effects in nickel-graphene nanocomposites. Philos. Mag. Lett. **93**, 196–203 (2013)
35. Muller, S.E., Santhapuram, R.R., Nair, A.K.: Failure mechanisms in pre-cracked Ni-graphene nanocomposites. Comput. Mater. Sci. **152**, 341–350 (2018)
36. Yazdandoost, F., Boroujeni, A.Y., Reza, M.: Nanocrystalline nickel-graphene nanoplatelets composite: superior mechanical properties and mechanics of properties enhancement at the atomistic level. Phys. Rev. Materials **1**, 076001 (2017)
37. Vardanyan, V.H., Urbassek, H.M.: Dislocation interactions during nanoindentation of nickel-graphene nanocomposites. Comput. Mater. Sci. **170**, 109158 (2019). https://doi.org/10.1016/j.commatsci.2019.109158
38. Vardanyan, V.H., Urbassek, H.M.: Strength of graphene-coated Ni bi-crystals: a molecular dynamics nano-indentation study. Materials **13**, 1683 (2020)
39. Shuang, F., Aifantis, K.E.: Dislocation-graphene interactions in Cu/graphene composites and the effect of boundary conditions: a molecular dynamics study. Carbon **172**, 50–70 (2021)
40. Shuang, F., Dai, Z., Aifantis, K.E.: Strengthening in metal/graphene composites: capturing the transition from interface to precipitate hardening. ACS Appl. Mater. Interfaces **13**, 26610–26620 (2021)
41. Zhang, S., Huang, P., Wang, F.: Graphene-boundary strengthening mechanism in Cu/graphene nanocomposites: a molecular dynamics simulation. Mater. Des. **190**, 108555 (2020)
42. Ma, Y., Zhang, S., Yunfei, X., Liu, X., Luo, S.-N.: Effects of temperature and grain size on deformation of polycrystalline copper-graphene nanolayered composites. Phys. Chem. Chem. Phys. **22**, 4741–4748 (2020)
43. Shuang, F., Aifantis, K.E.: Relating the strength of graphene/metal composites to the graphene orientation and position. Scripta Mater. **181**, 70–75 (2020)
44. Vardanyan, V.H., Urbassek, H.M.: Morphology of graphene flakes in Ni-graphene nanocomposites and its influence on hardness: an atomistic study. Carbon **185**, 660–668 (2021)
45. Mishin, Y., Farkas, D., Mehl, M.J., Papaconstantopoulos, D.A.: Interatomic potentials for monoatomic metals from experimental data and ab initio calculations. Phys. Rev. B **59**, 3393 (1999)
46. Stuart, S.J., Tutein, A.B., Harrison, J.A.: A reactive potential for hydrocarbons with intermolecular interactions. J. Chem. Phys. **112**, 6472–6486 (2000)
47. Huang, S.-P., Mainardi, D.S., Balbuena, P.B.: Structure and dynamics of graphite-supported bimetallic nanoclusters. Surf. Sci. **545**, 163–179 (2003)
48. Kelchner, C.L., Plimpton, S.J., Hamilton, J.C.: Dislocation nucleation and defect structure during surface indentation. Phys. Rev. B **58**, 11085–11088 (1998)
49. Plimpton, St.: Fast parallel algorithms for short-range molecular dynamics. J. Comput. Phys. **117**, 1–19 (1995). http://lammps.sandia.gov/

50. Stukowski, A.: Structure identification methods for atomistic simulations of crystalline materials. Model. Simul. Mater. Sci. Eng. **20**, 045021 (2012)
51. Stukowski, A., Bulatov, V.V., Arsenlis, A.: Automated identification and indexing of dislocations in crystal interfaces. Model. Simul. Mater. Sci. Eng. **20**, 085007 (2012)
52. Stukowski, A., Arsenlis, A.: On the elastic-plastic decomposition of crystal deformation at the atomic scale. Model. Simul. Mater. Sci. Eng. **20**, 035012 (2012)
53. Stukowski, A.: Visualization and analysis of atomistic simulation data with OVITO – the open visualization tool. Model. Simul. Mater. Sci. Eng. **18**, 015012 (2010). http://www.ovito.org/
54. Zhao, Y., Peng, X., Tao, F., Sun, R., Feng, C., Wang, Z.: MD simulation of nanoindentation on (001) and (111) surfaces of Ag-Ni multilayers. Physica E **74**, 481–488 (2015)
55. Tao, F., et al.: Molecular dynamics simulation of plasticity in VN(001) crystals under nanoindentation with a spherical indenter. Appl. Surf. Sci. **392**, 942–949 (2017)
56. Alhafez, I.A., Ruestes, C.J., Urbassek, H.M.: Size of the plastic zone produced by nanoscratching. Tribol. Lett. **66**, 20 (2018)
57. Olmsted, D.L., Foiles, S.M., Holm, E.A.: Survey of computed grain boundary properties in face-centered cubic metals: I. Grain boundary energy. Acta Materialia **57**, 3694–3703 (2009)
58. Jin, Z.-H., et al.: The interaction mechanism of screw dislocations with coherent twin boundaries in different face-centred cubic metals. Scr. Mater. **54**, 1163–1168 (2006)
59. Jin, Z.-H., et al.: Interactions between non-screw lattice dislocations and coherent twin boundaries in face-centered cubic metals. Acta Mater. **56**, 1126–1135 (2008)
60. Kulkarni, Y., Asaro, R.J.: Are some nanotwinned fcc metals optimal for strength, ductility and grain stability? Acta Mater. **57**, 4835–4844 (2009)
61. Kulkarni, Y., Asaro, R.J., Farkas, D.: Are nanotwinned structures in FCC metals optimal for strength, ductility and grain stability? Scripta Mater. **60**, 532–535 (2009)
62. Wang, J., Zhou, Q., Shao, S., Misra, A.: Strength and plasticity of nanolaminated materials. Mater. Res. Lett. **5**, 1–19 (2017)
63. Tsuru, T., Kaji, Y., Matsunaka, D., Shibutani, Y.: Incipient plasticity of twin and stable/unstable grain boundaries during nanoindentation in copper. Phys. Rev. B **82**, 024101 (2010)
64. Voyiadjis, G.Z., Yaghoobi, M.: Role of grain boundary on the sources of size effects. Comput. Mater. Sci. **117**, 315–329 (2016)
65. Hou, Z., Tian, Z., Liu, R., Dong, K., Aibing, Y.: Formation mechanism of bulk nanocrystalline aluminium with multiply twinned grains by liquid quenching: a molecular dynamics simulation study. Comput. Mater. Sci. **99**, 256–261 (2015)
66. Hou, Z.Y., Dong, K.J., Tian, Z.A., Liu, R.S., Wang, Z., Wang, J.G.: Cooling rate dependence of solidification for liquid aluminium: a large-scale molecular dynamics simulation study. Phys. Chem. Chem. Phys. **18**, 17461–17469 (2016)
67. Tavazza, F., Senftle, T.P., Zou, C., Becker, C.A., van Duin, A.C.T.: Molecular dynamics investigation of the effects of tip-substrate interactions during nanoindentation. J. Phys. Chem. C **119**, 13580–13589 (2015)
68. Misra, A., Hirth, J.P., Hoagland, R.G.: Length-scale-dependent deformation mechanisms in incoherent metallic multi-layered composites. Acta Mater. **53**, 4817–4824 (2005)
69. Gao, X., et al.: Mechanical properties and thermal conductivity of graphene reinforced copper matrix composites. Powder Technol. **301**, 601–607 (2016)
70. Zhang, D., Zhan, Z.: Strengthening effect of graphene derivatives in copper matrix composites. J. Alloy. Compd. **654**, 226–233 (2016)

Physical Modeling of Grinding Forces

F. Kästner[✉] and K. M. de Payrebrune

Institute for Computational Physics in Engineering, RPTU Kaiserslautern-Landau,
Kaiserslautern, Germany
felix.kaestner@rptu.de

Abstract. In order to address the increasing demands on precision in manufacturing, the prediction of various processes by model-based methods is increasingly becoming a key technology. With respect to this, the grinding process still reveals a lot of potential in terms of reliable predictions. In order to exploit this potential and to improve the understanding of the process itself, a physical force model is developed. Here, process-typical influencing factors, as well as commonly used cooling lubricants, are considered. In addition to the simulative effort for the actual model, basic experimental investigations have to be carried out. In single scratch tests, it has been found that process and deformation mechanisms such as rubbing, ploughing, and cutting of the material and also the pile-up of this material on both sides of the cutting grain are significantly involved in the development of forces. It also turned out that the resulting forces are greater when cooling lubricants are used and that the topographic characteristics of a scratch are also affected by them. For a realistic mapping of these effects within the force model, the deformation model, according to Johnson and Cook, and a discretization, according to Arbitrary Lagrangian-Eulerian, proved most suitable. For integrating the cooling lubricants, the Reynolds equation using a subroutine proves to be a suitable instrument. The challenge to complete the force model is combining the scratch and the Reynolds equation simulation.

1 Introduction

For the development of new technologies or the optimization of existing systems, an essential guarantor for this is high-precision manufactured components and tools. A key element for a precision-manufactured component is choosing the right manufacturing process and how well it is understood. The grinding process is one of the most important surface-finishing processes in the industry [1]. This is also due to the fact that grinding is indispensable for precision-manufactured high-performance components for major industries such as aerospace, the automotive industry, and the energy sector [2]. Grinding itself is a material-removing process in which geometrically undefined abrasive grains are used as tools [3]. Due to these irregularly shaped abrasive grains and the very high process speeds, grinding is a complex manufacturing technology [4]. While this technology has been used by humankind since ancient times, its complexity leads to the fact that the grinding process is still not adequately understood and researched to this day [5, 6]. Due to the complex abrasion mechanisms between abrasive grain and material,

© The Author(s) 2023
J. C. Aurich et al. (Eds.): IRTG 2023, *Proceedings of the 3rd Conference on Physical Modeling for Virtual Manufacturing Systems and Processes*, pp. 70–89, 2023.
https://doi.org/10.1007/978-3-031-35779-4_5

experimental investigations are extremely costly or hardly possible when considering an entire grinding wheel [7, 8]. Existing approaches to determining characteristics of the grinding process by experimental investigations are usually associated with very high costs and therefore are not economically lucrative [9, 10].

In order to avoid these expensive investigations, there exist already several approaches. For example, there are considerations to reduce the required experiments to individual aspects and to measure only the total forces during grinding [11, 12]. In addition to examining the entire grinding wheel, some considerations relate to a single abrasive grain. For instance, Nie et al. [13] have mathematically mapped an abrasive grain statistically and used it to describe the influence of cutting speed and cutting depth on the process forces. In addition, several further considerations attempt to describe the process using numerical or analytical approaches [11, 15]. Notably, the cooling lubricants commonly used in manufacturing are not included or play a minor part in most considerations [13, 14]. The latter influence the process itself and are essential to produce the required fine surface finish [15].

The simulation of a manufacturing process in a computer model has proven to be a suitable method to optimize such processes [16, 17]. Therefore, experiments are mainly needed to validate the model and do not have to be carried out for each constellation of test parameters. An additional advantage of such a model is the possibility of observing processes that are taking place inside the material, such as the stresses that occur. Depending on the complexity of the manufacturing process, it is easier or more difficult to represent these effects within a model. The grinding process reveals a high potential with respect to modeling approaches. This is also due to the fact that special removal mechanisms such as rubbing, ploughing and cutting are not yet sufficiently considered in models for grinding. The model developed here for the grinding process is intended to close the gap by considering and modeling deformation processes and process-required additives such as cooling lubricants. In this context, it is important to map the influence of the cooling lubricants and the mechanical removal mechanisms in detail. In particular, processes in the very small gaps between the tool and the material are of special complexity. For this purpose, the behavior in these areas is to be approached by employing the Reynolds equation.

2 Experimental Investigation

This section discusses the procedure and results of the experiments performed. The experiments carried out here are single scratch tests. In these tests, a diamond tip, which represents a grit of the grinding tool, is driven through the surface of a sample, which represents the material of the workpiece. The scratches created in this way are used for further investigations. Real experiments are important for developing a physical force model for several reasons.

On the one hand, they serve to compare the data from the grinding model with reality. The focus here is on the forces that occur, which in this case, are divided into normal and tangential forces. On the other hand, the material behavior of the samples used is investigated. Special attention is paid to process-typical effects such as rubbing, ploughing, and cutting. But also the effects, such as chip formation and the pile-up of the material. For the development of the physical force model, it is necessary to take all these effects into account and reproduce them since they also contribute significantly to force development.

2.1 Requirements for Performing Experiments

In order to simulate the pile-up effect, an appropriately suitable test rig is required. Therefore, this project uses a scratch test rig with longitudinally guided grains. This makes it possible to generate individual scratches in a reproducible manner. Consequently, the pile-up effect can also be reproduced and investigated. By using an entire grinding wheel, on the other hand, it is not possible to examine the pile-up effect for each scratch individually. In addition, it is possible to determine the force distributions individually for each abrasive grain using the single scratch test rig.

2.1.1 Test Rig for Scratch Tests

In order to generate reproducible and utilizable scratches on a sample surface, the test rig displayed in Fig. 1 shows its most important components. It is important to underline that the abrasive grain in the test rig used here is moved through the sample exclusively in a translational movement. This is also the major difference to test rigs with a rotating grinding wheel. To realize this translational movement, an elementary component of this test rig is the linear unit. The sample is clamped on this unit via a corresponding device and moves translationally through the indenter tip during the scratch test. The linear unit can also be used to set and perform the different scratch speeds (0 mm/s to 800 mm/s) for the tests. Here, too, stepless adjustment is possible, with the restriction that the drive

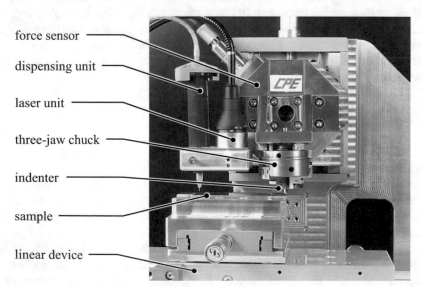

Fig. 1. Test rig with the main modules and functional elements to perform scratch tests in dry and lubricated conditions

used only works reliably up to a speed of 800 mm/s and is no longer precisely controlled beyond this threshold.

In order to ensure ideal fixation, it is necessary that the indenter does not twist during clamping (important when using pyramidal indenters) or does not move up or down (occurs while using a screw to press the indenter against a wall). A three-jaw chuck has proven to be a reliable clamping with low susceptibility to mistakes.

A confocal distance laser (CL-3000 series from Keyence) is used to set the required penetration depth of the indenter into the corresponding sample. The test rig is equipped with a dynamometer (type 9109AA from Kistler) to record the forces during a scratch test. This dynamometer records the tangential and normal forces.

In order to be capable of carrying out the scratch tests either dry, i.e., unlubricated, or lubricated with cooling lubricants, a dispensing unit (2000 series from Vieweg) for various liquids is integrated into the test rig. With the aid of this dispensing unit, the reference oil, here FVA2 and FVA3, is applied onto the sample directly in front of the indenter. As a sample material, aluminum alloy A2024-T351 (see Table 1) is used to represent ductile material.

Table 1. Material parameters of aluminum alloy A2024-T351 used as samples

Density ρ in kg/m^3	Young's Modulus E in GPa	Poisson's ratio ν [–]	Specific heat Cp in J/kgK^{-1}	T_{melt} in °C
2700	73	0.33	875	1793

2.1.2 Indenter

In order to get reproducible scratches, it is advisable to use indenters instead of real abrasive grains. The problem of low reproducibility results primarily from the complexity of placing the abrasive grains always in the same position or classifying their adjusted position. Indenters used for the experiments can be seen schematically in Fig. 2. The first two indenters (from left to right) are geometrically standardized, whereas the last indenter has an undefined geometry and therefore comes closest to the real abrasive grain, but again with the previously described problems in the usage.

Both the conical and the pyramidal indenters are available in different versions with regard to the angle of their tip. For the experiments carried out here, mainly conical indenters of 90° to 150° are used. The advantage of using conical indenters is their simple and rotationally symmetrical geometry, which makes the alignment of the indenter considerably easier than with the pyramidal indenter. When aligning the pyramidal indenter, paying attention to the orientation of the pyramid faces is always necessary. In Fig. 3, three different examples of conical indenters are shown. The first picture shows a microscope image of an indenter in a three-dimensional perspective, and the two following pictures show two indenters with different factory-specified angles of their tip. The angles were measured manually to check the production precision. The diamond tip and the carrier material of the indenter are also clearly visible based on the three-dimensional

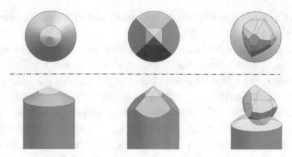

Fig. 2. Conical indenter (left); pyramidical indenter (center); indenter with real grit (right) used in single grit scratch experiments to measure grinding forces

image. The three-dimensional scan is required for the development of the force model. This scan enables the remodeling of the indenter in the finite element software used. The examination of the angle of the indenter tips, on the one hand, is necessary for quality control. On the other hand, they are required for the later documentation and simulation of the wear of the indenter tips.

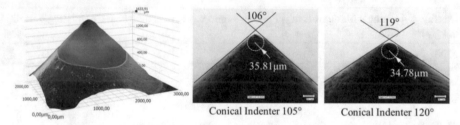

Conical Indenter 105°   Conical Indenter 120°

Fig. 3. 3D magnification of a conical Indenter (left); conical indenter with a factory-specified angle of 105° (center); conical indenter with a factory-specified angle of 120° (right)

2.2 Preparations for the Scratch Tests

Since the samples are manufactured in a metal processing facility, it is important to ensure they are free of production residues, such as lubricating oils and greases, before they are used in the test rig. Hence, they are cleansed of possible impurities with acetone in an ultrasonic bath. The procedure is the same for the respective indenter to be used. The cleaned samples are then fixed in the clamping device of the test rig, and the indenter is inserted into the jaw chuck. Due to the weight of the indenter (approximately 5 g), the tip of the indenter now lightly contacts the surface of the sample, and the chuck is then firmly tightened. The position of the sample to the indenter tip is set as scratch depth "zero" in the used LabView program. After the sample has been moved away from the indenter tip, the confocal distance laser is used to set the specified scratch depth for a specific test run. Moving the linear unit to the start position completes the setup for a scratch test.

2.3   Performing Scratch Tests in Dry Conditions

In order to exclusively investigate the effect of different scratching speeds, scratching depths, and indenter angles on the samples in relation to the normal and tangential force and the sample topography, scratching tests are first carried out in a dry environment. This approach ensures that only the pure interaction between the diamond tip of the indenter and the aluminum surface of the sample is investigated. The distribution of the forces and their changes under different test parameters allow essential conclusions to be drawn about the material parameters. These parameters are important for the grinding model to predict forces with this model in its final state. For this reason, a range of parameter constellations is tested for these scratch tests. Scratch depths from 50 μm to 250 μm, scratch speeds from 50 mm/s to 1000 mm/s, and indenter angles from 90° to 120° have turned out to be practicable constellations.

Regarding the selection of the scratch depth, it is important to consider that the deeper the scratch depth, the more reliable the results are and the less they scatter. This is because the indenter tip is slightly rounded. However, it is important not to scratch too deep since from a scratch depth of approximately 300 μm, the substrate material of the indenter can be partially involved in the formation of the scratch. The selected scratch depths are also subject to the production-related properties of the indenters used. Due to the rounded tip, a certain penetration depth is required to obtain error-free data for the simulation and its validation. Although smaller scratch depths are common in grinding, characterizing processes are scalable in most cases. After overhauling the test rig, it was decided not to use the maximum speed of the device due to technical control reasons and to use it only up to 800 mm/s.

Following the setting of the corresponding test parameters, the test is started via the implemented software of the test rig. After a pre-defined acceleration phase of the sample, which is mounted on the linear unit, the sample moves with a constant speed under the indenter in the scratching area and is scratched. Figure 4 shows a sample used for the scratch tests. On this sample three sets each with nine scratches can be seen. The same test parameters apply within each set. Therefore, all scratches within one area are repetition tests. The three sets differ in their scratch speed. The upper set with nine scratches generated at 50 mm/s feed rate, the middle set with nine scratches were generated at 400 mm/s feed rate and the lower set with nine scratches were generated at 750 mm/s feed rate.

50 mm/s

400 mm/s

750 mm/s

Fig. 4. Three sets of scratches with nine repetitions each on an aluminum sample, the sets differ in scratch speed. Test parameters: scratch speed 50, 400 and 750 mm/s; scratch depth 0.08 mm; conical indenter of 105°; lubricated with FVA2

The topographic properties of the scratched samples are then examined and evaluated with suitable optical equipment. For this purpose, 3D-capable microscopes such as the confocal microscope μSurf-explorer from NanoFocus or the digital microscope VHX of the 7000 series from Keyence are generally used. Based on the topography obtained in this way, conclusions can be drawn about the deformation behavior of the sample material. These are important to realistically simulate effects such as the pile-up of the displaced material.

The dynamometer integrated with the test rig directly records the force signals and saves them corresponding to the test parameters. Figure 5 shows a typical normal and tangential force distribution during a scratch test.

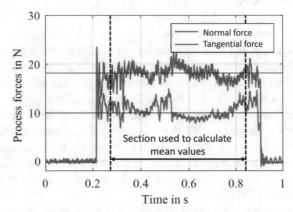

Fig. 5. Typical distribution of normal and tangential force during a scratch test. For the evaluation, only the forces without the edge areas are used to calculate the mean values. Here illustrated exemplary by the area within the dashed lines.

To avoid interfering edge effects where the indenter enters and leaves the material, 10% of the force signal after entering and before leaving are each ignored in the evaluation. Figure 5 illustrates which section of the force signal is used to calculate the mean values for further evaluation. Figure 5 also clearly shows that the normal forces in a scratch test are higher than the corresponding tangential forces. This property is also unaffected by the selected indenter angle and the set scratch speed, as shown in Fig. 6.

Furthermore, it can be seen in Fig. 6 that with increasing scratching speed, both the normal and tangential forces decrease. This trend has already been observed in previous publications [18, 19]. One reason for this behavior could be a temporary temperature increase in the cutting area, which reduces the flow stress in the area.

2.4 Performing Scratch Tests in Wet Conditions

In industrial production, grinding processes are almost exclusively done in combination with cooling lubricants, and it is also necessary to perform scratch tests with cooling lubricants. This is also important to determine whether, for example, the pile-up effect is increased or decreased by the influence of cooling lubricants. Another important point

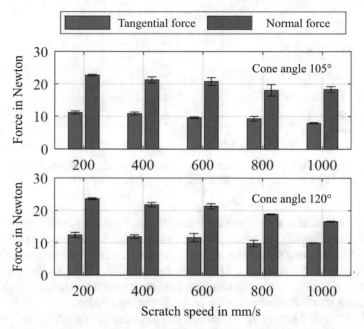

Fig. 6. Mean values of normal and tangential forces for different scratch speeds and two indenter angles with a scratch depth of 50 μm: cone angle of 105° (top); cone angle of 120° (bottom)

is to investigate the influence of cooling lubricants on the tangential and normal forces. This also raises the question of whether these forces increase or decrease.

The attached dispensing unit on the test rig enables the application of selected cooling lubricants under which the scratch test will be carried out. Since industrially used cooling lubricants generally contain various additives to adjust their properties, they are not suitable for basic research with regard to the scratch test. One of the reasons for this is that the added substances cannot always be determined qualitatively and quantitatively due to confidentiality. In addition, evaluating the tests is difficult when such additives are present since many processes occur in part at the molecular level. For this reason, reference oils (FVA2 and FVA3 from Weber Reference Oils) are used instead of real cooling lubricants for the tests carried out here. The two reference oils essentially differ in their viscosity. Here the reference oil FVA2 with 85 mm²/s at 20 °C has a significantly lower viscosity than the reference oil FVA3 with 300 mm²/s at 20 °C. The tests with reference oils are carried out the same way as those in a dry environment.

Based on the results obtained from dry and wet tests, it is then possible to detect and evaluate both analogies and differences. The first effect that attracts attention is that it can be confirmed that the normal forces are still larger than the tangential forces when using cooling lubricants. This can be seen in Fig. 7. Figure 7 also displays another important influence of the cooling lubricants. It can be seen that the normal and tangential forces under dry test conditions are smaller than those under wet test conditions. One possible reason for this may be the additional liquid phase that must be displaced by the indenter, which increases the force.

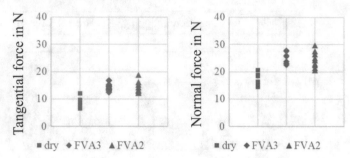

Fig. 7. Normal and tangential force for scratch tests under dry and wet conditions. Test parameters: scratch speed 50 mm/s; scratch depth 0.08 mm; conical indenter of 105°

Another remarkable aspect is that the deviation in the forces caused by the reference oil FVA2 (low viscosity) is slightly wider than that of the reference oil FVA3 (high viscosity). The different viscosities of the two reference oils can be used as an important clue to explain this phenomenon. The reference oil FVA3, with a higher viscosity than the reference oil FVA2, might absorb possible vibrations of the indenter due to its higher viscosity; as a result, a small deviation in the forces is recorded.

Another essential part of the realization of a physical force model is the understanding of the surface characteristics of the workpiece after a scratch test. Here, however, the focus of the investigation is less on the quality of the surface itself but more on the topography and characteristics of a single scratch. In exclusively dry tests, the topographical characteristics of a scratch are mainly influenced by the use of different indenter angles. Since a further component is contributed when cooling lubricants are used, it is important to find out how this additional component affects the scratch characteristics.

To examine the respective scratches, all scratches of a sample are measured optical. With the help of the digital microscope VHX 7000 from Keyence, the surface is recorded and converted into a three-dimensional image. These surface profiles are then exported as scatter plots and further processed in Matlab. During further processing, the data volume of the scatter plot is reduced from the data export. This is necessary for performance reasons only. Thus, a surface profile can be generated, as shown in Fig. 8.

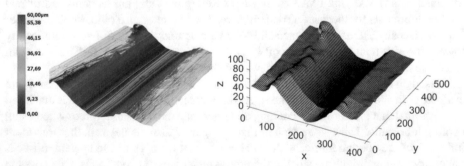

Fig. 8. Topography of a scratch magnified by using a digital microscope (left); recreated relief of such a scratch by using the exported csv data (right)

In a subsequent step, various profiles are extracted from such a relief (cf. Fig. 9), describing the scratch and its cross-section.

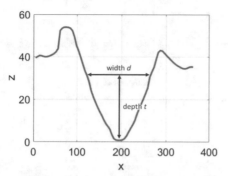

Fig. 9. Extracted profile from the relief of a scratch with the key values scratch width and scratch depth to evaluate the scratch topography. The extracted profile depicts the cross section of the scratch from Fig. 8 at the border line between the colored and grey areas. Values are displayed in μm.

The scratch width d and the scratch depth t have proven to be suitable values for comparing the topography of different scratches. Instead of using the values individually, the ratio d/t is used. The reason for this is that, for example, the scratch depth t changes depending on the scratch depth set before the start of the test. However, since the scratch always represents the negative image of the indenter used and the indenter always has the same height and corresponding width ratio, this ratio must also be present in the imprint. Detectable deviations of this ratio within a scratch profile are, therefore, due to process- or parameter-related influences. The values d and t used are determined by Matlab from the measured values of the digital microscope. Figure 10 shows the ratio d/t depending on the three environmental conditions dry, FVA3, and FVA2.

Figure 10 shows that the ratio d/t becomes smaller as soon as the tests are carried out with the reference oils. It is also notable that the reference oil FVA2 (low viscosity) differs from the reference oil FVA3 (high viscosity) with regard to this ratio. FVA2 always shows the larger ratio d/t. However, in wet condition the ratio is always smaller than in dry conditions, which indicates that the scratches become smaller when performed with cooling lubricants. A material buildup in front of the indenter without reference oils could cause this phenomenon. This buildup of material could cause an increase in material removal at the sides of the indenter. However, we cannot explain this phenomenon with absolute certainty at present. To identify the exact causes, further experimental setups designed for this purpose must be developed. For example, it is necessary to optically examine the cutting front during the scratch test.

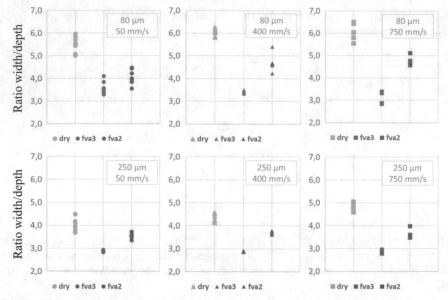

Fig. 10. Ratio of scratch width d to scratch depth t for velocities 50 mm/s and 400 mm/s and the scratch depth of 80 μm and 250 μm, dry and wet condition generated by a conical indenter with an angle of 105°

3 Development of the Grinding Model

For the development of the grinding model, the realistic behavior between abrasive grain and sample material has to be considered and implemented in the model. To develop a reliable force model, first of all, it is essential to simulate the material behavior of the sample in a physically precise manner. To accomplish this, it is necessary to take a more detailed consideration of simulation approaches and techniques. In the first instance, it is advisable to consider the indenter as rigid to focus on the sample material and its behavior. With this assumption transferred to the scratching problem, the represented abrasive grain is treated as wear-free and fracture-resistant.

3.1 Selection of the Suitable Material Model

To simulate the behavior of the material realistically, a suitable approach must be selected and implemented. The deformation model, according to Johnson and Cook (JC), is a candidate for this purpose. Using the JC model, the strain hardening of the corresponding material can be described analytically. Furthermore, the strain rate and temperature dependence of a material are also described. The material behavior, according to JC, is integrated as standard in most finite element method (FEM) programs. The von Mises stress σ can be calculated according to the JC model by using the equation

$$\sigma = \left[A + B(\varepsilon)^n\right]\left[1 + C\ln\left(\frac{\dot{\varepsilon}}{\dot{\varepsilon}_0}\right)\right]\left[\left(\frac{T - T_{room}}{T_{melt} - T_{room}}\right)^m\right]. \tag{1}$$

Here, A is the quasi-static yield stress, B is the modulus of strain hardening, n is the work hardening exponent, C is the strain rate sensitivity, and m defines the temperature sensitivity. Significant temperature development in the process of a single scratch is not expected. Due to the high thermal conductivity of the aluminum selected here, it is assumed that any process heat is immediately transported off the scratch area. Hence, temperature-relevant effects in the JC model are of negligible importance for the current state (Table 2).

Table 2. Johnson–Cook material parameters used for aluminum [21]

Initial yield strength	A	369	MPa
Strain hardening constant	B	684	MPa
Strengthening coefficient of strain rate	C	0.0083	–
Strain hardening coefficient	n	0.73	–
Thermal softening coefficient	m	1.7	–

The Crystal Plasticity Finite Element Simulation Method (CPFEM) has been considered an alternative method for describing material behavior. However, the comparison between the JC model and CPFEM carried out in this context does not indicate any significant advantage with respect to CPFEM. This is also due to the fact that the data used here for the CPFEM by [20] is still from its initial phase.

3.2   Discretization Approaches

Various approaches exist in continuum mechanics and are implemented in many finite element programs to simulate the motion of material points. In Abaqus, the used FE program for this study, the mesh-based approaches according to Lagrangian (LAG) and Arbitrary Lagrangian-Eulerian (ALE), and the mesh-free smooth particle hydrodynamics (SPH) approach are to be mentioned. Each approach is particularly well suited for certain problems, so comparing the three approaches was first carried out on a simplified 2D scratching process.

3.2.1   2D Discretization Benchmark

The computational effort differs significantly depending on the type of discretization and how fine the mesh is set. For the estimation of basic aspects of the individual approaches, the problem is therefore first considered on a two-dimensional level. For the preliminary study conducted here, we started with a tool rake angle with a positive value of $\gamma = 20°$ in dry conditions, similar to a turning process, for which experimental and simulated data are available to compare from the literature [21]. For the sample, aluminum alloy A2024 T351 was used. Figure 11 shows the FE output models used to evaluate the different discretization approaches. For the Lagrangian (LAG) and Arbitrary Lagrangian-Eulerian (ALE) approach, the sample is divided into three layers, with a sacrificial layer L2 separating the chip area L1 from the base material L3. When an element in the sacrificial layer L2 reaches a critical damage value, the elements are deleted and separated. The element size of the mesh-based approach is similar to the particle size of the mesh-free smooth particle hydrodynamics approach. No sacrificial layer is required here since separation can occur between any pair of particles when the cohesive bonds are no longer sufficiently large.

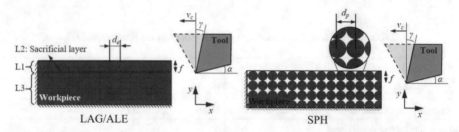

Fig. 11. 2D simulative output model for benchmarking of the LAG/ALE models (left) and SPH model (right), according to [22]

Figure 12 shows the results from the simulation and the experimental values. Based on this figure, it can be seen that, with respect to the tangential forces, all approaches are close to the experimental values with a deviation from the experimental value by a maximum of 1%. In contrast, if the normal force is observed, it is noticeable that all approaches have a deviation of about 40%. A possible reason for this large deviation could be the active element deletion in the mesh-based approaches. When the affected elements are deleted, they can no longer cause any force to be exerted on the tool. In practice, however, no material elements are deleted, and the remaining material builds up in front of the tool and leads to an increase in force. In the mesh-free approaches, a weighting function controls when a cohesive material bond is dissolved. The parameters used here may be the reason why the composite is dissolved earlier.

In contrast to the measured forces shown in Fig. 5 and Fig. 6, the tangential forces are much larger than the normal forces in the 2D simulation. Besides the already discussed effects of element deletion and weighting function on the forces, an important aspect is the effects of material deformation when rubbing, ploughing, and cutting occurs. Since these material deformations run in all directions, these effects can only be displayed reasonably in a 3D simulation.

3.2.2   Single Grit Scratch Model as 3D Approach

Even if investigations in two-dimensional form can provide values for approximate predictions, a three-dimensional approach is necessary for an overall analysis since only in this way boundary effects such as the pile-up of the material and the general material deformation effects like rubbing, ploughing, and cutting can be represented in a useful way. A simplified three-dimensional model of a scratch test is used to investigate the discretization approaches. The model height is 0.2 mm, the model depth is 0.5 mm, the model length is 1 mm and the spacing between the elements or particles are 0.003 mm. Figure 13 shows the stress distribution for the ALE and SPH approaches.

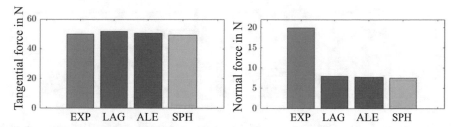

Fig. 12. Comparison of the tangential forces (left) and normal forces (right) of the discretizational approaches with the experimental results from [22] for the 2D orthogonal cutting model with tool rake angle $\gamma = 20°$, according to ref. [22]

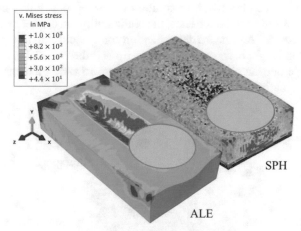

Fig. 13. Simulative output model for benchmarking LAG, ALE, and SPH in 3D. Already with an example of the distribution of the von Mises stress in the ALE and SPH models. Test parameters: model height 0.2 mm; model depth 0.5 mm; model length 1 mm; indenter geometry cone 105°; feed rate 200 mm/s; scratch depth 0.03 mm; element spacing for the LAG/ALE and particle spacing for the SPH 0.003 mm, respectively [22].

Figure 14 shows the simulated and experimentally determined values for the normal and tangential forces of the three approaches.

As can be seen from these results, the simulated tangential forces now show the measured tendency with smaller values than the normal forces. Additionally, the ALE approach agrees best with the values from the experiments. Therefore, the ALE approach will be used for the more detailed development of the physical force model.

3.3 Simulative Integration of the Cooling Lubricants

As experimental investigations have already shown that cooling lubricants have a detectable influence on the normal and tangential forces during scratching and that the topographical nature of a scratch also changes under their influence, the cooling lubricants must also be integrated into the grinding model. The inclusion of an additional

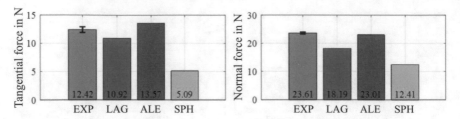

Fig. 14. Process forces of measurement and discretization approaches for a 3D single grit scratch (cutting speed $v_c = 200$ mm/s, depth of cut $a_p = 50$ μm that corresponds to $a_{p,\text{sim}} = 30$ μm, cone angle of $\gamma = 105°$), according to [22]

material turns out to be non-trivial. In addition to the reproducibility of the influence of cooling lubricants during scratch tests, the realizability via discretization approaches must also be discussed. A standard discretization for a liquid film in combination with the very small gap height between the indenter tip and the material is computationally almost impossible or even difficult to perform. Figure 15 shows the interspace that can be rated as problematic by classical FEM discretization.

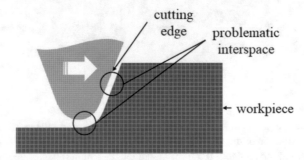

Fig. 15. Schematic illustration of the small gap height between the indenter tip and the material in which the lubricant is located

3.3.1 Basic Principle According to Reynolds Equation

After various solution methods for this problem have been considered, the approach, according to Reynold, is the most suitable solution. The Reynold equation describes and calculates pressure distribution problems of thin viscous fluid films in lubrication theory. The Reynolds equation can be described by

$$\frac{\partial}{\partial x}\left(\frac{h^3}{12\eta}\frac{\partial p}{\partial x}\right) + \frac{\partial}{\partial z}\left(\frac{h^3}{12\eta}\frac{\partial p}{\partial z}\right) = \frac{1}{2}\frac{\partial(U_2 - U_1)h}{\partial x} + (V_1 - V_2) + \frac{1}{2}\frac{\partial(W_2 - W_1)h}{\partial z}. \tag{2}$$

Here, h is the gap height between two plates, U and W are the respective velocities of the plates in the x and z directions, V is the velocity in the y direction, p is the pressure between the plates, and η is the dynamic viscosity of the fluid. In principle, the Reynolds equation converts a three-dimensional problem into a two-dimensional one. Applied to the problem of Fig. 15, it is no longer necessary to describe the interspace with a very fine mesh.

3.3.2   Implementation of the Reynold Equation by a User Element

With Abaqus, a direct implementation of the Reynolds equation is not possible without further effort. However, special subroutines can be integrated into Abaqus by scripts. Such subroutines are called user elements (UEL) and are used to apply the Reynolds equation to the problem under consideration. [23] has already programmed such a UEL to investigate and simulate plain bearings in Abaqus. With the help of this UEL, it is sufficient to discretize the liquid within the gap with only one element of thickness. The UEL converts this three-dimensional mesh to a two-dimensional layer, solves the Reynolds equation in it, and returns the results to the nodes of the three-dimensional sample. Therefore, the pressure distribution of a fluid within a very small gap can be calculated and simulated. Figure 16 shows an exemplary simulation of oil between two plates. In the simulation, the upper gray body represents a rigid planar plate, which is loaded initially with a pressure field in the negative y-direction. The lower element is designed as a user-defined deformable material. In this example, the material properties and the corresponding deformation behavior are assumed for the aluminum alloy A2024-T351. The element height is 0.04 mm, the element depth is 0.1 mm, the model length is 0.3 mm and the spacing between the elements are 0.0006 mm. The meshing in this element was done manually.

A parameterizable fluid is located between the two bodies. The force from the rigid body causes a pressure field in the fluid, which transmits a resulting force to the deformable body. The liquid can flow in the x-direction, but cannot flow in the z-direction due to the infinite expansion of the elements in this direction.

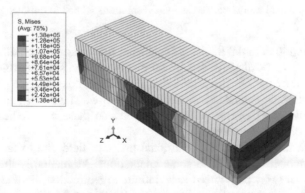

Fig. 16. Example of a simulation with the Reynolds equation integrated as user element (UEL) in Abaqus. Test parameters: deformable model height 0.04 mm; model depth 0.1 mm; model length 0.3 mm; gap height 0.005 mm; applied pressure 1 MPa; element spacing in height 0.0133 mm; element spacing in length 0.006 mm.

Figure 17 shows the numerical solution by the UEL compared with the analytical solution of the gap height over time, which slowly reduces due to the force applied by the upper body. The reduction in gap height over time is considered here as a reference point. The expression

$$h(t) = \sqrt{\frac{h_0^2 B^3 L\eta}{2p_p BLh_0^2 t + B^3 L\eta}} \quad (3)$$

is derived by transforming the Reynolds equation according to the gap height $h(t)$. Here, h is the gap height, B and L are the dimensions of the element, p_p is the applied pressure, and η is the dynamic viscosity. Based on the two solutions, it can be seen that they are identical. This proves that the UEL in Abaqus reliably computes the Reynolds equation.

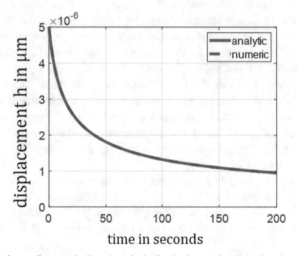

Fig. 17. Comparison of numerical and analytical solutions related to the change in gap height

A highly simplified initial model is considered first to adapt the UEL to the scratch test problem. Figure 18 displays this initial model and the stress development due to the pressure field of the oil film after solving the Reynolds equation. The gray body represents an infinitely extended indenter in z-direction. The lower part represents the aluminum sample as a deformable material. Between these two bodies is the liquid, which can flow off in x-direction.

Even though the results of the additional pressure field due to the fluid film are promising, the model must include some adaptations. Momentarily the sample (blue body) and indenter (gray body) need to be infinite in z-direction. For the further course of the project, the quasi-two-dimensional simplification will be removed step by step. In this way, the Reynolds equation will also be rendered implantable for a complete three-dimensional scratch test. This requires modifications to both the UEL and the modeling of the indenter tip and sample in Abaqus. Additionally, further considerations are required to simulate the flow of the reference oils between the indenter tip and the sample in the most efficient way. The Reynolds equation does not provide for flow through the gap. To solve this problem, a further UEL may be necessary.

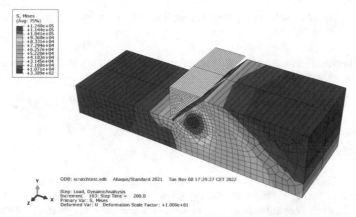

Fig. 18. Initial model for adapting the UEL to the scratch test problem

4 Conclusion

For the development of the grinding model presented here, two major aspects have been addressed in this report. On the one hand, the approach to develop the actual model and, on the other hand, the necessary basic experimental investigation. The experimental data are important for reproducing the material behavior and mapping the forces correctly in the simulation. Single scratch tests were considered here as the basis for the experimental investigation. These were carried out either dry, for the basic behavior, or wet, to consider cooling lubricants. The normal forces in each test parameter constellation are shown to be higher than the corresponding tangential forces. Moreover, it turned out that the resulting forces in scratch tests also depend on the deformation behavior of the sample material used. Thus, effects such as the pile-up of the material have a non-negligible influence on the normal and tangential forces. With the addition of cooling lubricants, the trend of higher normal force remains. However, it has been shown that both normal and tangential forces are generally higher when cooling lubricants are used. In addition to the influence of the forces during scratching, the topography of a scratch itself is also affected by the cooling lubricants. Thus, with the help of the scratch width and depth ratio, it can be seen that scratches produced under the influence of cooling lubricants are less wide.

For the development of the grinding model, the experiments revealed that, on the one hand, the consideration of cooling lubricants is important to obtain a realistic force model and, on the other hand, how different test parameters influence the normal and tangential forces. It became clear that preliminary observations in a two-dimensional simulation are only of limited value. In particular, the pile-up effect of the material and its influence on the forces can only be mapped realistically in a three-dimensional simulation. The influence of the cooling lubricants and the associated narrower scratches can only be correctly reproduced in a three-dimensional simulation. With regard to the cooling lubricants to be simulated, the employment of the Reynolds equation using a user element (UEL) has crystallized as a promising option. Here, problems in discretizing the very small gap between the indenter and sample can be avoided. The challenge is to couple the three-dimensional scratch test, with all its deformation aspects, with the Reynolds equation.

Acknowledgements. Funded by the Deutsche Forschungsgemeinschaft (DFG, German Research Foundation) – 252408385 – IRTG 2057.

References

1. Chatti, S., Laperrière, L., Reinhart, G., Tolio, T., et al.: CIRP Encyclopedia of Production Engineering. Springer, Heidelberg (2019). https://doi.org/10.1007/978-3-642-20617-7
2. Klocke, F., et al.: Abrasive machining of advanced aerospace alloys and composites. CIRP Ann. **64**, 581–604 (2015). King, R.I. and Hahn, R.S.: Handbook of modern grinding technology. Springer Science & Business Media, 2012
3. Klocke, F.: Fertigungsverfahren 2: Zerspanung mit geometrisch unbestimmter Schneide, vol. 6. Springer, Berlin (2018). ISBN 978-3-662–58091-2. https://doi.org/10.1007/978-3-662-533 10-9
4. Inasaki, I., Tönsho, H.K., Howes, T.D.: Abrasive machining in the future. CIRP Ann. **42**(2), 723–732 (1993)
5. Brinksmeier, E., et al.: Advances in modeling and simulation of grinding processes. CIRP Ann. **55**(2), 667–696 (2006)
6. Setti, D., Kirsch, B., Aurich, J.C.: Experimental investigations and kinematic simulation of single grit scratched surfaces considering pile-up behaviour: grinding perspective. Int. J. Adv. Manuf. Technol. **103**(1–4), 471–485 (2019). https://doi.org/10.1007/s00170-019-03522-7
7. Feng, B.F., Qi Cai, G.: Experimental study on the single-grit grinding titanium alloy TC4 and superalloy GH4169. Key Eng. Mater. **202**, 115–120 (2001)
8. Zahedi, A., Azarhoushang, B.: FEM based modeling of cylindrical grinding process incorporating wheel topography measurement. Procedia Cirp **46**, 201–204 (2016)
9. Mabrouki, T., Girardin, F., Asad, M., Rigal, J.-F.: Numerical and experimental study of dry cutting for an aeronautic aluminium alloy (A2024-T351). Int. J. Mach. Tools Manuf. **48**(11), 1187–1197 (2008)
10. Huang, S., Yu, X.: A study of grinding forces of SiCp/Al composites. Int. J. Adv. Manuf. Technol. **94**, 3633–3639 (2018)
11. Liu, Q., Chen, X., Wang, Y., Gindy, N.: Empirical modelling of grinding force based on multivariate analysis. J. Mater. Process. Technol. **203**(1–3), 420–430 (2008)
12. Nie, Zg., Wang, G., Jiang, F., Lin, Yl., Rong, Ym.: Investigation of modeling on single grit grinding for martensitic stainless steel. J. Central South Unive. **25**, 1862–1869 (2018)
13. Eder, S.J., Leroch, S., Grützmacher, P.G., Spenger, T., Heckes, H.: A multiscale simulation approach to grinding ferrous surfaces for process optimization. Int. J. Mech. Sci. **194** (2021)
14. Opoz, T.T.: Investigation of material removal mechanism in grinding: a single grit approach. Doctoral thesis, University of Huddersfield (2012)
15. Madaj, M., Píška, M.: On the SPH orthogonal cutting simulation of A2024–T351 alloy. Procedia Cirp **8**, 152–157 (2013)
16. Woldman, M., Van Der Heide, E., Tinga, T., Masen, M.A.: A finite element approach to modeling abrasive wear modes. Tribol. Trans. **60**(4), 711–718 (2017)
17. Wu, C., Guo, W., Wu, Z., Wang, Q., Li, B.: Ductility-oriented high-speed grinding of silicon carbide and process design for quality and damage control with higher efficiency. Int. J. Adv. Manuf. Technol. **105**(7–8), 2771–2784 (2019). https://doi.org/10.1007/s00170-019-04461-z

18. Fang, C., Yang, C., Cai, L., Zhao, Y., Liu, Z.: Predictive modeling of grinding force in the inner thread grinding considering the effect of grains overlapping. Int. J. Adv. Manuf. Technol. **104**(1–4), 943–956 (2019). https://doi.org/10.1007/s00170-019-03925-6
19. Lerra, F., Candido, A., Liverani, E., Fortunato, A.: Prediction of micro-scale forces in dry grinding process through a FEM — ML hybrid approach. Int. J. Precis. Eng. Manuf. **23**, 15–29 (2022)
20. Arrazola, P.: Modelisation numerique de la coupe: Etude de sensibilite des parametres d'entree et identification du frottement entre outil-copeau. Ph.D. thesis, L'École Centrale de Nantes, Nantes, France (2003)
21. Subbiah, S.: Some investigations of scaling effects in micro-cutting. Ph.D. thesis, Georgia Institute of Technology, Atlanta, GA, USA (2006)
22. Sridhar, P., Rodríguez Prieto, J.M., de Payrebrune, K.M.: Modeling grinding processes—mesh or mesh-free methods, 2D or 3D approach? J. Manuf. Mater. Process. **6**(5), 120 (2022)
23. Gradl, C.: Hydraulic stepper drive: conceptual study, design and experiments Doctoral thesis, Universität Linz, Dissertation (2017)

Modeling and Implementation of a 5G-Enabled Digital Twin of a Machine Tool Based on Physics Simulation

J. Mertes[1](✉), M. Glatt[1], L. Yi[1], M. Klar[1], B. Ravani[2], and J. C. Aurich[1]

[1] Institute for Manufacturing Technology and Production Systems, RPTU Kaiserslautern, Kaiserslautern, Germany
jan.mertes@rptu.de

[2] Department of Mechanical and Aerospace Engineering, UC Davis, Davis, CA, United States

Abstract. The cellular network standard 5G meets the networking requirements for different industrial use cases due to the advantages of low latency, high bandwidth, and high device density while providing a very good quality of service. These capabilities enable the realization of wireless digital twins (DTs), a key element of future cyber-physical production systems. DTs for prediction, monitoring, and control of machine tools need physical modeling as well as the bidirectional exchange of information between the digital and the physical world. 5G is a wireless communication technology with the potential to disruptively change industrial communication. 5G enables wireless, highly scalable, and flexible realization of even safety- and latency-critical connections. In this paper, a 5G enabled DT of a machine tool for process control, monitoring and simulation is developed and implemented. A bidirectional communication between the physical machine tool and the DT is realized via 5G. Moreover, process prediction is enabled based on physics simulation. Next to the physical modeling of the machine behavior, a 5G-capable interface between the input and output signals of the machine control system and the developed DT is implemented. Moreover, the DT is migrated in a wireless form to an edge server. Furthermore, the capabilities of the DT are demonstrated. Therefore, the architecture and implementation of the DT as well as its benefits and challenges are outlined.

1 Motivation

Modern manufacturing systems that incorporate digital technologies - the so-called cyber-physical production systems (CPPS) - seek to interconnect the digital and real world. For better flexibility, scalability, and reconfigurability, CPPS aim to connect individual cyber-physical systems and soften the hierarchical automation pyramid to decentralized distribution of computing units [1]. Due to increased demand for customized products, the requirements on the connectivity, functionality, flexibility, and intelligence of machine tools are also increasing [2]. Consequently, CPPS lead to complex manufacturing systems that are increasingly challenging to control and understand, due to many different decentralized, connected systems (e.g., machine tools, automated guided vehicles, human-machine-interfaces, etc.) with different characteristics, functionalities, and thus different heterogenous information technologies [3].

© The Author(s) 2023
J. C. Aurich et al. (Eds.): IRTG 2023, *Proceedings of the 3rd Conference on Physical Modeling for Virtual Manufacturing Systems and Processes*, pp. 90–110, 2023.
https://doi.org/10.1007/978-3-031-35779-4_6

Therefore, the concept of digital twins (DTs) enables and facilitates the operation of CPPS and is considered a prerequisite for providing model-based decision support and process control [4].

However, enhanced physical simulation of the manufacturing process, virtual models, and high connectivity of the involved sub-systems are required for full potential of DTs in manufacturing [5]. Especially for real-time monitoring, high requirements have to be met regarding the communication technology. Current wired solutions cannot achieve the required flexibility and scalability of the manufacturing systems, while wireless technologies cannot meet the required low latency and reliability in communication.

The 5G communication standard addresses these issues of insufficient flexibility, scalability, and low latency while maintaining high reliability [6]. 5G supports real-time capable, reliable, and wireless connectivity within a CPPS [7], thus providing the basis for wireless DTs with full functionality in manufacturing. However, currently no implementation of 5G-enabled DT for machine tools in manufacturing exists. Accordingly, this paper advances the authors' preliminary work [8] by developing an architecture for 5G-enabled DTs of machine tools based on physics simulation with different functionalities such as process prediction, monitoring, control, and diagnosis. Furthermore, the detailed implementation with different hardware and software components as well as benefits and challenges of the implemented architecture are outlined.

2   State of the Art

2.1   5G Communication Standard

5G - the fifth mobile communication standard - was introduced in 2018 and is standardized by 3rd Generation Partnership Project (3GPP) [9]. The 5G standard intends to improve the most important functions of mobile networks, with specific consideration of industrial requirements. The 5G network architecture consists of a centralized radio access network (RAN) with multiple remote radio reads and aggregating base band units. In addition, a 5G core network provides different management functionalities [10]. This leads to the following beneficial performance characteristics of 5G-based wireless communication [11]:

- Ultra-reliable low latency communication (**uRLLC**): user plane latency down to 1 ms and reliability of 99.9999%
- Massive machine type communication (**mMTC**): high device density with up to 10^6 devices per km^2
- Enhanced mobile broadband (**eMBB**): high bandwidth and data rates up to 20 Gbit/s

In particular, low latency combined with high reliability provide the required robustness for fast and safety-critical communication within manufacturing. However, these performance characteristics conflict with each other, implying that not all extreme values (minimum latency and reliability, maximum data rate, maximum number of devices) can be achieved simultaneously. Therefore, the method of network slicing is applied to implement different network layers with different characteristics in the same 5G network. Network slicing enables maintaining different communication requirements for different use cases [12]. However, 5G networks operated by telecommunication companies usually cannot be optimized regarding the individual, use-case-dependent demands. Therefore, so-called private networks are needed. Private networks are locally limited and operated independently by the individual organization. This enables high flexibility and reconfigurability of the communication network regarding the needed requirements [13].

In the future, it is highly likely that industrial communication based on 5G will be widely implemented and needed for modern manufacturing systems [14]. 5G shows better performance characteristics in comparison to other communication technologies such as WiFi 6. Due to high quality of service by using dedicated spectrum resources, 5G is especially well suited for safety-critical, low-latency use cases [15]. Because of its high performance characteristics as well as the ongoing standardization process for further improvements, 5G bears potential to be the communication platform for industrial automation, control, and holistic interconnectivity - even of safety-critical applications – within manufacturing systems [16]. For example, different control tasks offloaded in a wireless format [17, 18], flexible 5G-enabled human-machine interfaces with augmented reality [19, 20], and Industrial Internet of Things networks for monitoring and diagnosis [21, 22] show strong application potential.

5G offers the possibility to develop a scalable and flexible framework for the implementation of DTs in manufacturing. In addition, 5G can serve as the communication platform for advanced and challenging industrial communication. In particular, 5G communication architecture is suitable for real-time capable, wireless DTs for simulation, monitoring, and control of a machine tool.

2.2   Physics Simulation in Manufacturing

Physics simulation is a widely used tool for analyzing physical phenomena in a virtual world. Physics simulation is based on a physics engine, which is a software platform containing reusable resources to compute specific physical behavior of material bodies [23]. Due to the capability to model mechanical behaviors, kinematics, collisions, and other physics-related properties, physics simulations and engines have been widely used in manufacturing engineering [24]. Especially for flexible CPPS, gaming engines have a very high potential for the simulation of manufacturing on a macro scale (non-molecular levels). Gaming engines combine physical simulation (kinematics, dynamics, collision, etc.) with graphical elements and user interaction capabilities [25].

On the level of machine tools, physics simulations as well as gaming engines have been used to study the system dynamics of machine tools. For example, prior to the start of a manufacturing task, the physics simulation can be used to simulate the procedure of human operations at a workplace [26] or the machining processes to analyze the chip formulations [27]. During the manufacturing process, physics simulation can be used to mirror the manufacturing activities for process monitoring, such as the component building process of a 3D printer [28] or machine tool motion [29]. On the level of manufacturing systems, physics simulation has been used for the analysis the performance of transportation vehicles in the material flow [30], the design and optimization of a workspace layout [31], or the validation of a manufacturing process [32].

All these works have demonstrated the benefits and potentials of physics simulations and engines in manufacturing engineering. In assessing these works, it is observed that a number of different commercial or non-commercial physics engines are available for different purposes. Moreover, in terms of the programming platform as well as the development environment, it is found that most physics engines are either suitable for C++, C#, Python, JavaScript, or Lua [25]. Nevertheless, in terms of the interdisciplinary research considering 5G-enabled DTs with real-time data transmission between physics simulation and real system, no works has been found in the literature so far.

2.3   Digital Twin in Manufacturing

To manage the complexity of CPPS, digital representations of relevant processes and involved systems are needed [33]. Therefore, DT links simulation models to real systems. By utilizing cloud-based simulation capacities, available data (sensor or physics simulation based), and high interconnectivity, DT enables model-based decision support for the real system [4].

In current research, there exist many different definitions of DTs with different delimitation criteria such as the life-cycle phase (design phase, manufacturing phase, service phase, retire phase) of the DT or the level of integration and information flow (digital model, digital shadow, digital twin) [34].

This paper follows the DT definition as a composition of digital models (physical simulation or data-driven) to process information in real-time from the real system. In this manner, the manufacturing process can be monitored and controlled in real-time and profound decision support for the real system is possible [35]. In addition, DT need autonomous and bidirectional data transmission in an adequate time frame regarding the specific use case [36].

According to the initial concept of DT, which roots in aerospace, three major functionalities can be distinguished [37]:

- **Prediction** for pre-process simulation of the real system
- **Monitoring and control** for prediction, analysis, and direct interaction of the real system during the process
- **Diagnosis** for model-based analysis of unpredicted failures after the operation of the real system

For implementation of DTs in manufacturing systems, a connection between CPPS and DTs is needed. Therefore, digital models have to be linked to the real system utilizing physical models, real-time augmentation with data, and a reliable communication link. A recent study emphasizes that especially the data link and thus the communication system are very important for interoperability and scalability of DTs [38]. In addition - depending on the use case - there exist high requirements regarding bidirectional communication to control the real system based on the simulation-based instructions [39]. Due to low cycle times and safety criteria of machine tool control, very fast and reliable connectivity is mandatory for DTs of machine tools.

In a recent literature survey, a shift from conceptual and framework-oriented research regarding DTs towards implemented applications of DTs is identified [34]. It was highlighted that DTs can be used to simulate, analyze, and control a variety of different processes on different levels of the manufacturing system. These range from the factory planning and factory control level, to the process and machine level [34].

For simulating and planning on a factory level, Zhang et al. develop a DT-driven smart shopfloor for dynamic resource allocation [40]. Moreover, human-based production processes can be incorporated to the DT of the factory level for better optimization results regarding time and cost-efficiency [41]. Glatt et al. implement a DT based on physics simulation for optimizing material flows through the manufacturing system [4]. On the process and machine level, DT are mainly developed for monitoring, visualizing and optimizing machining processes [42–44]. However, not many implemented DTs with bidirectional information flow between machine tools and digital system have been reported. A brief overview of implemented DT of machine tools, is provided in [8].

Not many DT in manufacturing have been reported utilizing the potentials of 5G communication. For example, Groshev et al. developed and validated a 5G-enabled DT for the control of robotics [45]. They migrated the control of a robot arm to an edge server via 5G. In addition, 5G-enabled DT are also used for remote monitoring and operation of different robots and machines [46]. In preliminary work of the authors, an architecture for DT of machine tools with migrated computerized numerical control (CNC) to an edge server is developed for wireless real-time closed-loop control [19]. However, currently, there exist no implemented DT for machine tools that utilizes the potential of 5G communication standard. Moreover, there exists no wireless DT for latency-critical machine tool control in general. Current safety-critical and time-sensitive applications are still wired and the physical communication layer on a technological level is not sufficiently considered [47]. In a current literature review on DTs, Zeb et al. determined that edge-computing as well as 5G networks and future network technologies are needed for real-time capable, wireless DTs. However, research regarding in this area is still in its infancy [47].

In summary, existing approaches do not sufficiently address the needed data link or communication system between the digital and real world and thus neglect the associated benefits and opportunities for the design of DT. This connectivity is especially essential for DT of CNC machine tools, as fast response times are required. Current solutions cannot simultaneously allow flexibility and scalability while exploiting the full potential and functionalities of DT. Therefore, a generic architecture, and implementation of a DT of a machine tool based on physics simulation and utilizing 5G and edge computing is developed. This will involve both physics simulation and the integration of disruptive communication architectures to meet the requirements of a DT of machine tools in CPPS.

3   Modeling of the Architecture for 5G-Enabled Digital Twin

The DT is a link between the physical and virtual world in CPPS. Therefore, DTs are developed and operated in the virtual world with interdependencies of systems of the physical world. The resulting combination of the digital and real world leads to a comprehensive development project with a high degree of interdisciplinary within involved research fields [4]. To manage this complexity, the requirements for functionality, the overall architecture, and the physical model of the machine tool and its functions are outlined prior to the elaboration of the detailed DT implementation. The methodology is based on model-based systems engineering, which is used for the description of complex and multidisciplinary technical systems [48].

3.1 Objectives and Requirements

As a result of the analysis of the state of the art, our approach addresses several objectives for DT in manufacturing:

- **Prediction**: The DT should allow pre-process prediction of the kinematic and dynamic behavior of the machine tool. This enables the visualization and analysis of the real system prior the actual runtime. Moreover, the verification of the correct manufacturing program and prevention of collisions of the machine tool should be realized.
- **Monitoring/Control**: Monitoring and simulation of the state of the ongoing machining process are mandatory. In addition to the monitoring process, it should also be possible for the DT to control the machine tool in order to react immediately to monitored process anomalies. Therefore, direct control of the machine tool by the DT should be possible (e.g., emergency stop or change of feed overrides)
- **Diagnosis**: The analysis of the behavior of the machine tool after the operation should also be integrated to the DT framework. Based on the monitored information (e.g., position deviations, vibration data or downtimes), process optimizations can be implemented. Therefore, predictive maintenance and big data analytics will be possible.
- **Flexibility**: The demanded flexibility of the DT framework is manifested in two characteristics. On the one hand, the deployment of the DT should be independent of the used device and operating system. On the other hand, the implementation of different DT in the manufacturing system should be possible without high additional infrastructural efforts.
- **Comprehensibility**: To increase user acceptance, the DT framework should allow easy understandability for human operators.
- **Scalability**: Besides flexibility, the DT framework should enable scalability of the whole system. This allows both the addition of new DTs of different machine tools and the expansion of existing DTs of machine tools by adding new information (e.g., by sensors, actors, inputs/outputs (I/Os) of control units). This leads to the possibility of the implementation of many different DTs of different machine tools in a manufacturing system. Moreover, the possibility to expand functionalities and interconnected devices without high implementation effort is required.
- **Real-time capability and reliability**: To enable the real-time adapted monitoring and control of the machine tool by the DT, a real-time capable interface for data transmission is needed. In addition, the communication has to be reliable, even for safety-critical processes.

To achieve these goals, various technological requirements have to be met. These requirements can be divided into hardware, software, and communication requirements.

On the communication side, the data transmission between physical and digital world should be completely wireless to enable flexibility and scalability. Therefore, especially for monitoring and control of safety-critical processes, a low latency below 50 ms for monitoring and remote control and below 1 ms for offloaded motion control is required [49]. In addition, high reliability (>99.9999% or <0.5 min of downtime per year) is required. Furthermore, the communications infrastructure should be scalable and flexible to enable high connectivity for devices to be integrated into the system in the future. For this purpose, a network protocol should be utilized that enables machine-to-machine communication.

On the software side, a physical model of the machine tool is needed, integrated into an appropriate physics engine for simulation of kinematics and dynamics of rigid bodies (e.g., collision, movement). The pre-process simulation should be based on the interpretation of the programmable language used in the CNC of the machine tool (G-Code). Due to the needed flexibility, the physics engine should be platform independent and enable the deployment of mobile applications (Android and iOS). Furthermore, a graphical user interface (GUI) for visualization is needed to provide the required comprehensibility and interactivity. Next to the software requirements of the physics simulation, there is also the need for a real-time capable CNC software that has all stages of the motion planning process (path planning, trajectory generation, and trajectory tracking) implemented. Another requirement for the CNC to monitor and control the machine tool by the DT is the ability to read and write I/O with open, adaptable, and expandable interfaces for the needed flexibility.

The required hardware consists of a machine tool with closed-loop motion control. The machine tool is the physical counterpart of the DT and enables the wireless transmission of the needed data for simulation and prediction. Therefore, 5G modules to enable the 5G-capability of the CNC are also required. In addition, the transfer of encoder feedback from the motors is required. Next to that, an edge server near the production site is mandatory to achieve the required low latencies and reliability. For future real-time control of machine processes, offloading of computing processes to outsourced servers is not possible. Processing on a local edge-server near the production site is needed to ensure low latencies [47]. In addition, the edge server needs powerful computing resources to enable machine learning based diagnosis of the DT and adequate responsiveness for control.

3.2 System Architecture

As shown in Fig. 1, the overall system consists of three interacting sub-systems: the digital system (1), the communication system (2), and the real system (3). This architecture represents the progression of ongoing research and is based on preliminary work by the authors about 5G-enabled DTs for closed-loop machine tool control [8].

The *real system* consists of the machine tool to be operated with its physical components needed for operation (motors with motion controller, spindle, limit switches, etc.). In addition, the 5G-enabled CNC is a mandatory part of the real system to ensure the wireless communication between digital and real system. The CNC needs an adaptable interface to enable the transfer of I/O values of the machine tool via 5G to an edge server. In addition, workpiece features and the G-Code is transferred via 5G from the edge server to the CNC. In the real system, there are sensors integrated to provide information for further analysis and diagnosis of the machining process. Next to the sensors, there are different possible human-machine-interfaces (HMI) and mobile devices that

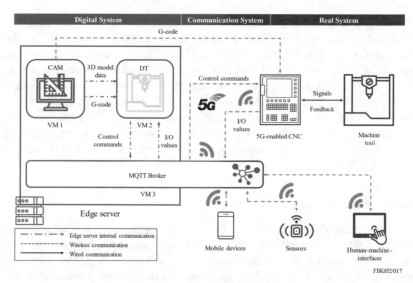

Fig. 1. Architecture of the overall system.

can be integrated via 5G. Due to the centralized MQTT broker and the 5G connection, visualization and further data processing can also occur decentralized on various devices.

The *communication system* consists of a 5G network and creates the link between the digital and real system. In this architecture, the 5G communication as part of the communication system is mandatory for functionality of the overall setup. 5G enables reliable, low-latency, and data-intensive interconnectivity for the different components of the system. The data flow via 5G is based on the MQTT protocol that enables high flexibility and scalability as well as machine-to-machine communication between different machines or computing units. The different I/O values are sent to the MQTT broker that is running on a virtual machine (VM 3) on the edge server. In addition, the commands computed by the DT (VM 2) on the edge server are sent to the CNC via MQTT. The detailed functionality of the interfaces is outlined in Sect. 4.

The *digital system* is offloaded in a wireless manner to a powerful edge server that is directly connected to the 5G core. It consists of different VMs with different types of functionalities. The first VM 1 contains supporting instances such as computer aided manufacturing (CAM). The 3D model data from the CAM software is transferred internally on the edge server to the DT on VM 2. In addition, the G-Code generated by the CAM software is sent via 5G to the 5G-enabled CNC unit. The G-Code is also sent to the DT to enable the pre-process prediction of the kinematics and dynamics of the machining processes based on G-Code. On the second VM 2 the DT for prediction, monitoring, control, and diagnosis is operated. The prediction functionality is achieved by interpreting machine specific G-Code received from VM 1. Based on this G-Code interpretation, the machining process is simulated. The simulation is based on a physics engine for simulating kinematics and dynamics. Next to the prediction mode, there is also a monitoring and control mode. The DT receives real-time data from the CNC unit via 5G. The data is transferred via the MQTT protocol that is lightweight and based on the publish and subscribe principle with a broker as intermediary. This MQTT broker is also on the edge server on VM 3. VM 3 orchestrates the entire 5G traffic of different devices connected (DT, CNC, sensors, mobile devices, HMI). The DT interprets the received I/O data from the CNC to monitor and visualize the machine tool during process. Therefore, a dynamic real-time image of the process and the involved components based on the real-time data is rendered. Another mode of the DT is the diagnosis mode. Additional sensor data, as well as data received from CNC (e.g., encoder feedback from the motors, power consumption, etc.) can be evaluated with statistical models as well as machine learning-based algorithms. A detailed description of the implementation and functionality of the DT is provided in Sect. 4.3.

3.3 Interactions and Information Flow

Based on the general model described in Sect. 3.2, the interactions and information flow between the different sub-systems - real, digital, and communication system – are outlined in the following. For better comprehensibility, the software and hardware utilized for implementing the DT is specified. However, for detailed implementation, refer to Sect. 4. All physical and digital systems of the implemented architecture are represented in an extended UML diagram (see Fig. 2). The top column of each box in this Figure describes the system affiliation and the needed specific soft- or hardware to implement the architecture. The input and output variables of the systems are represented by a " + " and "−" respectively, followed by a descriptive nomenclature, and, if reasonable, the file format. Furthermore, Fig. 2 illustrates the processes within the respective component to ensure further processing of the different variables and to forward them to the appropriate subsequent instance (e.g., ".publish();"). Additionally, the flow of information and utilized communication architecture between the individual sub-systems are illustrated.

The CAM system is based on Autodesk Fusion 360 (VM 1) and the MQTT broker is based on Eclipse Mosquitto (VM 3). For the prediction mode, the G-Code (.ngc) is sent within the edge server from Autodesk Fusion 360 (VM 1) to the Unity application (VM 2). In Unity, the G-Code is interpreted and visualized for predicting the kinematics and dynamics of the machine tool behavior (.predictBehavior();).

For monitoring, there exists a loop for continuous data transmission and interpretation. First, a G-Code is generated in Autodesk Fusion 360 (VM 1) that is sent via 5G to LinuxCNC. The G-Code is transferred from CAM to CNC (LinuxCNC) using the WebSocket protocol, which is based on TCP/IP to ensure no packet loss. There the machine is moved according to the G-Code and position feedbacks from the encoders of the machine tool are returned to LinuxCNC. The position data and other relevant information from the CNC unit are then sent to the MQTT Broker (VM 3: Mosquitto broker) on the edge server via 5G using the monitoring interface. The Unity application subscribes to the data (VM2: Python Subscribe) from the broker to monitor the actual machining process. Therefore, the current status and spatial position of the machine tool is visualized.

For control, the data transmission is triggered by the Unity application. When the need for action is identified by the operator, a command is triggered (e.g., emergency stop, change feed override, etc.) via a graphical interface of the DT. The information is published via 5G using MQTT protocol (VM 2: Python Publish) to the broker (VM 3) and subscribed, processed, and forwarded to LinuxCNC utilizing the action interface. LinuxCNC interprets and forwards the command to the machine tool, which is then executed. For feedback of successful outcome of control tasks, the DT receives the information of the current machine status via the continuous loop for monitoring.

Currently, the prediction and monitoring mode are two different functionalities of the DT. They will be merged in the future for automated process preparation. Moreover, data from the diagnosis function (e.g., sensor data) and automated process adjustments (e.g., automated emergency stop) are not implemented yet. However, some of the needed data is already transferred to the DT via 5G (e.g., encoder feedback).

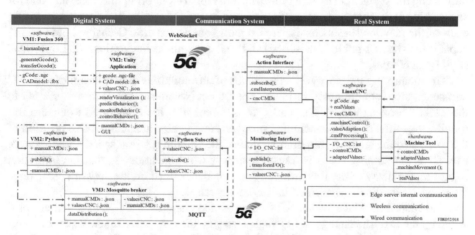

Fig. 2. Extended UML diagram of information flow.

4 Implementation

In the following sub-sections, the implementation of the system is outlined. In particular, the functionality of the DT with its different modes is elaborated. It is worth mentioning, that the implementation of 5G-enabled sensors and human-machine-interfaces - as a part of the architecture - is an ongoing area of research. It should be referred to Sect. 5 for further information.

4.1 Real System

The real system consists of a three-axis gantry milling machine and the CNC control unit based on LinuxCNC. The system has been developed to meet the needed requirements, especially regarding 5G connectivity. LinuxCNC   as an open source CNC for machine tools – enables a high degree of freedom regarding its configuration, manipulation, and connectivity. LinuxCNC allows simple definition of the hardware abstracted layer of the machine tool, allowing high flexibility in potential I/Os. It is possible to manually add virtual, software defined I/Os, which is needed for monitoring and controlling the real machine tool with the DT.

Another benefit of LinuxCNC is the presence of a dedicated Python interface [50] that allows reading, controlling, manipulating, and creating I/O signals. However, the Python interface does not provide a way to interact with the machine tool via network interface out of the box.

The developed interface to monitor and control the machine tool via network is implemented in LinuxCNC and thus also part of the real system. It consists of the aforementioned Python interface that is expanded with a MQTT publish and subscribe function. To ensure both operating modes – monitoring and control -, two interfaces are needed as can be seen in Fig. 2 (action interface and monitoring interface). One for publishing the status information of the machine tool (monitoring interface) and one for subscribing to the manual commands triggered by the DT (action interface). As a result, each of the new interfaces consist of two parts, one to for the 5G-enabled networking capabilities (MQTT publish/subscribe) and one to ensure the integration into LinuxCNC (Python interface).

The monitoring interface reads the values of different I/Os directly from LinuxCNC and sends them continuously in JavaScript Object Notation (JSON) format via 5G to the MQTT broker (VM 3). Currently, 44 different I/Os are transferred to ensure full monitoring of the ongoing machining process. Next to absolute and relative position of the axes, also the status of the limit switches, the spindle rotation speed and direction, as well as the homing status is transmitted. It is worth mentioning, that the limitation regarding the delay of the transmission speed is determined on the network side, since the publish function is performed in the sub ms range.

The control interface subscribes to pre-defined changes of I/O values that are provided by the DT. The change of I/O values triggers the respective functions (e.g., emergency stop, manual jogging), which then are interpreted and executed by LinuxCNC to control the physical machine tool. This enables the machine tool to be controlled via 5G by the DT on the edge server. Simultaneously, the change in machine tool status is captured by the monitoring interface and transmitted back to the DT, providing feedback of the control process.

4.2   Communication System

The communication system provides interconnectivity of all implemented devices and is completely based on 5G wireless technology. However, due to the absence of integrated, 5G-enabled systems for manufacturing, both software and hardware have to be adapted. The hardware structure of the system is illustrated in Fig. 3.

The 5G connectivity of LinuxCNC is ensured by retrofitting 5G capability with a 5G-Gateway based on a Raspberry Pi 4B with a M2 5G-module from Quectel[1]. The Raspberry Pi is operated with OpenWRT, a Linux-based operating system that was developed specifically for network routing.

On the software side, MQTT is used as a communication protocol between DT and LinuxCNC. Due to the MQTT broker as the central orchestration unit of the data

---

[1] Naming of specific manufacturers is done solely for the sake of completeness and does not necessarily imply an endorsement of the named companies nor that the products are necessarily the best for the purpose.

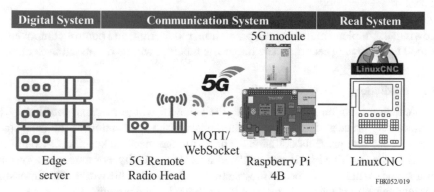

Fig. 3. Implemented communication system based on preliminary work [8].

transferred, the protocol with its publish and subscribe functions enables high flexibility and scalability within the manufacturing system. In addition, MQTT enables triggering of commands (e.g., machine stop) when the connection of any publishing or subscribing client is interrupted. The MQTT broker operates with quality-of-service level of 0 for fastest transmission speeds. It is worth mentioning that just one air interface is involved, which reduces the overall latency. The subscription of the data by the DT is handled edge server internal. Next to the MQTT protocol, the WebSocket protocol is utilized for transmitting the G-Code between the CAM system and LinuxCNC.

4.3   Digital System

The digital system consists of the three virtual machines: one supporting instance based on Microsoft Windows with the CAM system (VM 1: Autodesk Fusion 360), the MQTT broker running on Ubuntu (VM 3: Mosquitto Broker), and the DT with its interfaces running on Ubuntu (VM 2: Unity application). The detailed interaction and information flow of the instances involved is outlined in Sect. 3.3. In the following, the functionality of VM 1 and VM 3 will not be outlined in this paper.

The main part of the digital system is the DT that is shown in Fig. 5. It should ensure the prediction, wireless monitoring, and control, as well as the diagnosis of the real machining process. The DT is based on Unity, a gaming engine based on the C# programming language. Unity enables the required independency of the operating system, physics simulation, deployment as mobile application, and visualization of the DT. For example, due to the publish and subscribe architecture of MQTT and the 5G communication standard, the DT can also be deployed on tablets or different computers in the manufacturing system – even simultaneously. Therefore, the devices need a 5G interface and the interfaces for subscription or publishing.

The diagnosis function is currently under development. Therefore, a smartphone with 5G-capability as well as 5G-enabled IoT sensors are implemented into the process for further data analysis. Due to its flexibility and communication based on 5G, the developed architecture in Sect. 3.2 enables simple integration of different sensors and actors to the DT. Moreover, the edge server enables fast processing of data analysis algorithms and low-latency communication due to geographical proximity. However, as described in Sect. 3.3, the diagnosis function is not fully implemented yet. In the following, the implementation of the prediction, monitoring, and control components is outlined the following sections. The diagnosis function will be discussed in Sect. 5.

4.3.1 Prediction

The prediction function should simulate the behavior of the machine tool before the manufacturing process starts. The function enables the verification of the correctness of the G-Code and prevents collisions. Therefore, the machine specific G-Code (.ngc format) for the machine tool is analyzed and simulated in Unity. For the simulation environment, the spatial dimensions of the machine tool as well as the workpiece dimensions are integrated, and physical characteristics are added to the models.

The kinematic and dynamic behavior of the machine tool based on the G-Code interpretation is simulated by adding equations of motion (translational movement and acceleration). In addition, the acceleration parameters of the stepper motors and mass of the machine tool are integrated into the simulation. The process flow of the G-Code simulation is shown in Fig. 4.

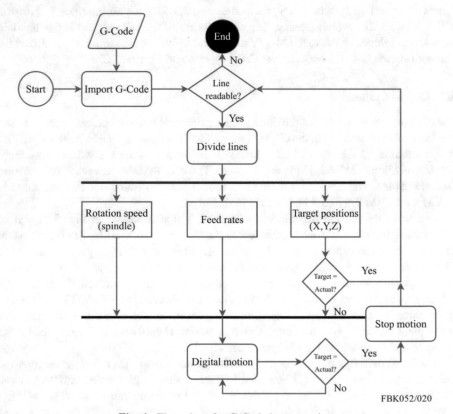

FBK052/020

Fig. 4. Flow chart for G-Code interpretation.

As it is depicted in this Figure, the G-Code is completely imported to the DT (Unity application) and the loop for the G-Code simulation starts. At the beginning of each loop, the interpretation of the current line of the G-Code is checked. If this is true, the new line is split into three parts with different available information. If the line is not interpretable, the simulation ends. The data required for the movement of the machine are the target positions (X,Y,Z) of each axis, the feed rates of each axis, as well as the rotation speed and direction of the spindle. This information is simulated, and the digital motion occurs. However, first the target and actual position of each axis are compared to neglect redundant calculations of the trajectories of the axes. For low computational intensity, the feed rates and spindle control inputs are directly processed for simulation. The resulting simulated digital motion of the machine tool is stopped when the target position is reached. Next to the automated stop, the operator has the ability to stop the simulation via the GUI.

4.3.2 Monitoring and Control

Fig. 5. DT during manufacturing process.

The monitoring and control function enables wireless anomaly detection of the manufacturing process. Encoder feedback is transferred to the DT to simulate deviations of the actual and target position. The inaccuracies can then be visualized. Furthermore, manual control such as emergency stop or jogging is enabled for direct process manipulation. In Fig. 5 the GUI of the DT, the visualized machining process, and the real process is shown. The visualization is based on the CAD model of the machine tool and the workpiece. The DT has four different operation modes (stand by, streaming, manual, G-Code) and five different viewpoints with different angles and distances to the work piece that the operator can switch through. In addition, information about the current tool can be manually added. The current position of the global coordinate system, the position of the used relative coordinate system (relative system 1 (G-Code 54) is used in Fig. 5), and the current spindle speed in rounds per minute is also displayed in the GUI. Moreover, the current values of the transmitted I/Os can be monitored in the lower half of the GUI.

For monitoring, the I/O values of the CNC unit – currently 44 different I/Os – are transferred via 5G to the DT. Each set of I/Os generates a new digital state of the machine tool that contains spatial dimensions, and different information about the system status. In addition, the target positions, actual positions, and the acceleration of the axes are transferred for further diagnosis of process accuracy. Each set of points generates the digital motion of the DT according to the information transferred via 5G. The detailed information flow is illustrated in Fig. 2.

To enable the monitoring function, the information from the real system which is published to the MQTT broker has to be subscribed and interpreted by the DT. For this, Eclipse PAHO is utilized. The data is written into a JSON format that is continuously accessed and interpreted by the DT. To simulate material removal of the work piece, mesh manipulation is integrated based on Clipper2 [51].

For control mode, commands are generated by human input via the GUI of the DT. This allows direct control of the process and direct response to process anomalies. The manual commands are published edger server internal to the MQTT broker. The transferred commands trigger the action interface of the real system, which interprets them and forwards the operation task to the CNC unit for execution. Feedback on the success of the control tasks is then provided by the monitoring function of the DT. Currently jogging and emergency stop is integrated. However, further control functions such as spindle control or starting a homing process will be implemented.

4.4   Benefits and Challenges

The benefits of the implementation of the architecture and thus the DT can be derived by addressing the requirements in Sect. 3.1 and are summarized in different categories below:

- **Manageability of system complexity:** The architecture enables prediction, monitoring, control, and diagnosis of manufacturing processes. Due to this functionality, the complexity of CPPS can be managed and the operator's understanding of manufacturing processes in an interconnected manufacturing system prior to, during, and subsequent to the process is ensured. In addition, the GUI provides an understanding of the current status of the machine as well as of the full manufacturing process, resulting in better comprehensibility for humans. For machine tools, processes can be simulated, and the G-code can be verified for correctness. Therefore, defective parts, waste, and thus costs can be reduced. Moreover, the implemented DT can be integrated into the framework of 5G-enabled HMIs (see another preliminary work of the authors [19]).

- **Flexibility and Scalability:** Due to the wireless realization of the DT by utilizing 5G communication technology and the possibility for deployment as mobile applications, the architecture enables high flexibility of the manufacturing system. Different machinery or process equipment can be implemented to the DT regardless of their physical location. The resulting scalability is also facilitated by the utilization of the MQTT protocol. This allows the DT to run simultaneously on different devices (computers, tablets, smartphones) independent of the operating system, without issues regarding data integrity and availability. For the machine tool of the implemented DT, this means that it can be moved flexibly within the manufacturing system without connectivity problems and thus functional losses. Moreover, further information can be implemented easily for example by adding 5G-enabled sensors.

- **Transferability:** Due to the performance characteristics of 5G, the DT is transferable for different processes. The DT of a CNC machine tool can be seen as a benchmark use case for uRLLC. The required performance characteristics for the communication technology for machine tool control are one of the highest within manufacturing systems [49]. Thus, if 5G shows sufficient performance for the developed DT, 5G also enables real-time monitoring and control of various manufacturing processes. In addition, the adaptability of the communication system enables technology transfer capability towards upcoming communication standards (5G+, 6G, etc.) due to the use of modularized network interfaces.

- **Robustness:** The 5G communication standard allows reliable and fast communication. Therefore, even safety-critical processes can be offloaded to the edge server. In addition, MQTT supports the triggering of individual commands (e.g., machine stop) when the connection is interrupted.

- **Cost savings:** By offloading computational units wirelessly to the edge server, space savings are realized, and less wiring infrastructure is needed. The implemented setup at our institute leads to around 56.7% less wiring infrastructure. Due to less wiring, cost savings in double-digit euro range are possible [52]. Furthermore, the centrality of the edge server and consequently the central software deployment leads to less maintenance effort and thus costs. Moreover, the prediction and diagnosis functions of the DTs for machine tools leads to cost savings due to less defective parts and less downtimes of the machine tool.

Next to the benefits of the 5G-enabled DT, there exist also challenges during implementation. Reading and manipulating I/Os requires an open CNC unit with appropriate interfaces for external control by third party programs. Currently, no manufacturer of common industrial grade CNC units delivers this capability. Reading and overwriting - especially of safety critical and motion control related I/Os - is not possible. Therefore, LinuxCNC is utilized, which is a very comprehensive CNC software. However, LinuxCNC requires a lot of machine-specific configuration and expertise for implementation.

Another issue could be the centralized MQTT broker connected to the 5G core. When scaling up the publish and subscribe operation (adding more DTs in the manufacturing system), the broker could be overloaded with traffic orchestration.

Another challenge is the complexity of the operation of a private 5G network. As shown in [19], the communication performance depends on different configurable characteristics of the 5G network. If set up correctly, 5G is a good basis for wireless DT but currently the operator needs high expertise for the integration of 5G networks into the CPPS. Furthermore, there are also high costs in the five – to low six-figure range for building, setting up and operating a private 5G network.

5 Summary and Outlook

In this paper an architecture for 5G-enabled DT for prediction, monitoring, and diagnosis is developed. The framework is implemented by developing a 5G-enabled DT of a machine tool. The DT is completely wirelessly migrated to a central high-performance edge server, which is facilitated by 5G campus networks. To simulate and control the machining process in real-time, reliable and low-latency communication is needed. 5G is the wireless communication technology that meets the strict communication requirements for DTs of machine tools and thus has potential for disruptive change of industrial communication.

This work is the first implemented use case of 5G-enabled DTs for bidirectional control of machine tools in manufacturing and extends previous research by the authors regarding this topic. The implementation demonstrates many advantages, e.g., less wiring efforts in manufacturing, centralized maintenance and management, better comprehensibility of complex CPPS as well as high scalability, flexibility and reconfigurability of the DT. In addition, the presented framework provides transferability for further DTs of different assets and expandability of the current implementation. Real-time capable DT of a machine tool can be a benchmark for uRLLC. Therefore, the architecture is suitable for the majority of processes in CPPS.

Currently, the DT of the machine tool is expanded by integrating sensor data of 5G smartphones and 5G-enabled IoT sensors. For this reason, vibration data from the workpiece is captured and machine learning algorithms are utilized for analyzing status and predicting anomalies of the machining process. The sensor data will also be implemented to the monitoring and control mode to enable better anomaly detection and inline process control. Moreover, 5G-enabled mobile HMI are implemented that operates the DT as an edge device. For diagnosis, there is also encoder feedback transferred to the DT via 5G. This enables the detection of anomalies in the machine tool movement.

Regarding performance evaluation of the overall monitoring and control system, different experiments are currently being conducted with different 5G network configurations and traffic load on the network. The end-to-end latencies, jitter and deviation will be measured between DT and CNC unit. Moreover, the benchmark use case will be further defined and evaluated, for example regarding the monitoring accuracy of the DT in comparison to a wired solution or the resulting manufactured part quality with offloaded process control. First, experiments show that the private 5G network at TU Kaiserslautern meets the requirements for monitoring and manual control with latencies below 10 ms ($\mu = 6.8$ ms; $\sigma = 3.6$ ms with ~2600 × 44 data points sent) and low jitter (~2.5 ms). The full results with the description of the experimental design will be published in future research.

Acknowledgement. The authors would like to thank the Deutsche Forschungsgemeinschaft (DFG, German Research Foundation) for its financial support in the context of the IRTG 2057 (funding code: 252408385) and the Federal Ministry of Digital and Transport (BMDV) for its financial support in the context of the 5 × 5G strategy (funding code: VB5GFKAISE).

References

1. Monostori, L., et al.: Cyber-physical systems in manufacturing. CIRP Ann. **65**(2), 621–641 (2016). https://doi.org/10.1016/j.cirp.2016.06.005
2. Mourtzis, D.: Machine tool 4.0 in the era of digital manufacturing. In: Proceedings of the 32nd European Modeling & Simulation Symposium, 16–18 September 2020. CAL-TEK srl, pp. 416–29 (2020)
3. Wang, L., Shih, A.J.: Challenges in smart manufacturing. J. Manuf. Syst. **40**(1) (2016). https://doi.org/10.1016/j.jmsy.2016.05.005
4. Glatt, M., Sinnwell, C., Yi, L., Donohoe, S., Ravani, B., Aurich, J.C.: Modeling and implementation of a digital twin of material flows based on physics simulation. J. Manuf. Syst. **58**, 231–245 (2021). https://doi.org/10.1016/j.jmsy.2020.04.015
5. Qi, Q., et al.: Enabling technologies and tools for digital twin. J. Manuf. Syst. **58**, 3–21 (2021). https://doi.org/10.1016/j.jmsy.2019.10.001
6. Navarro-Ortiz, J., Romero-Diaz, P., Sendra, S., Ameigeiras, P., Ramos-Munoz, J.J., Lopez-Soler, J.M.: A survey on 5G usage scenarios and traffic models. IEEE Commun. Surv. Tutor. **22**(2), 905–929 (2020). https://doi.org/10.1109/COMST.2020.2971781
7. Cheng, J., Chen, W., Tao, F., Lin, C.-L.: Industrial IoT in 5G environment towards smart manufacturing. J. Ind. Inf. Integr. **10**, 10–19 (2018). https://doi.org/10.1016/j.jii.2018.04.001
8. Mertes, J., Glatt, M., Schellenberger, C., Klar, M., Schotten, H.D., Aurich, J.C.: Development of a 5G-enabled digital twin of a machine tool. Procedia CIRP **107**, 173–8 (2022). https://doi.org/10.1016/j.procir.2022.04.029
9. Penttinen, J.T.J.: 5G Second Phase Explained: The 3GPP Release 16 Enhancements. Wiley, Hoboken (2021)
10. Alawe, I., Ksentini, A., Hadjadj-Aoul, Y., Bertin, P.: Improving traffic forecasting for 5G core network scalability: a machine learning approach. IEEE Netw. **32**(6), 42–49 (2018). https://doi.org/10.1109/MNET.2018.1800104
11. ETSI. 5G (2022). https://www.etsi.org/technologies/5g. Accessed 28 Nov 2022
12. Foukas, X., Patounas, G., Elmokashfi, A., Marina, M.K.: Network slicing in 5G: survey and challenges. IEEE Commun. Mag. **55**(5), 94–100 (2017). https://doi.org/10.1109/MCOM.2017.1600951
13. Federal Ministry for Economic Affairs and Energy: Guidelines for 5G Campus Networks – Orientation for Small and Medium-Sized Businesses, Berlin (2020)
14. Cheng, J., Yang, Y., Zou, X., Zuo, Y.: 5G in manufacturing: a literature review and future research. Int. J. Adv. Manuf. Technol. (2022). https://doi.org/10.1007/s00170-022-08990-y
15. Oughton, E.J., Lehr, W., Katsaros, K., Selinis, I., Bubley, D., Kusuma, J.: Revisiting wireless internet connectivity: 5G vs Wi-Fi 6. Telecommun. Policy **45**(5), 102127 (2021). https://doi.org/10.1016/j.telpol.2021.102127

16. Mourtzis, D., Angelopoulos, J., Panopoulos, N.: Smart manufacturing and tactile internet based on 5G in industry 4.0: challenges, applications and new trends. Electronics **10**(24), 3175 (2021). https://doi.org/10.3390/electronics10243175
17. Girletti, L., Groshev, M., Guimaraes, C., Bernardos, C.J., de La Oliva, A.: An intelligent edge-based digital twin for robotics. In: 2020 IEEE Globecom Workshops, 7 December 2020–11 December2020, Taipei, Taiwan, pp. 1–6. IEEE (2020)
18. Kropp, A., Schmoll, R.-S., Nguyen, G.T., Fitzek, F.H.P.: Demonstration of a 5G multi-access edge cloud enabled smart sorting machine for industry 4.0. In: 2019 16th IEEE Annual Consumer Communications & Networking Conference (CCNC 2019), 11 January2019–14 January 2019, Las Vegas, NV, USA, Piscataway, NJ, pp. 1–2 (2019). IEEE
19. Mertes, J., et al.: Evaluation of 5G-capable framework for highly mobile, scalable human-machine interfaces in cyber-physical production systems. J. Manuf. Syst. **64**, 578–593 (2022). https://doi.org/10.1016/j.jmsy.2022.08.009
20. Siriwardhana, Y., Porambage, P., Liyanage, M., Ylinattila, M.: A survey on mobile augmented reality with 5G mobile edge computing: architectures, applications and technical aspects. IEEE Commun. Surv. Tutor. **1** (2021). https://doi.org/10.1109/COMST.2021.3061981
21. Chandra Shekhar Rao, V., Kumarswamy, P., Phridviraj, M.S.B., Venkatramulu, S., Subba Rao, V.: 5G enabled industrial Internet of Things (IIoT) architecture for smart manufacturing. In: Reddy, K.A., Devi, B.R., George, B., Raju, K.S. (eds.) Data Engineering and Communication Technology. LNDECT, vol. 63, pp. 193–201. Springer, Singapore (2021). https://doi.org/10.1007/978-981-16-0081-4_20
22. Chettri, L., Bera, R.: A comprehensive survey on Internet of Things (IoT) toward 5G wireless systems. IEEE Internet Things J. **7**(1),16–32 (2020). https://doi.org/10.1109/JIOT.2019.2948888
23. Gregory, J.: Game Engine Architecture. CRC Press, Boca Raton (2019)
24. Hofmann, D., Reinhart, G.: Raising accuracy in physically based simulations through scaling equations. J. Comput. Inf. Sci. Eng. **13**(4) (2013). https://doi.org/10.1115/1.4025590
25. Zarco, L., Siegert, J., Schlegel, T., Bauernhansl, T.: Scope and delimitation of game engine simulations for ultra-flexible production environments. Procedia CIRP **104**, 792–7 (2021). https://doi.org/10.1016/j.procir.2021.11.133
26. Vasilopoulos, G., Vosniakos, G.-C.: Preliminary design of assembly system and operations for large mechanical products using a game engine. Procedia CIRP **104**, 1395–400 (2021). https://doi.org/10.1016/j.procir.2021.11.235
27. Yang, X., Marx, T., Zimmermann, M., Hagen, H., Aurich, J.C.: Virtual reality animation of chip formation during turning. AMR **223**, 203–11 (2011). https://doi.org/10.4028/www.scientific.net/AMR.223.203
28. Yi, L., Glatt, M., Ehmsen, S., Duan, W., Aurich, J.C.: Process monitoring of economic and environmental performance of a material extrusion printer using an augmented reality-based digital twin. Addit. Manuf. **48**, 102388 (2021). https://doi.org/10.1016/j.addma.2021.102388
29. Abdul Kadir, A., Xu, X., Hämmerle, E.: Virtual machine tools and virtual machining—a technological review. Robot. Comput. Integr. Manuf. **27**(3), 494–508 (2011). https://doi.org/10.1016/j.rcim.2010.10.003
30. Glatt, M., Kull, D., Ravani, B., Aurich, J.C.: Validation of a physics engine for the simulation of material flows in cyber-physical production systems. Procedia CIRP **81**, 494–499 (2019). https://doi.org/10.1016/j.procir.2019.03.125
31. Messi, L., de Soto, B.G., Carbonari, A., Naticchia, B.: Spatial conflict simulator using game engine technology and Bayesian networks for workspace management. Autom. Constr. **144**, 104596 (2022). https://doi.org/10.1016/j.autcon.2022.104596
32. Serpa, Y.R., Nogueira, M.B., Rocha, H., Macedo, D.V., Rodrigues, M.A.F.: An interactive simulation-based game of a manufacturing process in heavy industry. Entertain. Comput. **34**, 100343 (2020). https://doi.org/10.1016/j.entcom.2020.100343

33. Uhlemann, T., Lehmann, C., Steinhilper, R.: The digital twin - realizing the cyber-physical production system for industry 4.0. Procedia CIRP **61**, 335–40 (2017). https://doi.org/10. 1016/j.procir.2016.11.152
34. Liu, M., Fang, S., Dong, H., Xu, C.: Review of digital twin about concepts, technologies, and industrial applications. J. Manuf. Syst. **58**, 346–361 (2021). https://doi.org/10.1016/j.jmsy. 2020.06.017
35. Kunath, M., Winkler, H.: Integrating the digital twin of the manufacturing system into a decision support system for improving the order management process. Procedia CIRP **72**, 225–31 (2018). https://doi.org/10.1016/j.procir.2018.03.192
36. Negri, E., Fumagalli, L., Macchi, M.: A review of the roles of digital twin in CPS-based production systems. Procedia Manuf. **11**, 939–948 (2017). https://doi.org/10.1016/j.promfg. 2017.07.198
37. Shafto, M., et al.: Draft modeling, simulation, information technology & processing roadmap (2010)
38. Autiosalo, J., Vepsalainen, J., Viitala, R., Tammi, K.: A feature-based framework for structuring industrial digital twins. IEEE Access **8**, 1193–1208 (2020). https://doi.org/10.1109/ACC ESS.2019.2950507
39. Tao, F., Cheng, J., Qi, Q., Zhang, M., Zhang, H., Sui, F.: Digital twin-driven product design, manufacturing and service with big data. Int. J. Adv. Manuf. Technol. **94**(9–12), 3563–3576 (2017). https://doi.org/10.1007/s00170-017-0233-1
40. Zhang, H., Zhang, G., Yan, Q.: Dynamic resource allocation optimization for digital twin-driven smart shopfloor. In: ICNSC 2018, 27 March 2018–29 March 2018, Zhuhai, Piscataway, NJ, pp. 1–5 (2018). IEEE
41. Nikolakis, N., Alexopoulos, K., Xanthakis, E., Chryssolouris, G.: The digital twin implementation for linking the virtual representation of human-based production tasks to their physical counterpart in the factory-floor. Int. J. Comput. Integr. Manuf. **32**(1), 1–12 (2019). https:// doi.org/10.1080/0951192X.2018.1529430
42. Zhu, Z., Liu, C., Xu, X.: Visualisation of the digital twin data in manufacturing by using augmented reality. Procedia CIRP **81**, 898–903 (2019). https://doi.org/10.1016/j.procir.2019. 03.223
43. Sharif Ullah, A.M.: Modeling and simulation of complex manufacturing phenomena using sensor signals from the perspective of Industry 4.0. Adv. Eng. Inform. **39**, 1–13 (2019). https:// doi.org/10.1016/j.aei.2018.11.003
44. Soares, R.M., Câmara, M.M., Feital, T., Pinto, J.C.: Digital twin for monitoring of industrial multi-effect evaporation. Processes **7**(8), 537 (2019). https://doi.org/10.3390/pr7080537
45. Groshev, M., Guimaraes, C., La Oliva, A., de, Gazda R.: Dissecting the impact of information and communication technologies on digital twins as a service. IEEE Access **9**, 102862–102876 (2021). https://doi.org/10.1109/ACCESS.2021.3098109
46. Isto, P., Heikkilä, T., Mämmelä, A., Uitto, M., Seppälä, T., Ahola, J.M.: 5G based machine remote operation development utilizing digital twin. Open Eng. **10**(1), 265–272 (2020). https://doi.org/10.1515/eng-2020-0039
47. Zeb, S., Mahmood, A., Hassan, S.A., Piran, M.J., Gidlund, M., Guizani, M.: Industrial digital twins at the nexus of NextG wireless networks and computational intelligence: a survey. J. Netw. Comput. Appl. **200**, 103309 (2022). https://doi.org/10.1016/j.jnca.2021.103309
48. Sinnwell, C., Krenkel, N., Aurich, J.C.: Conceptual manufacturing system design based on early product information. CIRP Ann. **68**(1), 121–124 (2019). https://doi.org/10.1016/j.cirp. 2019.04.031
49. Ho, T.M., et al.: Next-generation wireless solutions for the smart factory, smart vehicles, the smart grid and smart cities (2019). https://arxiv.org/pdf/1907.10102. Accessed 08 Feb 2023
50. LinuxCNC. Python interface (2022). https://linuxcnc.org/docs/devel/html/config/python-int erface.html. Accessed 14 Nov 2022

51. Johnson, A.: Clipper2 - polygon clipping and offsetting library (2022). https://github.com/AngusJohnson/Clipper2. Accessed 28 Oct 2022

52. Ahrend, U., Aleksy, M., Berning, M., Gebhardt, J., Mendoza, F., Schulz, D.: Sensors as the basis for digitalization: new approaches in instrumentation, IoT-concepts, and 5G. Internet Things **15**, 100406 (2021). https://doi.org/10.1016/j.iot.2021.100406

A Human-Centered Framework for Scalable Extended Reality Spaces

V. M. Memmesheimer[✉] and A. Ebert

Human Computer Interaction Lab, Department of Computer Science, RPTU
Kaiserslautern-Landau, Kaiserslautern, Germany
v.memmesheimer@rptu.de

Abstract. Mixed and Virtual Reality technologies have been assigned considerable potential to support training and workflows in various domains. However, available solutions are subject to scalability limitations which evoke temporal and cognitive efforts that outweigh the technology's intrinsic potential and prevent their application in profit-making, real-world settings. Addressing these issues, we developed a framework for Scalable Extended Reality (XR^S) spaces following a human-centered design process. To this end, we derived abstract high-level use cases which exploit key benefits of Mixed and Virtual Reality technologies and can be combined with each other to describe specific low-level use cases in many domains. Based on the defined high-level use cases, i.e., design and development of physical items, training, teleoperation, co-located and distributed collaboration, we specified functional and non-functional requirements and developed a framework design solution that implements multidimensional scalability enhancements: Multiple on-site and off-site users can access the XR^S space through customized Mixed or Virtual Reality interfaces and then reference or manipulate real or virtual scene components. Thereby, full scalability regarding options of interaction is provided through the integration of a robotic system that allows off-site users to manipulate real scene components on site. Eventually, the framework's applicability to different use cases is demonstrated in theoretical walkthroughs.

1 Introduction

Extended Reality (XR) technologies provide a versatile tool to inspect and interact with three-dimensional virtual elements in physical or completely computer-generated environments. Throughout the last decade XR technologies have been assigned considerable potential to support training and workflows in many different domains such as in the automotive and aerospace industry, healthcare, interior design, factory layout planning, at construction sites, for maintenance and repair tasks as well as for co-located and distributed collaboration in general. Even within each of these domains, XR technologies can support multiple different workflows and tasks. For instance, in the automotive industry, XR can be employed for early design and engineering reviews of virtual prototypes to save material costs as well as for training the teleoperation of machines and providing support by remote experts.

© The Author(s) 2023

J. C. Aurich et al. (Eds.): IRTG 2023, *Proceedings of the 3rd Conference on Physical Modeling for Virtual Manufacturing Systems and Processes*, pp. 111–128, 2023.
https://doi.org/10.1007/978-3-031-35779-4_7

While the field of applications for XR technologies is vast, its actual profit-making implementation in real-world scenarios is still limited, due to the use case driven development of XR applications resulting in solutions that are limited in terms of hardware, tasks, and number of users. As such, frequent system adaptations are required which evoke temporal and cognitive efforts that outweigh XR's actual potential and hinder its application in the real world.

Seeking to address these issues, this paper presents a framework for Scalable Extended Reality spaces that provides scalability between different degrees of virtuality, different devices, and different numbers of users. The development of the framework followed a human-centered design process: We first defined its context of use by high-level use cases that exploit XR's key benefits and can be combined with each other to describe specific low-level use cases. Next, we defined functional and non-functional requirements, based on which we developed a framework design solution. Eventually, theoretical walkthroughs are provided to demonstrate its applicability to different use cases.

2 Background

2.1 Terminology

Extended Reality (XR) is currently being used as an umbrella term for environments that are encompassed by the so-called Reality-Virtuality Continuum that was introduced by Milgram et al. [1]. It ranges from reality (i.e., physical environments) to *Mixed Reality (MR)*, to completely computer-generated virtual environments (i.e., *Virtual Reality, VR*). Initially, MR encompassed physical scenes augmented with virtual components (*Augmented Reality, AR*) as well as virtual scenes augmented with real components (*Augmented Virtuality, AV*). Due to both its complex implementation and its limited use cases, AV failed to materialize such that the terms AR and MR are now often used synonymously. The term MR, as used by us in this paper, encompasses AR as physical scenes with pure virtual overlays but also expands to more complex environments that virtually augment reality while considering the scene's physical constraints.

XR applications can be accessed with a variety of technologies. First, *head-mounted displays (HMDs)* can be employed for both, MR and VR applications. Especially for MR applications, *handheld displays (HHDs)* such as smartphones or tablets can be used as well. Besides these mobile accesses to XR, there also exist projection-based setups: The term *Spatial Augmented Reality (SAR)* is commonly being used for spaces that augment physical scenes with virtual elements that are projected directly into the scene. On the other hand, *CAVE* [2] systems provide a projection-based access to VR environments.

This paper is focused on the development of a human-centered framework that provides mobile access points to MR and VR scenes, i.e., MR-HMDs, VR-HMDs, and MR-HHDs. In this context, we distinguish between on-site and off-site users: *On-site users* access the system from the actual working environment. This can be the site at which a real machine to be operated is located, at which already existing physical parts of a prototype are located, or at which co-located collaborators are located. *Off-site users* on the other hand access the system from a different location. These can be persons that are operating a machine remotely as well as remote experts that join the system as distributed collaborators in virtual replications of the on-site environment. Furthermore, we distinguish between real and virtual components: We use the term *real component* to refer to a physically existing object and the term *virtual component* to exclusively virtual objects (i.e., there is no physically existing counterpart for this object). In this context, it is important to note that real components may be displayed as virtual replications off site. On-site users may interact with exclusively virtual components and with real components. Off-site users may interact with exclusively virtual components as well as with real components through interaction with their virtual replications. Thereby, we use the terms *static* and *dynamic* to indicate if scene components are meant to change their position or orientation during the session.

2.2 Developing Collaborative Extended Reality Applications

A major impediment to the development and application of XR technologies in collaborative real-world settings concerns the variety of hardware producers and the platform-specific requirements. While game engines like Unity and Unreal support the development of XR applications for different operating systems, these platforms still require the integration of different APIs. Seeking to reduce this fragmentation, OpenXR was introduced as a cross-platform API. Furthermore, real-time collaboration between different XR applications requires low-latency, wireless communication between the respective devices. For example, Photon Engine can be employed to handle communication between multiple clients as it offers SDKs for various platforms. Despite the existence of these commercial solutions, backend development is still far from straightforward. Previous research [3–5] presented architectures and frameworks that combine the different available solutions that seek to facilitate the backend development of collaborative, multi-device XR applications.

Apart from issues related to multi-platform development, network communication, and calibration, the application of XR technologies in real-world, profit-making settings is also impeded by inadequate user interfaces which opens further fields of research. To enable collaboration between on-site collaborators in MR scenes and off-site collaborators in VR scenes, the remote collaborator needs to be provided with a detailed virtual replication of the on-site environment in real time. Existing replication techniques such as 360-degree videos [6], RGB-D cameras [7], or light fields [8] however provide different quality-latency tradeoffs. Another important research topic concerns the semantic segmentation of these virtual replications [9] which is required to make single components of the replication referenceable. While most XR devices provide out of the box interaction techniques, plenty of research is being conducted to enhance their usability. Previous research has for example focused on reducing fatigue while performing in-air gestures to interact with virtual components displayed through a MR-HMD [10]. For MR-HHDs, device-based interaction that maps a HHD's movement to virtual objects has been proposed as an alternative to fatigue-prone touch based interaction techniques that require the device to be held with one hand [11]. In the context of collaborative settings, further research has been conducted regarding adequate visualizations of remote collaborators [12].

The framework presented by Pereira et al. [4] was implemented with Unity and supposed to support collaborative interior design. Multiple clients can connect to a server through the integration of Photon Engine which handles communication and maintains synchronization between a shared and several device-specific scenes. While their framework provides access points for VR-HMDs and MR-HHDs, it is focused on the interaction with virtual components only and does not provide interaction with virtually replicated parts of the physical scene. While they give users the option to highlight points of interest, their proposed avatar visualization is likely to produce visual clutter as the number of users increases. Furthermore, different interaction paradigms were implemented for object manipulation with VR-HMDs (motion controllers) and MR-HHDs (touch input), which may complicate switching between access points for the user.

Kostov and Wolfartsberger [3] presented a proof-of-concept application that supports collaborative training for engine construction. Their application is accessible through a VR-HMD, MR-HMD, MR-HHD, and a desktop PC. Communication between the clients and synchronization of the environment is handled through Unity's networking library. Again, the virtual replication of physical scenes remains unaddressed. In contrast to Pereira et al. [4], Kostov and Wolfartsberger [3] highlight the challenges related to the different device-specific input paradigms. Seeking to reduce the complexity for both developers and users they implement the same button-based user interface for all devices. By clicking the different buttons with the device specific input modality, the users can manipulate the virtual objects according to predefined increments. This interaction approach however limits flexibility and disregards the 3D nature of XR technologies.

In contrast to [3, 4], the framework presented by García-Pereira et al. [5] provides the VR-HMD user with a static virtual reconstruction of the physical environment that is scanned in advance. Physical markers are used to align the orientation of the virtual and the physical world. Further access points are provided through a desktop PC and a HHD that can display both the VR and MR scene. To develop the XR applications and set up the server, Unity and Node.js were employed. Again, different device-specific interaction techniques were provided to interact with virtual components: an external sensor was attached to the HMD to capture hand gestures, the HHD offered a touch-based interface, and the desktop application responded to mouse clicks. While more advanced avatars showing the user's point of view and a hand ray were generated for each access point, the problem of potentially occurring visual clutter remains unaddressed.

A detailed review of literature related to the research topics mentioned above as well as a future research agenda can be found in [13]. While the investigation of all these agenda items and the development of a total solution is far beyond the scope of a single research paper, consideration must be given in advance to how results from independent research in the different fields need to be integrated in the end. Addressing this issue, we build up on the general concept of scalable XR presented in [13] and develop a human-centered framework that considers scalability between the number of users, the degree of virtuality, and the type of device.

3 XR^S Framework: Basic Concept

XR technologies have been assigned considerable potential to decrease temporal and cognitive efforts as well as material costs in many domains. However, existing XR applications are limited to single use cases, specific hardware, or two collaborators.

This lack of scalability causes large overheads of time and cognitive resources when switching between devices or applications, such that reduced efficiency outweighs the potential awarded to XR technologies and impedes their application to real-world settings. Addressing these issues, we present a framework for Scalable Extended Reality spaces as introduced in [13]. The development of our framework followed the human-centered design approach as defined in ISO 9241-210: First, we specified the context of use by deriving abstract high-level use cases that exploit XR's key benefits from specific use cases. The following steps in the human-centered design process were completed based on these high-level use cases, i.e., specifying requirements, design solutions, and evaluation.

3.1 Scalable Extended Reality (XR^S)

Seeking to address these scalability limitations and to increase XR's application in the real world, we introduced the term Scalable Extended Reality (XR^S) as a concept for XR spaces that provide multidimensional scalability enhancements (see Fig. 1). Firstly, they should scale between different degrees of virtuality, i.e., from completely virtual spaces to spaces with single physical elements that are augmented with multiple virtual elements to physical scenes that are augmented with single virtual elements. Secondly, they should scale between different devices, i.e., the space should be accessible via HHDs and HMDs. Lastly, XR^S is supposed to scale between different numbers of users, i.e., from single users to multiple possibly distributed collaborators. As such, XR^S spaces could serve as highly flexible, long-time training or working environments [13].

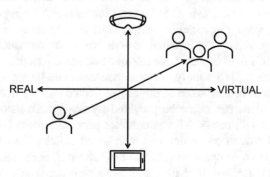

Fig. 1. Scalable Extended Reality (XR^S) Concept; modified replication from [13].

3.2 Context of Use

To understand and specify the context of use, we derived abstract high-level use cases that exploit XR's key benefits from various XR applications that have been proposed by previous research.

3.2.1 XR's Fields of Application

Throughout the last decade, research on potential application areas for XR technologies has been conducted and revealed promising use cases in different fields. Since XR technologies provide a more intuitive way to inspect and interact with three-dimensional elements than conventional desktop applications, they have been assigned considerable potential to support design and engineering reviews in different domains. For instance, Wolfartsberger [14] developed and evaluated a VR system to support the design review of power units, Gong et al. [15] developed a multi-user VR application that allows globally distributed users to cooperate in an automotive design review task, and Kaluza et al. [16] integrated methods from visual analytics in a MR application to support decision-making in automotive life cycle engineering. Another promising field of application for XR technologies is factory layout planning such as presented by Gong et al. [17] who developed a VR system for factory layout planning that seeks to facilitate the modeling process and to improve decision-making through more accessible visual representations. XR applications may be used in a similar way for supporting interior design. For instance, Vazquez et al. [18] developed a MR tool that provides scale-accurate virtual augmentations with virtual furniture. Furthermore, XR technologies may find application at construction sites for safety training in VR scenarios as presented by Wu et al. [19] as well as for supporting workers in monitoring and documentation tasks with virtual augmentations such as presented by Zollmann et al. [20]. A detailed review of potential aerospace applications of VR technologies was given by Pirker [21]: Use cases listed in the paper include training in simulations, teleoperation of remote machines, testing, design reviews, collaboration, and remote assistance. Further interesting fields of application for XR technologies can be found in healthcare. The literature review conducted by Sadeghi et al. [22] revealed several interesting XR applications in the context of cardiothoracic surgery, including surgical planning, training in virtual simulators, and intraoperative guidance. For instance, in assistive scenarios visual information augmenting the surgeon's field of view can be scaled and placed according to the surgeon's personal preferences [23]. Similarly, maintenance tasks can be supported by XR technologies as relevant information can be directly projected into the worker's field of view [24]. These instructions can either be provided by the system automatically or by a remote collaborator. Ultimately, XR technologies provide a powerful tool for supporting collaboration between co-located and distributed collaborators in more complex, 3D tasks that cannot be completed via 2D desktop sharing. For instance, Bai et al. [25] presented a system that allows to share a local working space with a remote collaborator who can deliver support in terms of visual cues that augment the local worker's field of view.

3.2.2 XR's High-Level Use Cases

As summarized in Sect. 3.2.1, the application of XR technologies is deemed beneficial in many domains. To develop our framework, we abstracted these specific use cases and grouped them into the following five high-level use cases. By implication, the framework can then be adapted to any low-level use case that can be described by the blueprints of the high-level use cases or any combination of them.

XR technologies can be applied to support the training of complex or safety-critical tasks as virtual environments reduce safety issues, such that less supervision is required, and trainees can practice the task independently and more often. Virtual training environments can be easily set up multiple times such that accessibility to the training environment is improved and multiple trainees may practice a task at the same time. Furthermore, XR technologies can be used to support both co-located and distributed collaboration. In co-located scenarios, multiple collaborators may be provided with a customized access to the XR space. As such they can be provided with the level and representation of information that fits their responsibilities, experience, and personal preferences. Distributed collaborators can join this collaborative session in a virtual replication of the scene. In both cases, collaborators can be provided with awareness cues displaying the other collaborators' locations and activities. Like the use case of distributed collaboration, XR applications can be applied for teleoperating machines and robots. Depending on how far the worker is away from the actual working place, he or she can be provided with a MR or VR scene in which the machine can be operated through virtual user interface components that allow reviewing the effect of a command in a virtual simulation prior to actual execution. Another promising field of application for XR technologies concerns the design and development of physical items. Temporal and financial costs of prototyping can be reduced by the integration of virtual components.

3.2.3 Dependencies Between XR's Key Benefits and High-Level Use Cases

The relevance of the high-level use cases described in Sect. 3.2.2 can be explained by their exploitation of XR's key benefits. The benefits exploited by each high-level use case and possible combinations of high-level use cases are displayed in Fig. 2.

Training scenarios are supported by the seamless integration of real and virtual elements that allow to display virtual augmentations in the exact right time and place. Hence, trainees do not have to shift their focus between multiple sources and can keep concentrating on their actual task. Since these virtual elements can be quickly modified, the trainee can be provided with the degree of virtuality and level of information that is needed. Similarly, in collaborative scenarios each user may be provided with a customized access to the XR space as virtual augmentations can be easily modified. To facilitate communication and prevent misunderstandings between collaborators, virtual elements that display user activities can be seamlessly integrated in the co-located collaborator's field of view, whereas distributed collaborators may be provided with real-time virtual replications of the on-site environment. Depending on a teleoperator's location he or she may be provided with a virtual replication of the on-site environment or with virtual augmentations that are seamlessly integrated in the physical scene. The design and development process of physical items also benefits from the digital nature of virtual prototypes which can be modified quickly and thus accelerate workflows and save material costs. As the physical item evolves, the degree of virtuality can decrease, and the existing physical parts of the prototype can be seamlessly augmented with virtual elements. Depending on the specific use case, decision-making and quality control can further be supported by the in-context visualization and analysis of data from relevant digital twins or sensors.

Apart from the independent implementation of each of these high-level use cases, they can also be combined with each other. For instance, design and development tasks can be performed by single users, together with co-located or distributed collaborators, or both. Furthermore, teleoperating a machine can be a task that is subject to a training scenario. At the same time, teleoperation of robotic arms may be implemented to allow distributed collaborators to remotely manipulate physical objects on site (see Sect. 5.7). Training tasks can be set up in collaborative scenarios, too. In that way the trainee can be supported by distributed or co-located collaborators that watch the trainee completing the task and may intervene if necessary.

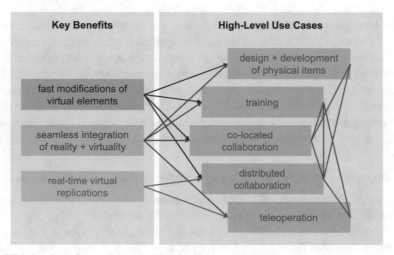

Fig. 2. XR's key benefits exploited by high-level use cases (red, green, and orange arrows on the left) and possible combinations of high-level use cases (blue lines on the right).

4   XR^S Framework: Requirements

Based on the identified high-level use cases as listed in Sect. 3.2.2, we specified functional and non-functional requirements for a human-centered XR^S framework.

4.1   Functional Requirements

The functional requirements of the XR^S framework concern the hardware and technology that is employed to access the XR^S space, the interaction modalities the users are provided with, and the visualization of real and virtual components as well as of the users' locations and activities.

Access

RQ 1 On-site users can access the XR^S space via a MR-HMD or a MR-HHD.

RQ 2 Off-site users can access the XR^S space via a VR-HMD.

Interaction

RQ 3   On-site users can reference real components.
RQ 4   On-site users can reference virtual components.
RQ 5   On-site users can manipulate real components.
RQ 6   On-site users can manipulate virtual components.
RQ 7   Off-site users can reference real components.
RQ 8   Off-site users can reference virtual components.
RQ 9   Off-site users can manipulate real components.
RQ 10 Off-site users can manipulate virtual components.

Visualization

RQ 11 Each collaborator sees where the other collaborators are.
RQ 12 Each collaborator sees what the other collaborators do.
RQ 13 Off-site users are provided with a virtual replication of static real components.
RQ 14 Off-site users are provided with a virtual replication of dynamic real components.
RQ 15 On-site users are provided with visual representations of virtual components that are seamlessly integrated into the physical scene.
RQ 16 Off-site users are provided with visual representations of virtual components that are seamlessly integrated into the virtual scene.

4.2 Non-functional Requirements

The application of such a system in real-world settings further requires the maintenance of usability across different system configurations.

RQ 17 Users can intuitively switch between devices.
RQ 18 Users can intuitively switch between degrees of virtuality.
RQ 19 Usability is maintained with an increasing number of collaborators.
RQ 20 The interaction techniques for manipulating and referencing virtual elements provide high usability.

5 XR^S Framework: Design Solution

Based on the functional and non-functional requirements specified in Sect. 4, we developed a framework design solution for XR^S spaces (see Fig. 3) that incorporates the following system features.

5.1 Access Points and Data – RQs 1, 2, 17, 18

Our framework provides three different access points to the XR^S space: On-site users can access the XR^S space either via MR-HMDs or via MR-HHDs and off-site users can access the XR^S space via VR-HMDs. Firstly, a virtual replication of the static components is generated. This builds the basis for the VR scene through which off-site users can access the XR^S space. Next, virtual replications of dynamic real components as well as of on-site and off-site collaborators are created and added to the VR scene. On the other side, virtual replications of off-site users are integrated into the MR scene. Furthermore, exclusively virtual components are added to both the MR and VR scene. Throughout the session, clients read and write data from and to a database storing information about each user, real and virtual component.

5.2 Subscribing to Collaborators – RQs 11, 12, 19

As the number of collaborators increases, usability may decrease as adding visualizations of each collaborator's location and activity to the scene is likely to produce visual clutter and confusion. To prevent these issues and allow collaborators to keep concentrating on their actual task, they should only be provided with the information needed for task completion. To this end, each user can individually subscribe to visual representations of the other collaborators. The database holds information about each user's id, role (i.e., on-site user or off-site user), activity (i.e., referencing or manipulating objects), position and orientation in space, and the individual subscriptions to other collaborators' locations and activities. For each collaborator, the database stores references to a set of collaborators whose location should be represented as an avatar and whose activities should be represented by visual cues such as hand pointers and gaze rays. In contrast to off-site users, on-site users cannot subscribe to avatars of other on-site users. These references can be set prior to the collaborative session and updated through the users during the session. Users can subscribe and unsubscribe to visual representations of other collaborators' locations and activities through different modalities that may be based on context-menus, speech recognition, or direct interaction (e.g., looking or pointing at avatars to activate cues). As such, the corresponding references in the database will be updated accordingly and the collaborators' visual representation will be adapted individually for each user.

5.3 Visualizing Static Scene Components – RQ 13

We refer to scene components as static if they are not meant to change their position or orientation during the session (e.g., the room in which on-site collaborators are located). To provide scalability between degrees of virtuality, off-site users should be provided with a virtual replication of these static real components. As described in [13], physical scenes can be virtually replicated with different techniques that differ in terms of their quality-latency trade-off. Hence, static scene components that require few to no updates during the collaborative session should be replicated with techniques providing the highest quality at the cost of high latencies. If the static scene components need to be referenceable, the virtual replication needs to be semantically segmented.

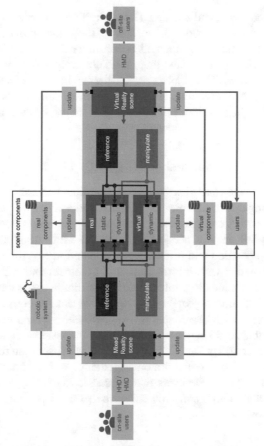

Fig. 3. The Human-Centered Scalable Extended Reality (XR^S) Framework.

5.4   Visualizing Dynamic Scene Components – RQs 14, 15, 16

Dynamic scene components on the other hand refer to real and virtual components whose position and orientation is meant to change a lot throughout the session. We use the term dynamic scene components for objects that physically exist on site, for their virtual replication, and for exclusively virtual objects. In this context it should be noted that while users themselves could be considered as dynamic components, we consider them separately in the next chapter because they have more properties that may change during the session than objects. Physically existing components do not have to be visualized for on-site but for off-site users. In contrast to static scene components, they are manipulable and hence the position and orientation of their virtual counterpart must be updated more often. The generation of the virtual replications of these dynamic scene components depends on each physical object's properties which are manipulated during the session. For instance, if only the position and orientation is manipulated, the object can be replicated in advance and integrated into the off-site collaborator's virtual environment such that during runtime, only the position and orientation of the object must be tracked, exchanged, and updated. However, if the object's appearance itself will be manipulated during the session, the effects of these manipulations on the object's appearance must be tracked and updated accordingly. The visualization of dynamic virtual components (i.e., components that are exclusively virtual in both the MR and VR scene) is less complex: The object's properties can be stored in a database which is updated in line with user interactions. As such, each client can read the same information about the object (e.g., its position, orientation, size, status, or current owner) from the database.

5.5   Visualizing User Location and Activity – RQs 11, 12

To support both, co-located and distributed collaboration, users should be provided with information regarding their collaborators' location and activity when needed. While co-located collaborators that are using MR devices can naturally see each other, remote collaborators using VR devices need to be provided with virtual replications of their collaborators whose position and orientation is updated in real time. Similarly, on-site users need to be provided with information about their remote collaborators. Besides information on the user's location, information regarding a user's current activity (e.g., referencing or manipulating objects) can also be relevant for both co-located and distributed collaborators. Hence, each user is provided with a visual representation of the other collaborators that corresponds to the user's individual subscriptions that are stored in the database as described in Sect. 5.2 . As such, off-site users can for example activate avatar representations of their collaborators together or without hand pointers and gaze rays. The same holds for on-site users that subscribe to off-site collaborators. On the contrary, on-site collaborators can only subscribe to hand pointers and gaze rays for other on-site collaborators.

5.6   Referencing Scene Components – RQs 3, 4, 7, 8

To support co-located and distributed collaboration, collaborators need to be able to reference scene components. This means they should be able to execute an action (e.g., pointing at a scene component) such that this scene component is highlighted for all collaborators that subscribed to the user referencing the component. For instance, this can be implemented by adapting the object's visual representation (e.g., by changing the color of a virtual component or a real component's virtual replication in VR or by augmenting real components with virtual overlays in MR) or by playing 3D audio. Referencing an object can also be interpreted as the selection of an object which can be followed by a manipulation.

5.7   Manipulating Dynamic Scene Components – RQs 5, 6, 9, 10

Our framework focuses on the manipulation of virtual and real components in terms of translation and rotation. To let on-site and off-site collaborators manipulate virtual components, appropriate input techniques are required such that the updated position and orientation of the virtual component can be computed, and its visual representation can be adapted accordingly. The manipulation of real components in the on-site environment, especially for off-site collaborators, is more complex. To this end, we propose the integration of a robotic system, that allows collaborators to remotely translate or rotate real components on site: First, off-site collaborators can manipulate the virtual replication of the real component. As soon as the off-site collaborator confirms the manipulation, the updated position and orientation of the virtual replication is sent to the robot application which automatically computes the necessary motion planning for the robot to adapt the corresponding real component's position and orientation respectively. A similar approach could also be useful for on-site users that want to manipulate large or heavy objects: Physical objects could be augmented with virtual overlays that are manipulated by the on-site user to control the robot. In the case of a training session, the connection with the robotic system can be disabled prior to the start of the session.

5.8   Scalable Interaction Techniques – RQs 17, 18, 20

To provide scalability between different devices and degrees of virtuality and allow users to intuitively switch between them, scalable interaction techniques for referencing and manipulating virtual components or virtual replications of real components are needed. These interaction techniques must scale between all access points to the XR^S space (i.e., MR-HHDs, MR-HMDs, and VR-HMDs). In this context we refer to an interaction technique to be scalable, if switching between access points is possible without large overheads of cognitive efforts that are required to relearn and re-adapt to the system. In other words, users should be able to switch between access points intuitively. While existing interaction techniques rely on different input paradigms (e.g., in-air gestures for MR-HMDs, touch for MR-HHDs, controllers for VR-HMDs), scalable interaction techniques should be based on similar input paradigms that provide high learnability and memorability. Despite the relevance of this question, it has been addressed by very few research papers so far. For example, Kostov and Wolfartsberger [3] implemented a cross-device button-based user interface for VR-HMDs, MR-HMDs, and MR-HHDs. However, the selection of these buttons still relies on the device-specific input modality and disregards advanced, spatial interaction paradigms. Apart from enhanced scalability, the interaction techniques themselves should provide high usability as defined in ISO 9241-11. The development of such scalable interaction paradigms is a complex and interesting topic for future research. The full scalability regarding options of interaction between on-site and off-site users as implemented in our framework builds the basis for a future integration of such scalable interaction techniques.

6 XR^S Framework: Walkthrough

As a first evaluation of our framework this chapter provides a theoretical walkthrough for two use cases that can be described by a combination of the high-level use cases introduced in Sect. 3.2.2.

6.1   Collaborative Prototyping

The design and development of physical items such as cars usually involves co-located as well as distributed collaborators from different fields. Throughout the design and development process, each of these collaborators has different tasks and responsibilities that require different levels of information. The framework presented in this paper can help providing these collaborators with this information exactly when and where it is needed.

At the beginning of the design and development process, during the ideation phase, users may be immersed in a completely virtual environment to collect ideas and develop a proof of concept. If no physical parts of the product exist yet, there is no on-site environment and all users join the XR^S space in a VR scene in which they can reference and manipulate dynamic virtual components (i.e., the components of the product to be developed). Each collaborator may subscribe to other collaborators to receive information regarding this collaborator's location and activity. In each frame the position, orientation, and activity of each user is written to the corresponding database and retrieved from the collaborators according to their subscriptions.

As soon as the first physical parts of the product exist, co-located collaborators that are located on site may switch to a MR scenario in which the physically existing real components are augmented with virtual components that display the missing parts. In iterative prototyping stages, different virtual configurations of the missing parts can be tested out and reviewed according to key parameters that may be generated in real time by integrated digital twins. To do so, on-site users can reference and manipulate the real and virtual components as described by the framework. Collaborators that are located off site can join the XR^S space through the same VR scene as before. To this end, a virtual replication of the static real components is generated which builds the basis for the VR scene. Properties of dynamic real components are tracked, and their virtual replication is integrated into the VR scene and updated together with the exclusively virtual components that are continuously tracked as well. Off-site users can then reference and manipulate both virtual replications of real components and exclusively virtual components as described in the framework. Like in the exclusively virtual space in the beginning, both on-site and off-site users may subscribe to each other to receive information about each other's location and activities.

The degree of virtuality decreases throughout these iterative prototyping stages until the physical prototype is only augmented with single virtual components. The implementation of scalable interaction techniques for referencing and manipulating objects allows users to intuitively switch between degrees of virtuality (which may change continuously from virtual to real as the product evolves or abruptly if on-site users become off-site users or vice versa). Considering real-world settings, it is very likely that one person is involved in multiple design and development processes of different products at once. As the current stage of development may differ between these products, this person may have to switch between devices and degrees of virtuality multiple times per day. Furthermore, on-site users may become off-site users depending on their location which again requires switching between devices and degrees of virtuality. Our framework implements multidimensional scalability enhancements to reduce the temporal and cognitive efforts of switching between these technologies and allows users to keep focusing on the actual task.

6.2 Training and Teleoperation

To guarantee the correct and safe execution of complex and dangerous tasks workers need to undergo appropriate training in advance. However, access to the corresponding machines may be limited as these are in use or occupied by other trainees. Furthermore, especially dangerous tasks may require supervision by experts whose availability is limited. Transferring these training sessions to virtual scenarios can reduce the necessity of supervisors as safety-critical parts are eliminated. On top of that, the training of multiple trainees can be parallelized as virtual environments can be replicated, given that enough hardware is available.

The application of our proposed framework allows trainees to start in a VR scene in which they are provided with a virtual replication of the actual working environment and can practice operations by manipulating virtual replications. In this case, the interaction with virtual replications should not be executed in the real working environment. As the trainees make learning progress, they may switch to MR scenarios where they can practice operations in the actual working environment while safety-critical parts may still be virtualized. Through the implementation of scalable interaction techniques that rely on the same interaction paradigms as in the VR scene, they can concentrate on the actual task (i.e., the operation to be practiced). Once training is accomplished and they move on to the operation in the real world, they can still be provided with visual overlays that display in-context information in the beginning.

At the same time, the novice worker can request help from a remote expert which is located off site and joins the XR^S space in a VR scene. Again, this VR scene can be generated by the virtual replication of static real components which is then augmented in real time with virtual replications of dynamic real components and exclusively virtual components. Depending on the specific task, the remote expert's interactions with virtual replications may be executed on site. Thus, remote experts may act as teleoperators that manipulate virtual replications of a machine in VR to command a robot or machine on site to execute the operation in the real world. The application of our framework allows to design similar user interfaces for remote assistance and teleoperation – tasks that are likely to be completed by the same person. As such, persons that act as both remote experts and teleoperators benefit from the multidimensional scalability enhancements that provide them with a highly memorable user interface.

7   Conclusion

This paper presents a human-centered framework for XR^S spaces that implements multi-dimensional scalability enhancements regarding different degrees of virtuality, different devices, and different numbers of users. As such, we contribute to the list of highly relevant research topics presented in [13] and seek to foster the application of XR technologies in profit-making, real-world use cases which is currently impeded by overheads of cognitive and temporal efforts that outweigh the potential inherent in XR technologies.

The presented framework provides three access points for on-site and off-site users that can reference and manipulate both real and virtual components. To this end, off-site collaborators are provided with a virtual replication of the on-site environment that integrates virtual replications of real components and exclusively virtual components to a VR scene which can be accessed with a VR-HMD. On-site users can access the XR^S space with MR-HMDs or MR-HHDs. The visualization of scene components is handled through a database that stores information about each user, real and virtual component. While all this information is accessible for all users, not all users may need all this information for effective task completion. Hence, users may subscribe to their collaborators individually to obtain the needed level of information. Furthermore, our framework provides full scalability regarding interactivity through the integration of a robotic system which allows remote users to manipulate real components on site. This scalability is of high relevance, as users should be able to switch between degrees of virtuality and devices depending on their role and location without losing options for interaction. In the future, the framework may also be extended with scalable interaction techniques that rely on similar input paradigms for referencing and manipulating real and virtual components through all access points.

The framework was developed based on five high-level use cases that exploit XR's key benefits: Design and development of physical items, training, teleoperation, co-located and distributed collaboration. Since these high-level use cases serve as blueprints that can be combined with each other to describe specific low-level use cases in the real world, the framework can by implication be used by collaborators as well as by single users in many different domains. Hence, our framework provides the foundation required to implement specific XR^S applications which will be part of our future work.

Acknowledgments. The authors thank Bahram Ravani for many valuable discussions. Funded by the Deutsche Forschungsgemeinschaft (DFG, German Research Foundation) – 252408385 – IRTG 2057.

References

1. Milgram, P., Takemura, H., Utsumi, A., Kishino, F.: Augmented reality: a class of displays on the reality-virtuality continuum. In: Das, H. (ed.) Telemanipulator and Telepresence Technologies, Bellingham, WA, USA, vol. 2351, pp. 282–292 SPIE (1995). https://doi.org/10.1117/12.197321
2. Cruz-Neira, C., Sandin, D.J., DeFanti, T.A.: Surround-screen projection-based virtual reality: the design and implementation of the CAVE. In: Proceedings of the 20th Annual Conference on Computer Graphics and Interactive Techniques, pp. 135–142. (1993). https://doi.org/10.1145/166117.166134
3. Kostov, G., Wolfartsberger, J.: Designing a framework for collaborative mixed reality training. Procedia Comput. Sci. **200**, 896–903 (2022). https://doi.org/10.1016/j.procs.2022.01.287
4. Pereira, V., Matos, T., Rodrigues, R., Nóbrega, R., Jacob, J.: Extended reality framework for remote collaborative interactions in virtual environments. In: 2019 International Conference on Graphics and Interaction (ICGI), pp. 17–24 (2019). https://doi.org/10.1109/ICGI47575.2019.8955025
5. García-Pereira, I., Gimeno, J., Pérez, M., Portalés, C., Casas, S.: MIME: a mixed-space collaborative system with three immersion levels and multiple users. In: 2018 IEEE International Symposium on Mixed and Augmented Reality Adjunct (ISMAR-Adjunct), pp. 179–183 (2018). https://doi.org/10.1109/ISMAR-Adjunct.2018.00062

6. Piumsomboon, T., Lee, G.A., Irlitti, A., Ens, B., Thomas, B.H., Billinghurst, M.: On the shoulder of the giant: a multi-scale mixed reality collaboration with 360 video sharing and tangible interaction. In: Proceedings of the 2019 CHI Conference on Human Factors in Computing Systems (CHI 2019), paper 228, pp. 1–17 (2019). https://doi.org/10.1145/3290605.3300458
7. Lindlbauer, D., Wilson, A.D.: Remixed reality: manipulating space and time in augmented reality. In: Proceedings of the 2018 CHI Conference on Human Factors in Computing Systems (CHI 2018), paper 129, pp. 1–13 (2018). https://doi.org/10.1145/3173574.3173703
8. Mohr, P., Mori, S., Langlotz, T., Thomas, B.H., Schmalstieg, D., Kalkofen, D.: Mixed reality light fields for interactive remote assistance. In: Proceedings of the 2020 CHI Conference on Human Factors in Computing Systems (CHI 2020), paper 162, pp. 1–12 (2020). https://doi.org/10.1145/3313831.3376289
9. Huang, S.-S., Ma, Z.-Y., Mu, T.-J., Fu, H., Hu, S.-M.: Supervoxel convolution for online 3D semantic segmentation. ACM Trans. Graph. (TOG) **40**(3), 1–15, Article no. 34 (2021). https://doi.org/10.1145/3453485
10. Brasier, E., Chapuis, O., Ferey, N., Vezien, J., Appert, C.: ARPads: mid-air indirect input for augmented reality. In: 2020 IEEE International Symposium on Mixed and Augmented Reality (ISMAR), pp. 332–343 (2020). https://doi.org/10.1109/ISMAR50242.2020.00060
11. Goh, E.S., Sunar, M.S., Ismail, A.W.: 3D Object manipulation techniques in handheld mobile augmented reality interface: a review. IEEE Access **7**, 40581–40601 (2019). https://doi.org/10.1109/ACCESS.2019.2906394
12. Piumsomboon, T., et al.: Mini-me: an adaptive avatar for mixed reality remote collaboration. In: Proceedings of the 2018 CHI Conference on Human Factors in Computing Systems (CHI 2018), paper 46, pp. 1–13 (2018). https://doi.org/10.1145/3173574.3173620
13. Memmesheimer, V.M., Ebert, A.: Scalable extended reality: a future research agenda. Big Data Cogn. Comput. **6**(1), 12:1–12:17 (2022). https://doi.org/10.3390/bdcc6010012
14. Wolfartsberger, J.: Analyzing the potential of virtual reality for engineering design review. Autom. Constr. **104**, 27–37 (2019). https://doi.org/10.1016/j.autcon.2019.03.018
15. Gong, L., et al.: Interaction design for multi-user virtual reality systems: an automotive case study. Procedia CIRP **93**, 1259–1264 (2020). https://doi.org/10.1016/j.procir.2020.04.036
16. Kaluza, A., Juraschek, M., Büth, L., Cerdas, F., Herrmann, C.: Implementing mixed reality in automotive life cycle engineering: a visual analytics based approach. Procedia CIRP **80**, 717–722 (2019). https://doi.org/10.1016/j.procir.2019.01.078
17. Gong, L., Berglund, J., Fast-Berglund, Å., Johansson, B., Wang, Z., Börjesson, T.: Development of virtual reality support to factory layout planning. Int. J. Interact. Des. Manuf. (IJIDeM) **13**(3), 935–945 (2019). https://doi.org/10.1007/s12008-019-00538-x
18. Vazquez, C., Tan, N., Sadalgi, S.: Home studio: a mixed reality staging tool for interior design. In: Extended Abstracts of the 2021 CHI Conference on Human Factors in Computing Systems (CHI EA 2021), Article no. 431, pp. 1–5 (2021). https://doi.org/10.1145/3411763.3451711
19. Wu, H.-T., Yu, W.-D., Gao, R.-J., Wang, K.-C., Liu, K.-C.: Measuring the effectiveness of VR technique for safety training of hazardous construction site scenarios. In: 2020 IEEE 2nd International Conference on Architecture, Construction, Environment and Hydraulics (ICACEH), pp. 36–39 (2020). https://doi.org/10.1109/ICACEH51803.2020.9366218
20. Zollmann, S., Hoppe, C., Kluckner, S., Poglitsch, C., Bischof, H., Reitmayr, G.: Augmented reality for construction site monitoring and documentation. Proc. IEEE **102**(2), 137–154 (2014). https://doi.org/10.1109/JPROC.2013.2294314
21. Pirker, J.: The potential of virtual reality for aerospace applications. In: 2022 IEEE Aerospace Conference (AERO), pp. 1–8 (2022). https://doi.org/10.1109/AERO53065.2022.9843324
22. Sadeghi, A.H., et al.: Current and future applications of virtual, augmented, and mixed reality in cardiothoracic surgery. Ann. Thorac. Surg. **113**(2), 681–691 (2022). https://doi.org/10.1016/j.athoracsur.2020.11.030

23. Al Janabi, H.F., et al.: Effectiveness of the HoloLens mixed-reality headset in minimally invasive surgery: a simulation-based feasibility study. Surg. Endosc. **34**(3), 1143–1149 (2020). https://doi.org/10.1007/s00464-019-06862-3

24. Siltanen, S., Heinonen, H.: Scalable and responsive information for industrial maintenance work – developing XR support on smart glasses for maintenance technicians. In: Proceedings of the 23rd International Conference on Academic Mindtrek (AcademicMindtrek 2020), pp. 100–109 (2020). https://doi.org/10.1145/3377290.3377296

25. Bai, H., Sasikumar, P., Yang, J., Billinghurst, M.: A user study on mixed reality remote collaboration with eye gaze and hand gesture sharing. In: Proceedings of the 2020 CHI Conference on Human Factors in Computing Systems (CHI 2020), paper 423, pp. 1–13 (2020). https://doi.org/10.1145/3313831.3376550

A Holistic Framework for Factory Planning Using Reinforcement Learning

M. Klar[1]([⊠]), J. Mertes[1], M. Glatt[1], B. Ravani[2], and J. C. Aurich[1]

[1] Institute for Manufacturing Technology and Production Systems (FBK), RPTU
Kaiserslautern, Kaiserslautern, Germany
matthias.klar@rptu.de
[2] Department of Mechanical and Aerospace Engineering, UC Davis, Davis, CA, USA

Abstract. The generation of an optimized factory layout is a central element of the factory planning process. The generated factory layout predefines multiple characteristics of the future factory, such as the operational costs and proper resource allocations. However, manual layout planning is often time and resource-consuming and involves creative processes. In order to reduce the manual planning effort, automated, computer-aided planning approaches can support the factory planner to deal with this complexity by generating valuable solutions in the early phase of factory layout planning. Novel approaches have introduced Reinforcement Learning based planning schemes to generate optimized factory layouts. However, the existing research mainly focuses on the technical feasibility and does not highlight how a Reinforcement Learning based planning approach can be integrated into the factory planning process. Furthermore, it is unclear which information is required for its application. This paper addresses this research gap by presenting a holistic framework for Reinforcement Learning based factory layout planning that can be applied at the initial planning (greenfield planning) stages as well as in the restructuring (brownfield planning) of a factory layout. The framework consists of five steps: the initialization of the layout planning problem, the initialization of the algorithm, the execution of multiple training sets, the evaluation of the training results, and a final manual planning step for a selected layout variant. Each step consists of multiple sub-steps that are interlinked by an information flow. The framework describes the necessary and optional information for each sub-step and further provides guidance for future developments.

1 Introduction

Machine learning approaches such as Reinforcement Learning (RL) have the potential to solve problems in various manufacturing-related areas, leading to improvements in terms of time, cost, and quality [1]. One promising application field for RL is factory layout planning [2], which is a central element of the factory planning process [3]. Existing literature on the subject mainly focuses on the principal utilization [2] and the scalability of the RL-based planning approach [4]. However, the potential to support planners in the early stage of layout planning is highlighted without presenting a holistic framework that covers the corresponding application steps and needed information. This paper

J. C. Aurich et al. (Eds.): IRTG 2023, *Proceedings of the 3rd Conference on Physical Modeling for Virtual Manufacturing Systems and Processes*, pp. 129–148, 2023.
https://doi.org/10.1007/978-3-031-35779-4_8

addresses this research gap by defining requirements that have to be met by a factory planning approach. These requirements build the basis for the framework developed for RL-based factory layout planning.

The paper is structured as follows: First, an introduction to factory layout planning (Subsect. 2.1) and the existing planning approaches (Subsect. 2.2) is presented. Subsection 2.3 introduces RL and its functionality, followed by the research gap in Sect. 3. Section 4 presents the requirements as well as the framework developed followed by an evaluation of the framework. Section 5 consists of a conclusion and a summarization of future developments.

2 State of the Art

2.1 Introduction to Factory Layout Planning

Factory layouts have to be adapted according to changing and influential factors such as technological innovations, shortened product life cycles, and changing market conditions [5]. Consequently, factory layout planning is frequently performed to ensure an economical, efficient, flexible, and versatile production under consideration of all external and internal characteristics [6].

Factory layout planning problems can be divided into new planning (greenfield planning) and the restructuring (brownfield planning) of a factory. The early stage of layout planning mainly focuses on the positioning of functional units (e.g. machines) in a given space without considering all planning details [7]. The result of the early stage can be a block layout that predefines the main characteristics of the future factory. It builds the starting point for further detailed manual planning steps in the later phase of layout planning. Consequently, the early planning phase is of high importance and it needs to be ensured that the initial solution quality is as high as possible [3].

In the early stage of layout planning, a variety of different and partially conflicting goals must be considered. Examples are the minimization of transportation costs, transportation intensities, throughput times, locked capital in the form of unused inventory stocks, and area demand while maximizing the transparency of the material flow, machine utilization, and supply readiness. Furthermore, boundary conditions such as the floor-bearing capacity, safety requirements, and media supply and disposal have to be considered [8]. In the case of brownfield planning, it must also be ensured that the restructuring costs do not exceed the positive effects of adapting the existing factory to new circumstances [9].

Among the multiple goals, the material flow related target variables have a predominant position. The material flow can be defined as the interconnection of all operations in the processing, and distribution process of material goods within defined areas of manufacturing. Thus, the material flow describes the linking between functional units and incorporates not only transport processes but also storage and handling processes [10]. The optimization complexity of the material flow and the corresponding material flow devices, such as conveyors, storage, and picking technology, is increased by multiple influential factors, such as [11]:

• Increase in product complexity and variant diversity

- Increase in the complexity of production networks
- Smaller batch sizes
- Shorter product life cycles

Therefore, multiple partially dynamically changing interdependencies must be considered; for example, the influence of order-related fluctuations on interlinked processes. Without considering dynamic effects in the planning phase, disturbances in the material flow can only be detected and eliminated during the operation phase. However, costly adaptions in the operation phase can be avoided by anticipative planning. Thus, the material flow in a manufacturing system can only be planned sufficiently if dynamic effects are already considered during layout planning [12]. Hence, discrete event material flow simulations (DES) are commonly used to analyze and optimize the material flow and its dynamic and stochastic characteristics. The central element of a DES is the simulation model, which is defined as a digital representation of the system behavior and its interdependencies. The DES allows to perform simulation experiments with different material flow configurations to analyze its influence regarding the throughput times or sensitivity to disruptions. This allows to design a robust and well-performing material flow that leads to reduced operational costs [11].

Summarized, a holistic planning approach needs to incorporate multiple planning objectives, the necessary boundary conditions, and DES to appropriately consider the dynamic effects in the material flow.

2.2 Approaches for the Early Phase of Factory Layout Planning

The existing factory planning approaches can be divided into manual and computer-aided. **Manual planning** requires expertise and can be characterized as a creative problem-solving process [12]. The starting point is usually an initial layout that was generated without consideration of detailed boundary conditions, such as geometrical restrictions (ideal layout). The cycle method, according to *Schwerdtfeger*, is an exemplary approach that can be used to generate the initial layout [13]. Another example is the Computerized Relative Allocation of Facilities (CRAFT) algorithm [14]. Furthermore, Sankey diagrams, which visualize the transportation intensity between functional units, can be used to generate the initial layout. Manual planning aims at transferring this initial layout into multiple layout configurations that satisfy the building-related boundary conditions of the factory (real layout) [13]. This process is often time-consuming due to the high variety of different positioning options. Consequently, only a limited number of factory layouts are generated in practice, and it can be assumed that the optimal layout configuration is difficult to find. The complexity and time consumption of generating one layout variant further increases in the case of multi-objective optimization [15]. Technologies such as virtual reality (VR) or augmented reality (AR) can be used to support the manual planning task. VR and AR applications can be interlinked with a DES which allows to analyze material flow properties in detail [16]. However, building the VR/AR planning environments is again time and resource-consuming on its own, which limits its industrial applicability [17].

Computer-aided approaches are used to support the planner in the early phase of layout planning by generating several layout variants. These approaches consider building-related boundary conditions and objective functions in the form of equations, which are aggregated to obtain an optimization problem that is called the facility layout problem [18]. However, since not all planning details are considered in the facility layout problem, a manual detailed planning step is required in the later phase of layout planning [3]. Computer-aided approaches can be further divided into exact and approximative planning approaches.

Exact planning approaches are built upon mathematical formulations such as the Branch-and-Bound method and calculate the optimal solution for a given problem. This is, however, computationally expensive since the facility layout problem is categorized as NP-hard. Consequently, such exact approaches are only suitable for small exemplary problem sizes [19].

Approximative planning approaches are used to overcome the computational problem by generating optimum-near solutions. Commonly used solving methods (meta-heuristics) are the genetic algorithm (GA), simulated annealing (SA), or the large adaptive neighborhood search [20]. The majority of the existing approaches are applied to a single objective optimization problem. An example is presented by *Lin et. al*, who use a GA to optimize a layout according to the resulting transportation costs [21]. *Chen et al.* optimize a layout according to the work-in-progress effects using SA [22]. Besides, novel techniques use quantum annealing to minimize the transportation intensity for different problem sizes within seconds [23].

Even though these approaches can generate valuable solutions, a single objective optimization seems inappropriate for the facility layout problem since usually multiple objectives must be considered at the same time. Only a few approaches allow the optimization under consideration of multiple objectives [20]. *Guan et al.* use a particle swarm approach to optimize multiple workshops regarding the transportation distance, the number of workshops, and the workshop floor utilization [24].

Most such approaches rely on the optimization of analytically formulated objective functions that are defined by using assumptions and simplifications such as optimizing according to the transportation intensity. Dynamic performance metrics of the material flow, such as the throughput time, are often neglected [25] which can lead to a mismatch between a generated solution and the real behavior of the manufacturing system. Consequently, a DES is a suitable basis for multi-objective optimization instead of analytically described objectives [12].

Our previous work highlights the potential of an RL-based factory planning approach [2, 4, 26]. RL-based factory layout planning is capable to solve problems with varying sizes by learning the metrics of the problem [4]. Furthermore, RL has been successfully used in other disciplines in combination with simulation environments [27] which also bears the potential for an RL and simulation-based factory planning approach. However, to extract the full potential of RL-based factory layout planning, a holistic framework is needed that describes the application phases, the information flow, and all sub-steps in detail.

2.3 Introduction to Reinforcement Learning

Besides unsupervised and supervised learning, RL is a subclass of machine learning algorithms. These algorithms have the ability to learn about complex relationships between different parameters by extracting patterns from data. While supervised and unsupervised learning require existing training data, RL approaches generate their training data within the training process [28]. Figure 1 visualizes the structure of an RL approach, which consists of an agent and the environment. As depicted in Fig. 1, the agent interacts with its environment at timestep t by selecting an action A_t based on the current state S_t. The environment receives the action and transitions to the next state S_{t+1}. Furthermore, a reward R_t is returned to the agent. The agent aims at learning a strategy (policy) that maximizes the accumulated reward over an episode. Consequently, the reward is used to incentivize a certain behavior [29].

A variety of alternative agent architectures exist. They can be categorized into value-based, policy-based, and hybrid (Actor-Critic) approaches according to their action selection and policy improvement process. Value-based approaches either estimate the value of selecting an action in the current state (action-value function) or the value of being in a state (state-value function). In contrast, policy-based approaches map the state to a probability distribution that is used directly for the action selection process [30].

One of the most commonly used architectures is the Double Deep Q Learning Agent (DDQN) which is categorized as a value-based approach. This agent is characterized by a stable and converging training behavior. Since two artificial neural networks (ANN) build the basis of the approach, DDQN can be used for problems with large state and action space representations. However, it only allows the selection of actions from a discrete action space [31]. The DDQN Agent uses a replay buffer to store its experience in form of transitions. One transition consists of the following elements:

- State S_t
- Action A_t
- Next State S_{t+1}
- Reward R_t

A subset of all transitions stored in the experience replay buffer is sampled between each step to train the agent by changing the weights of the DDQN networks according to the loss function [31]. The training process ends after a defined number of episodes, a certain training duration, or after reaching a performance threshold. Within the training process, the trade-off between exploration and exploitation is of special importance. The agent uses exploitation if the actions are selected according to the policy while exploration describes a behavior that results in a deviation from the policy. Only by defining high exploration rates, it is possible to experience novel state-action combinations which might help to overcome local minima. However, reaching meaningful states is only possible if the policy is exploited [29].

Continuous actions require different architectures, such as the Proximal Policy Optimization Agent, which can be categorized as a hybrid approach [32]. Hybrid approaches aim at combining the positive effects of both value- and policy-based approaches. Precisely, a hybrid approach consists of two agents the actor and the critic. The actor selects an action according to its policy and is categorized as policy-based. The policy is updated according to the feedback of the value-based critic. Consequently, both approaches are combined, leading to increased performances [33].

Besides, action masking methods can be applied to ensure that all actions that are selected are valid. Consequently, the agent must not learn the difference between valid and invalid actions which reduces the number of selectable actions and improves the learning behavior [34].

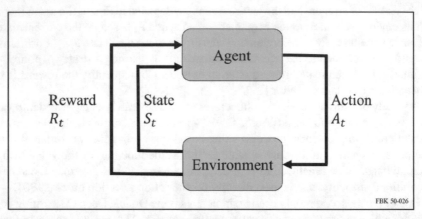

Fig. 1. Conventional reinforcement learning architecture that consists of an agent and its environment. The agent interacts with the environment by selecting actions based on the current state and reviews a reward as a feedback sign.

RL has several potential application fields and has been successfully applied to manufacturing-related problems. Examples are robot arm path planning in human-robot collaboration environments [35], job shop scheduling [36], or RL as a part of an energy management system that reduces energy-related costs in a manufacturing system [37]. Furthermore, RL has the ability to reach superior performance in complex decision-making problems with dynamic influences compared to existing approaches or manual methods. The reason for this superiority is the ability to extract and learn the patterns of a problem with only limited information. Hence, only the reward signal is necessary to develop a strategy compared to classic optimization problems and solving strategies that are based on defined behavior rules [38].

There are three advantages of the RL technique that can be summarized compared to standard metaheuristics. First, the agent can learn a strategy without defining rules for each situation. This would allow an optimized behavior even in situations that deviate from the normal case. Second, the agent can develop novel strategies that exceed the existing problem-solving knowledge. Third, the problem patterns are stored in the model of the ANN. Consequently, the same model can be reused again to solve a similar problem which is not possible with standard metaheuristics such as GA, or SA.

3   Research Gap

The advantages described in the previous section can be valuable in the context of factory layout planning for the following reasons. Factory layout planning is characterized as a decision-making process with a high combinatorial complexity that requires a high solution quality to reduce the operational costs to a minimum. The complexity of solving this problem is increased since dynamic effects due to changing interdependencies between functional units have to be incorporated which can be performed most appropriately using a DES. RL has demonstrated in other application fields that it is suitable to solve such complex decision-making processes since it is capable of learning the (often non-linear) problem metrics. Furthermore, it was successfully used in combination with simulation environments for problems that can only hardly be described by distinct rules.

Previous research also highlights the match between the challenges of the facility layout problem and the capabilities of RL. However, they mainly focused on the principal utilization [2] while ensuring the scalability of the existing approach [4].

However, the application perspective remains undiscussed. It is unclear which information is needed in order to use a RL-based factory planning approach, and how this information is processed.

This paper aims at closing this gap by improving the applicability of an RL-based factory layout planning approach with the development of a holistic framework that is divided into five steps. Within the framework, the information flow and all sub-steps are described in detail starting at the initialization phase of the approach up to the utilization of the results.

4 Framework for Factory Layout Planning Using Reinforcement Learning

4.1 Requirements

The first development step for the application of the RL method involves the definition of requirements for a holistic factory layout planning approach from a general perspective. These requirements define the boundary conditions for application purposes and influence the required information basis as well as the needed expertise to use an automated planning approach. The requirements are then transferred to an RL-related framework.

Requirement 1: Greenfield and Brownfield Planning of Factory Layouts
Factory planning can be divided into brownfield planning (restructuring) of an existing factory and greenfield planning of a new factory. Holistic approaches consider greenfield planning as a special case of brownfield planning without considering the restructuring effort. Hence, the problem and its boundary conditions must be modeled accordingly to allow the application to both cases.

The majority of existing planning approaches focus only on greenfield planning. However, brownfield planning shouldn't be neglected since minor changes in the factory layout do not require an entirely new planning but a slight variation of the existing layout. Furthermore, brownfield planning is performed more frequently than greenfield planning. A holistic approach should incorporate both planning cases and allow for greenfield as well as for brownfield planning. Even though, the characteristics associated with brownfield planning and the considered optimization objectives differ slightly from those in greenfield planning case.

Requirement 2: Multi-objective Optimization
Within layout planning, multiple partially conflicting planning objectives can be considered. Consequently, a holistic planning approach should allow the selection and prioritization of multiple optimization objectives at the same time. They can be divided into

analytic and dynamic optimization objectives. Dynamic planning objectives, such as the throughput time, should be incorporated using a DES. However, not all objectives can be handled by a simulation. Analytic objectives can be calculated without simulation such as the area demand. Layout planning problems have a heterogeneous character and require the consideration of different objectives. Consequently, a holistic approach should be able to consider one or multiple analytic and dynamic objectives that apply to a large variety of existing planning scenarios. Existing approaches often only use a single objective for optimization. Furthermore, the minority includes simulation results in the optimization phase.

Requirement 3: Practical Relevance of the Planning Results (Degree of Abstraction)
The applicability of the approach is directly connected to the degree of abstraction. A planning approach should generate results that are close to reality. To achieve this, diverse boundary conditions must be considered. Examples can be the floor-bearing capacity or the consideration of media supply and disposal. Furthermore, the modeling approach defines, whether functional units can be placed freely or only in defined positions. Modeling the facility layout problem as an open field layout allows a high degree of freedom since more placement options are available compared to a single-row planning problem.

The degree of abstraction differs significantly in the existing approaches. The majority of the approaches model the problem as an open-field problem but only consider the width and length of the functional units. However, boundary conditions such as the floor-bearing capacity or availability of media supply are often neglected.

Requirement 4: Scalability
The scalability requirement ensures the industrial applicability of a layout planning approach. As described in Sect. 2, layout planning requires high-quality solutions to reduce the operational costs. However, high solution quality should be combined with a reasonable computation time to ensure a fast planning process. Ensuring scalability with a rising number of functional units is a central challenge in the development of automated planning approaches.

The scalability of existing automated approaches differs depending on the used algorithm. RL seems to be a valuable tool to overcome this problem since it has demonstrated in other application fields that it is capable of solving problems with a large degree of combinatorial complexity while ensuring high-quality results.

Requirement 5: Accessibility
The initialization of computer-aided automated planning approaches can be challenging since mathematic formulations for optimization objectives and boundary conditions have to be defined. This process requires a large degree of expertise. Consequently, the planning approach should support the planner by providing suitable recommendations to reduce the manual modeling effort.

Requirement 6: Comprehensibility of the Planning Results
The last requirement is especially important for computer-aided, automated planning approaches. In contrast to manual planning approaches, computer-aided approaches

lack comprehensibility since the optimization process is to some degree characterized by probabilistic influences. This effect is even more pronounced in the case of an RL-based planning approach since a machine learning model can be considered a black box model. However, it is important to generate comprehensible and trustable results. Consequently, a holistic planning approach should be able to provide additional information regarding the main influencing factors as well as the degree of uncertainty in the solution. This will increase the trustability and supports the applicability of an automated planning approach. Conventional metaheuristics also lack comprehensibility since the solution process is characterized by probabilistic processes such as mutations for the GA. RL, in contrast, offers the potential to analyze the solution strategy (policy) if suitable explainable RL methods are applied.

4.2 Description of the Framework

The described requirements build the basis for the framework depicted in Fig. 2. As indicated in this Figure, the framework consists of five sequentially executed steps.

Step 1: Initialization of the Layout Planning Problem

In the first step, all major characteristics of the layout planning problem are defined. First, the planning step (greenfield or brownfield planning) must be specified. It directly influences the initialization of the layout characteristics and optimization objectives since brownfield planning requires the existence of an initial layout while greenfield planning doesn't.

The 4 dynamic and 5 analytic optimization objectives are summarized in Table 1. They can be considered individually or in an arbitrary combination, depending on the problem characteristics and strategic goals of the company. The target variables have a heterogeneous character and can thus be combined as desired. An exception is the transport intensity and the throughput time. In the case of mass production that uses conveyor belts for material transportation, an optimization regarding the throughput time or the transportation intensity can lead to similar results. In these cases, only one of the two objectives should be considered simultaneously. In the brownfield planning case, the objective of similarity is available and a mandatory requirement. Similarity to the existing layout helps to reduce the restructuring effort. As mentioned in Sect. 2, restructuring is only profitable if the restructuring costs do not exceed the positive effects of the optimization. Consequently, the objective of similarity should always be combined with an additional objective to generate a layout that slightly differs from the existing solution.

The layout characteristics influence the available optimization objectives. The layout characteristics summarize the geometrical, and boundary-related properties of the layout. They can be divided into mandatory, and optional conditions. A mandatory condition is the geometrical 2D shape of the layout. Optional properties are information regarding existing media supply, floor bearing capacity, and height information. In the best case, all optional properties are available, since optimization regarding media supply and disposal is only possible if this information is provided. If the information is not provided and optimization regarding the throughput time or transportation intensity is still possible.

Consequently, the available information density of the layout characteristics will directly influence the practical relevance of the planning results (Requirement 3).

The layout characteristics are also interlinked with the characteristics of the functional units. The functional units must provide the same information density as the layout, and vice versa. Hence, the 2D shape of each functional unit is mandatory information, while additional information will only increase the practical relevance of the solution. Furthermore, the following general, process-related information must be provided:

- Identification number or name of each functional unit
- Functional unit type: warehouse or processing unit
- Maximum product storage capacity of each functional unit
- Initial storage inventory at the start of the simulation of each functional unit
- Information per processing mode:

    o Processing time
    o Setup time
    o Input-Output relationship

Finally, the material handling system has to be defined. This involves information regarding the means of transportation, external supply, and the considered jobs that can be defined using forecasting methods for future orders based on historical data.

For the means of transportation, the following information must be provided:

- Identification number or name of the material handling systems
- Loading capacity
- Loading and unloading duration
- Speed

The external supply delivers products to defined functional units, such as warehouses. Consequently, the amount of goods and the frequency of delivery need to be defined. The job-related information provides the number of goods that should be produced and the corresponding production steps. Furthermore, control-related information must be provided, which involves the scheduling sequence of the jobs as well as a control strategy for the material flow, for example, a push or pull strategy.

At the end of step 1, all characteristic information of the layout planning problem is defined. In the case of brownfield planning, this also involves an existing layout and information about additional or missing functional units for the new layout configuration.

Step 2: Initialization of the RL-Algorithm

The second step builds upon the defined characteristics of the first step. The agent of the RL approach consists of two ANNs with identical architecture that are designed to process the layout, material flow, and functional unit characteristics in order to select the position of each functional unit. The starting point for action selection is the current state of the layout combined with additional information regarding the material flow characteristics. Based on the provided information in step one, the state representation provides a different degree of information density. Mandatory information is the material flow characteristics aggregated in form of a transportation matrix, and the information about occupied and free space in the layout. The latter information can be encoded by

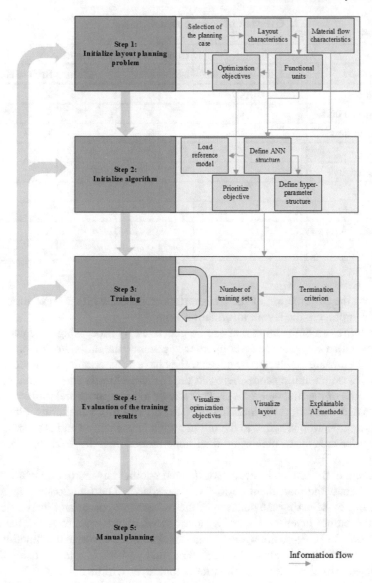

Fig. 2. Framework for RL-based factory layout planning structured as 5 sequentially executed steps containing loops to adjust earlier steps according to the results of later steps.

dividing the layout into a grid with n positions and transferring the positional information to a matrix. The additional but optional information (media, height, and floor-bearing capacity) can be transferred in a similar way. The input layer must fit the layout size and the amount of available information. One possible way is to use a graph neural network

Table 1. Overview of optimization objectives

	Type		Goal		Planning case	
Planning objective	Analytic	Dynamic	Min	Max	Greenfield	Brownfield
Throughput time		X	X		X	X
Utilization functional unit		X		X	X	X
Utilization material flow		X		X	X	X
Traffic congestion		X	X		X	X
Transportation intensity	X		X		X	X
Media supply	X		-	-	X	X
Floor-bearing capacity	X		-	-	X	X
Clarity of the material flow	X			X	X	X
Similarity	X			X		X

(GNN) to process the information of the transportation matrix, while a convolutional neural network can process the layout information (Fig. 3).

The GNN requires an embedding that can be generated by processing the material flow connections with a single dense layer in order to generate the initial node representation. Functional units are connected by edges if they have a material flow connection. Furthermore, pairwise attention scores between the functional units can be computed based on the outputs of the first dense layer. This allows to incorporate the influence of connected nodes and will reflect the interdependence between functional units. Finally, the result of this aggregation is added to the initial node representation and is processed by an additional dense layer to obtain the final node embeddings for each functional unit.

The architecture of the hidden layers can be defined without any restriction while the output layer represents the placement options (action space). The functional units are placed sequentially by selecting the position of the bottom left corner and its rotation. Consequently, the output layer consists of n_x neurons which are used to select the x-position, n_y neurons for the y-position, and 4 additional neurons to rotate the functional unit up to three times with an angle of 90° each. The final position is selected using the estimated Q-values. The n_x Q-values of the first output layer will be compared and the neuron with the highest value defines the x-position of the functional unit (y-position respectively). The rotation is obtained by a similar logic. The first neuron implies a rotation of 0° followed by 90° for the second neuron up to 270° for the last one. All subparts of the action are combined to define the final action that should be executed.

Afterwards, the hyperparameters of the ANN and the agent have to be defined. This includes the optimizer, learning rate, activation functions, and agent-related parameters such as the exploration strategy. An alternative for that sub-step is the usage of a reference model that was already trained to solve a similar layout planning problem. Consequently, it stores information about the solving process, which can be used to solve another problem without training the model from scratch (transfer learning). The weights of

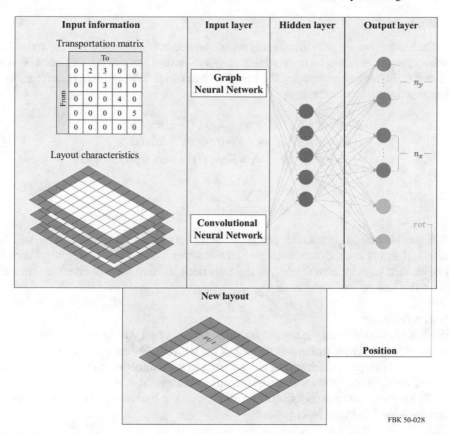

Fig. 3. Placement process and corresponding information. The input information is processed by a Graph neural network and a convolutional neural network to obtain an action from the output layer.

the model can be used completely if the entire ANN structure of the pre-trained model matches the structure of the new ANN. If that isn't the case, only parts can be reused. Using a pre-trained model can lead to faster and better results, depending on the model quality and the problem characteristics (Requirement 4: Scalability).

Finally, the reward function (R_t) is defined based on the following optimization objectives:

- Throughput time (R_{TT})
- Utilization of functional units (R_{UFU})
- Utilization of the material flow entities (R_{UMF})
- Traffic congestion (R_{TC})
- Transportation intensity (R_{TI})
- Media supply (R_{MS})
- Floor-bearing capacity (R_{FBC})
- Clarity of the material flow (R_{CMF})

- Similarity (R_S)

Each objective i is prioritized using weight w_i according to Eqs. 1 and 2. Within the environment, it needs to be ensured that each sub-reward reaches only values between -1 and 0. Better values will lead to rewards closer to 0. Without this transformation, prioritization will be ineffective.

$$
\begin{aligned}
R_t = {} & w_1 * R_{TT}(t) + w_2 * R_{UFU}(t) + w_3 * R_{UMF}(t) \\
& + w_4 * R_{TC}(t) + w_5 * R_{TI}(t) + w_6 * R_{MS}(t) \\
& + w_7 * R_{FBC}(t) + w_8 * R_{CMF}(t) + w_9 * R_S(t)
\end{aligned}
\tag{1}
$$

$$
\sum_{i=1}^{9} w_i = 1
\tag{2}
$$

If the planner knows which prioritization is appropriate only one model will be trained. If that is not the case, multiple prioritizations and corresponding weights need to be defined leading to multiple training sets. Each training set will result in one final layout that will be evaluated in step 4.

Step 3: Training

For each configuration of the reward function, a training set will be initiated which will lead to different layout variants at the end of the training. Before the training starts, the termination criterion, which can be either linked to the number of training episodes, the training duration, or a reward threshold, needs to be defined.

Within the training process, all improvements of the best-known solution are stored containing the following information:

- Current episode and training duration
- Total reward and all sub-rewards
- Weight of the prioritization
- Position of all functional units (actions)

The changes in the training loss, the average reward, and the individual reward values can be observed, for example, using a tensorboard or self-build visualization tools. A stable and converging behavior is a sign of a successful training process as depicted in Fig. 4. If the training process isn't stable a change in the ANNstructure or a change of the hyperparameters might lead to a stabilization.

Step 4: Evaluation of the Training Results

The evaluation process aims at selecting one layout alternative for further detailed manual planning steps. The planner will be supported by different visualization techniques. If multiple layouts are generated and multiple conflicting objectives are considered, a visualization of the corresponding pareto frontier leads to valuable insights into the problem characteristics. By analyzing the pareto frontier, the trade-off between the objectives can be evaluated. Furthermore, different solutions can be selected in order to visualize the corresponding layout (Fig. 5). The most suitable layout will be selected based on the preferences and the strategic goals.

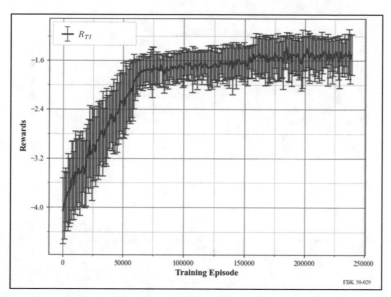

Fig. 4. Exemplary reward curve of a training process that stabilizes after 15000 training episodes

A second sub-step focuses on the field of explainable artificial intelligence (ExAI) and helps to increase the comprehensibility of the planning results (Requirement 6). The ExAI methods will provide information about the overall training behavior. The methods can ensure, that the reward function is aligned with the overall optimization objective. Furthermore, it is beneficial to evaluate the uncertainty in the placement process and to analyze in which situations a conflict in positioning occurred (two or more functional units should be placed in the same position). This information can be extracted using policy summarization methods, which analyze the placement behavior of the RL model. Besides the feature relevance explains which input information has the biggest influence on the decision-making process. This information might also help to identify bottlenecks and positions of higher relevance in the layout. At the end of step 4, the most favorable layout alternative is selected as a starting point for step 5.

The loops in the framework allow to repeat parts of the sequence if the planning results are not satisfying. Changes in the ANN structure in step 2 can lead to a change in the training behavior and new prioritizations might lead to new layouts for the evaluation step. Furthermore, lengthening the training duration can lead to additional improvements in the solution quality.

5 Step 5: Manual Planning

The manual planning step is the final step of the framework and focuses on the detailed planning tasks for the selected layout variant of step 4. First, the final position of each functional unit needs to be defined. The explainability methods can highlight conflicting goals and uncertainties to support this process. Afterwards, the functional units are

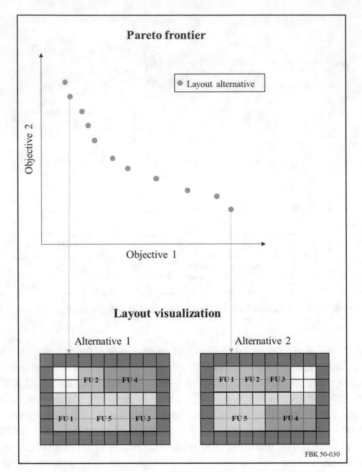

Fig. 5. Exemplary pareto frontier. Each datapoint corresponds to a different layout alternative. Two exemplary layouts are visualized

planned in detail, incorporating lighting conditions, emission limits, ergonomic conditions, and industrial safety to improve the workplace from an employee perspective. The detailed planning step can be supported by AR or VR technology, which will support the planner by designing the workspace according to the considered goals.

5.1 Evaluation of the Framework

In this Subsection, the framework and the corresponding RL-based planning approach are evaluated under consideration of the requirements defined in Subsect. 4.1. The placement process of the functional units is modeled in a way that allows for both a planning a new factory layout and restructuring an existing one. The only difference occurs regarding the input information and the reward function. Brownfield planning requires an initial layout configuration, and the reward function must contain the similarity reward. Consequently, Requirement 1 is satisfied. Furthermore, the framework allows to select and prioritize one or multiple objectives at the same time, which can be calculated based on a simulation or analytic formulation, which satisfies Requirement 2. However, until now, no simulation-based approach has been published.

The practical relevance of the planning results (Requirement 3) can only be fully ensured if the mandatory and optional information is provided. Furthermore, the free placement process (open field layout) leads to a higher level of practical relevance compared to commonly used single- and multi-row planning approaches. Furthermore, if only dynamic objectives are considered, the number of simplifications is smaller, since the DES allows a more realistic representation of the manufacturing system than a pure evaluation based on analytic formulations.

The scalability requirement can be ensured using an action masking method as introduced in [4] which prevents the selection of invalid actions, makes the learning phase faster, and makes the approach applicable for problems of different sizes.

Requirement 5 (accessibility) is improved by this research by presenting the sequential configuration process and highlighting the mandatory and optional information basis. The process can be further improved by developing a graphical user interface that supports a planner within the initialization and evaluation process.

The last requirement (comprehensibility of the planning results) is still an open research topic. The presented framework contributes to this open research question by providing a first overview of the valuable information that can be generated by an ExAI method. However, the ANNs of the agent can be considered as black box models. Consequently, reaching full transparency is challenging. Nevertheless, the methods mentioned in step 4 can increase the trustability and provide the necessary comprehensibility to apply the framework to industrial applications.

6 Conclusion and Outlook

Factory layout planning is an important but time-consuming process. Recently developed RL-based planning approaches have the potential to support the planner in the early planning phase by generating optimized layout variants. This paper contributes to the recent developments by presenting a holistic framework that increases the applicability of the RL-based approach by presenting the necessary steps and the underlying information flow that are required to utilize the layout planning potential. Consequently, the framework can be used as guidance on how to initialize, train, and evaluate the results of an RL-based factory layout planning approach. The framework is developed based on six layout planning-related requirements. It consists of five steps: the initialization of the layout planning problem, the initialization of the algorithm, multiple training sets, the evaluation of the training results, and the manual planning step for the selected layout variant. Each step consists of multiple sub-steps that are interlinked by an information flow. The framework describes for each sub-step which information is needed and further distinguishes between mandatory and optional information.

Furthermore, the framework provides guidance for further developments. Future work will consequently, focus on developing an discrete event material flow simulation and the necessary interfaces to integrate it into the environment of an RL approach. The development aims at including the dynamic optimization objectives presented in this framework and further allows to validate the framework in detail. Besides, the scalability of multiple optimization objectives needs to be investigated since the complexity increases compared to the existing single objective investigations. Furthermore, explainable artificial intelligence methods will be developed to provide insights into the training process. For example, a policy summarization method can be used to increase the trustability by providing information about the reward structure, the placement uncertainties, and the alignment of short-term rewards with the overall optimization objective.

Acknowledgements. This research was funded by the Deutsche Forschungsgemeinschaft (DFG, German Research Foundation) – 252408385 – IRTG 2057.

References

1. Mayr, A., Kißkalt, D., Meiners, M., Lutz, B., Schäfer, F., Seidel, R., et al.: Machine learning in production – potentials, challenges and exemplary applications. Procedia CIRP **86**, 49–54 (2019). https://doi.org/10.1016/j.procir.2020.01.035
2. Klar, M., Glatt, M., Aurich, J.C.: An implementation of a reinforcement learning based algorithm for factory layout planning. Manufacturing Letters **30**, 1–4 (2021). https://doi.org/10.1016/j.mfglet.2021.08.003
3. VDI 5200. Factory planning (2011). Accessed
4. Klar, M., Hussong, M., Ruediger-Flore, P., Yi, L., Glatt, M., Aurich, J.C.: Scalability investigation of double deep q learning for factory layout planning. In: Procedia CIRP, vol. 107, pp. 161–166 (2022). https://doi.org/10.1016/j.procir.2022.04.027
5. Westkämper, E., von Briel, R.: Continuous improvement and participative factory planning by computer systems. CIRP Ann. **50**(1), 347–352 (2001). https://doi.org/10.1016/S0007-8506(07)62137-4
6. Warnecke, H.-J.: Organisation, Produkt, Planung, 3rd edn. Springer, Berlin (1995)
7. Stephens, M.P., Meyers, F.E.: Manufacturing Facilities Design and Material Handling. Purdue University Press, West Lafayette (2013)
8. Drira, A., Pierreval, H., Hajri-Gabouj, S.: Facility layout problems: a survey. Annu. Rev. Control. **31**(2), 255–267 (2007). https://doi.org/10.1016/j.arcontrol.2007.04.001
9. Benjaafar, S., Heragu, S.S., Irani, S.A.: Next generation factory layouts: research challenges and recent progress. Interfaces **32**(6), 58–76 (2002). https://doi.org/10.1287/inte.32.6.58.6473
10. VDI 3330. Costs of material flow (2007). Accessed
11. VDI 3633. Simulation of systems in materials handling, logistics and production - Fundamentals (2014). Accessed
12. Tompkins, J., White, J.A., Bozer, Y.A.: Facilities Planning, 4th edn. Wiley, Hoboken (2010)
13. Grundig, C.-G.: Fabrikplanung: Planungssystematik, Methoden, Anwendungen, 5th edn. Hanser, München (2014)
14. Buffa, E.S., Armour, G.C., Vollman, T.E.: Allocating facilities with CRAFT. Harvard Bus. Rev. **42**, 136–59 (1964)
15. Şahin, R., Niroomand, S., Durmaz, E.D., Molla-Alizadeh-Zavardehi, S.: Mathematical formulation and hybrid meta-heuristic solution approaches for dynamic single row facility layout problem. Ann. Oper. Res. **295**(1), 313–336 (2020). https://doi.org/10.1007/s10479-020-03704-7

16. Doil, F., Schreiber, W., Alt, T., Patron, C.: Augmented reality for manufacturing planning. In: Kunz A, Deisinger J, editors. Proceedings of the workshop on Virtual environments 2003 - EGVE '03; 22.05.2003 - 23.05.2003; Zurich, Switzerland. New York, USA, ACM Press, pp. 71–6 (2003)

17. Nee, A., Ong, S.K., Chryssolouris, G., Mourtzis, D.: Augmented reality applications in design and manufacturing. CIRP Ann. **61**(2), 657–679 (2012). https://doi.org/10.1016/j.cirp.2012.05.010

18. Kusiak, A., Heragu, S.S.: The facility layout problem. Eur. J. Oper. Res. **29**(3), 229–251 (1987). https://doi.org/10.1016/0377-2217(87)90238-4

19. Amaral, A.R.: On the exact solution of a facility layout problem. Eur. J. Oper. Res. **173**(2), 508–518 (2006). https://doi.org/10.1016/j.ejor.2004.12.021

20. Hosseini-Nasab, H., Fereidouni, S., Fatemi Ghomi, S.M.T., Fakhrzad, M.B.: Classification of facility layout problems: a review study. Int. J. Adv. Manuf. Technol. **94**(1–4), 957–977 (2017). https://doi.org/10.1007/s00170-017-0895-8

21. Lin, Z., Yingjie, Z.: Solving the Facility Layout Problem with Genetic Algorithm. In: 2019 IEEE 6th International Conference on Industrial Engineering and Applications (ICIEA), 12.04.2019 - 15.04.2019; Tokyo, Japan, pp. 164–168. IEEE (2019)

22. Chen, C., Tiong, L.K.: Using queuing theory and simulated annealing to design the facility layout in an AGV-based modular manufacturing system. Int. J. Prod. Res. **57**(17), 5538–5555 (2019). https://doi.org/10.1080/00207543.2018.1533654

23. Klar, M., Schworm, P., Wu, X., Glatt, M., Aurich, J.C.: Quantum annealing based factory layout planning. Manuf. Lett. **32**, 59–62 (2022). https://doi.org/10.1016/j.mfglet.2022.03.003

24. Guan, C., Zhang, Z., Liu, S., Gong, J.: Multi-objective particle swarm optimization for multi-workshop facility layout problem. J. Manuf. Syst. **53**, 32–48 (2019). https://doi.org/10.1016/j.jmsy.2019.09.004

25. Robinson, S.: Simulation: The Practice of Model Development and Use, 2nd edn. Palgrave Macmillan, Basingstoke (2014)

26. Klar, M., Langlotz, P., Aurich, J.C.: A framework for automated multiobjective factory layout planning using reinforcement learning. Procedia CIRP **112**, 555–560 (2022). https://doi.org/10.1016/j.procir.2022.09.099

27. Osinski, B., Jakubowski, A., Ziecina, P., Milos, P., Galias, C., Homoceanu, S. et al.: Simulation-based reinforcement learning for real-world autonomous driving. In: 2020 IEEE International Conference on Robotics and Automation (ICRA), 31.05.2020–31.08.2020, Paris, France, pp. 6411–6418. IEEE (2020)

28. Goodfellow, I., Bengio, Y., Courville, A.: Deep learning. The MIT Press, Cambridge, Massachusetts (2016)

29. Sutton, R.S., Barto, A.: Reinforcement learning, second edition: An introduction. The MIT Press, Cambridge (2018)

30. Wiering, M., van Otterlo, M.: Reinforcement Learning. Berlin, Heidelberg (2012)

31. van Hasselt, H., Guez, A., Silver, D.: Deep Reinforcement Learning with Double Q-learning (2015)

32. Schulman, J., Wolski, F., Dhariwal, P., Radford, A., Klimov, O.: Proximal policy optimization algorithms (2017)

33. Han, M., Zhang, L., Wang, J., Pan, W.: Actor-critic reinforcement learning for control with stability guarantee. IEEE Robot. Autom. Lett. **5**(4), 6217–6224 (2020). https://doi.org/10.1109/LRA.2020.3011351

34. Kanervisto, A., Scheller, C., Hautamaki, V.: Action space shaping in deep reinforcement learning. In: 2020 IEEE Conference on Games (CoG), 24.08.2020–27.08.2020; Osaka, Japan, pp. 479–486. IEEE (2020)

35. Wu, X., Yi, L., Klar, M., Hussong, M., Glatt, M., Aurich, J.C.: Intelligent robotic arm path planning (IRAP2) framework to improve work safety in human-robot collaboration (HRC) workspace using deep deterministic policy gradient (DDPG) algorithm. In: Kim, KY., Monplaisir, L., Rickli, J. (eds) Flexible Automation and Intelligent Manufacturing: The Human-Data-Technology Nexus. FAIM 2022. LNME, pp 179–187. Springer, Cham (2023). https://doi.org/10.1007/978-3-031-18326-3_18

36. Zhao, F., Jiang, T., Wang, L.: A reinforcement learning driven cooperative meta-heuristic algorithm for energy-efficient distributed no-wait flow-shop scheduling with sequence-dependent setup time. IEEE Trans. Ind. Inf. 1–12 (2022). https://doi.org/10.1109/TII.2022.3218645

37. Yi, L., Langlotz, P., Hussong, M., Glatt, M., Sousa, F.J., Aurich, J.C.: An integrated energy management system using double deep q-learning and energy storage equipment to reduce energy cost in manufacturing under real-time pricing condition: a case study of scale-model factory. CIRP J. Manuf. Sci. Technol. **38**, 844–860 (2022). https://doi.org/10.1016/j.cirpj.2022.07.009

38. Silver, D., Hubert, T., Schrittwieser, J., Antonoglou, I., Lai, M., Guez, A., et al.: A general reinforcement learning algorithm that masters chess, shogi, and Go through self-play. Science **362**(6419), 1140–1144 (2018). https://doi.org/10.1126/science.aar6404

Simulation-Based Investigation of the Distortion of Milled Thin-Walled Aluminum Structural Parts Due to Residual Stresses

D. Weber[1]([✉]), B. Kirsch[1], C. R. D'Elia[2], B. S. Linke[2], M. R. Hill[2], and J. C. Aurich[1]

[1] Institute for Manufacturing Technology and Production Systems, RPTU Kaiserslautern, Gottlieb-Daimler-Str., 67663 Kaiserslautern, Germany
daniel.weber@rptu.de

[2] Department of Mechanical and Aerospace Engineering, University of California, One Shields Avenue, Davis, CA 95616, USA

Abstract. Nowadays, aluminum components in aircraft are mainly found in the form of thin-walled monolithic structural parts of the internal fuselage and the wings as spars and ribs [1]. This is because these components have excellent material properties for lightweight applications, such as a high strength-to-weight ratio and good corrosion resistance [2]. A typical manufacturing process to produce such structural components is milling. For these weight-optimized, monolithic components, up to 95% of the material is removed by machining [3]. The challenge with these thin-walled structural components, which are up to 14 m long, is that part distortion can occur because of the manufacturing-specific process chain [4]. Residual stresses due to machining and upstream processes such as forming, and heat-treatments are known to be the key factor for causing those distortions [5].

In this research the effect of the residual stresses, the machining strategy, the part topology and the geometry, including the wall-thickness, on distortion were investigated experimentally, and simulatively by validated virtual models based on the finite-element method. Those models can then be used to predict the distortion. At the end distortion minimization techniques were derived.

1 Introduction

Aluminum, with an annual production of 67,2 Mt in 2021, is the most common metal and the third most common element in the earth [6, 7]. Starting from the mineral bauxite one obtains aluminum by first refining it to alumina (Al_2O_3) via the Bayer process and then smelting the alumina by the Hall-Héroult process [8]. Aluminum alloys are used as structural materials in numerous fields such as automotive, aerospace, installation and apparatus, electrical engineering, food technology, chemical industry, and the optical industry [9]. Particularly in aerospace engineering, a high proportion of aluminum is required, especially in the form of thin-walled monolithic structural components. For example, these components are used in aircrafts as internal fuselage structures, wing ribs and spars, as well as window and crown frames [1]. These integral structural components, which are made of rolled sheets, extruded sections or forgings, can be up to 14 m long

© The Author(s) 2023
J. C. Aurich et al. (Eds.): IRTG 2023, *Proceedings of the 3rd Conference on Physical Modeling for Virtual Manufacturing Systems and Processes*, pp. 149–169, 2023.
https://doi.org/10.1007/978-3-031-35779-4_9

[4]. Thin-walled structural components are typically used because they have excellent material properties for lightweight applications, such as a high strength-to-weight ratio and good corrosion resistance [2]. For example, the mass fraction of aluminum alloys used in the B747, B757, B767, B777, A300, A380, A340, and A320 aircraft models is between 60 and 80% of the total weight [10, 11]. In these weight-optimized, monolithic components, the largest proportion of the material is removed by machining [5]. The information in the literature varies from a maximum machining volume of 80% [12] to 90% [5] and up to 95% [3] of the total volume of the semifinished product. A typical manufacturing process to produce thin-walled aluminum structural components is milling. The challenge with these monolithic thin-walled structural components is that part distortion can occur because of the manufacturing-specific process chain [4]. It is known that those distortions are caused by residual stresses (RS), which are defined as the locked in stresses being in equilibrium. They exist in materials and structures independent of the presence of any external loads [13]. Those RS typically are divided into machining induced residual stress (MIRS) and initial bulk residual stress (IBRS) due to upstream processes such as rolling, casting, and especially heat-treatments [5]. The equilibrium of the RS is disturbed due to the material removal process and the application of MIRS. Once the part is released from its clamping fixture, distortion is the result of the re-equilibrium of the RS [14]. Although the semifinished products typically undergo a process to relief the high IBRS due to quenching, e.g. by controlled compression or stretching, part distortion remains as a problem [15]. The MIRS can be found in the boundary layer of the parts and are the result of plastic deformations during machining [5]. The analysis of the effects of the cutting parameters and the tool on the MIRS was subject of research [5]. Consensus is that high mechanical loads result in compressive MIRS [16]. It was found that an increase of the feed per tooth or the cutting edge radius and a decrease of the corner radius led to greater, in terms of amount, and deeper compressive MIRS [17]. Investigations on the variation of the cutting speed did not show a clear influence on the MIRS at all: For example, Perez et al. [18] and Denkena et al. [19] noticed higher MIRS with increasing cutting speed, whereas Rao et. al. [20] and Tang et. al. [21] observed lower MIRS.

Above-mentioned research has in common that no repeated RS measurements on different or even the same sample(s) were carried out. Only limited or even no statistical confidence was present. Furthermore, there is a lack of studies that measure the MIRS of milled samples with multiple RS measuring techniques.

The literature review by Aurrekoetxea et al. highlighted that both RS types lead to the distortion of thin-walled aluminum components [22]. It is known that with decreasing wall-thickness the effect of MIRS on distortion increases. However, different critical values for material thicknesses were found to determine when MIRS dominate compared to IBRS [5].

Besides analytical models, which are based on the plate bending theory, numerical simulation models, typically finite element method (FEM) models, offer the possibility to predict the distortion due to known RS for complex part geometries. Mostly two approaches have predominantly prevailed in the literature: The RS were applied to the final machined part geometry [4, 23, 24] or the material removal process was modelled via element deletion techniques [25–27]. First approach provides faster results but leading potentially to a reduced accuracy [22]. However, the mutual influence of both RS types on distortion is still not fully understand. The entire RS, including MIRS on all surfaces and effects like machining-induced shear stresses were mostly not considered, although e.g. second have been shown to induce a torsional moment and thus influence the part distortion [28].

There are different ways to obtain the MIRS needed as an input for the distortion models. By RS measurements a data-basis can be built for different machining conditions varying in tools and process parameters. Using empirical regression models is a possibility to extend the data-basis within the process space. Furthermore, analytical and numerical models were used to predict the MIRS. With increasing computing power, as well as the availability of commercial FEM programs, numerical models replaced analytical and empirical ones in manufacturing, since the latter are usually only valid for a limited process space (tool, workpiece, cutting parameter combination) [29]. However, a comparative study by Jawahir showed that FEM models of different research groups, which modeled the same real-world process, led to large discrepancies in the simulated MIRS [29]. Possible sources of error are inappropriate simplifications, wrong assumptions, improper modeling of the boundary conditions, numerical run-out errors or discretization errors [30].

Two main categories of distortion control due to RS were identified in the literature review by Li et al. [5]: The postcorrection is characterized by processing the finished part with thermal or mechanical loads, like peen forming, laser heat treatments or other stress relieve techniques. The precontrol techniques on the other hand improve the machining conditions in a way that mainly the magnitude of RS is reduced, or a change of their distribution is aimed for [5]. Besides adjusting the cutting parameters to induce less MIRS, controlling the cutting sequence [31] and changing the process strategy (applying different MIRS on different machined surfaces [26], subsequent cutting steps to remove the boundary layer containing high MIRS [32], changing the milling direction [33]) led to a beneficial shift of the MIRS distribution and therefore reduced the distortion. The beneficial modification of the distribution of the IBRS was realized by changing the position of the part in the raw material [24, 34–38].

In summary both IBRS and MIRS lead to part distortions, which can be reduced by considering their magnitude and distribution in a favorable way. However, their mutual impact on part distortion and the potential of deriving compensation techniques are still not fully understood yet.

This research article is a summary to provide a holistic overview of our research in the field, where most of the content has already been published elsewhere ([39–43, 46–48, 51]). The novelty of this research consists of the holistic consideration of the combined influence of the part geometry, including its topology, and the machining strategy and distinguishing the dominating RS type for analyzing the part distortion; including investigations on the effect of IBRS on MIRS, the comparison of different RS measurement techniques and the repeatability of the RS and distortions.

2 Methodology

Figure 1 provides an overview of the concept developed to understand the residual stress induced distortion of milled thin-walled aluminum structural parts, predict the distortion using FEM simulations, and derive methods to minimize the distortion [39]. It contains a combination of experiments and simulation models. The experiments serve as a validation for each simulation model. As Fig. 1 illustrates, each, IBRS and MIRS, was analyzed individually before investigating the superposition of both to understand their fundamental principles and effects on part distortion. The structure of the experiments was divided into two main parts with regard to the target parameter to be investigated: RS investigations and distortion investigations. In the former, both RS types were characterized (Sect. 3.1 and 3.2) and the influence of the machining parameters, clamping strategy and tool type on the MIRS in the workpiece was identified (Sect. 3.2) [40]. This includes measuring the MIRS with various techniques [41] and a repeatability study [40]. In the distortion investigations, different thin-walled geometries were manufactured, and their distortion was determined to investigate the influence of the IBRS, the MIRS (Sect. 3.3) and their superposition on the part distortion (Sect. 3.4) [42]. The developed FEM models predict the part distortion (Sect. 4.1) based on the information about MIRS (Sect. 4.2), resulting from different machining parameters, tools, and the IBRS. The part geometry, including its topology (angle of stiffeners, wall thickness, size, complexity), and the machining strategy were varied to analyze their effect on the part distortion and to highlight possibilities to minimize the distortion (Sect. 5) [43].

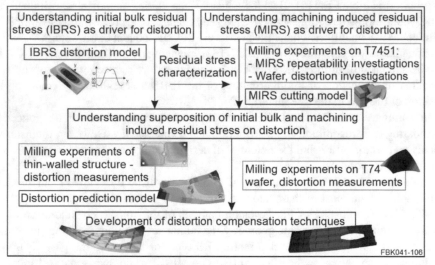

Fig. 1. Concept to minimize distortion of milled thin-walled aluminum structural parts due to residual stresses acc. to [39]

3   Experiments

3.1   Initial Bulk Residual Stress Characterization

To understand the effect of the IBRS on part distortion, different IBRS configurations were examined. Workpieces made of a rolled plate of the high strength aerospace aluminum alloy 7050 were investigated in two conditions: First 7050-T74, which is a solution heat treated and quenched material, containing high IBRS and secondly the 7050-T7451 state, which was in addition stress relieved by stretching and therefore containing low IBRS [44]. All blocks used for machining experiments (see Sect. 3.2 and 3.3) were cut from the original rolled plate into individual samples measuring 206 mm (x-direction: longitudinal rolling L), 102 mm (y-direction: short transverse ST) and 28.5 mm (z-direction: longitudinal transverse LT) (see Fig. 2a). The low IBRS were measured via the slitting method using wire electric discharge machining and the high IBRS via a variation of the slitting method, the so-called cut mouth opening displacement method, from which a 2D IBRS map was deduced [45]. The measurements showed that the low IBRS had a maximum of about 20 MPa (see Fig. 2b), whereas the high IBRS reached from -150 MPa to 100 MPa, which is a significant fraction of the material strength (see Fig. 2c) [45]. The quenched samples had a paraboloid spatial distribution of normal RS with a significant directionality. The stress state near the center is nearly uniaxial tension with σ_{xx} much larger than σ_{yy}. Whereas near the upper and lower boundaries the stress is nearly uniaxial compression (σ_{xx} compressive and σ_{yy} near zero) [45]. The distortion caused by those IBRS is discussed in Sect. 3.3.

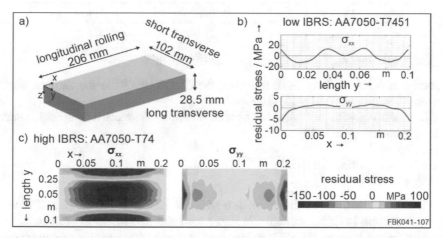

Fig. 2. Initial bulk residual stresses for low (b) and high stress configuration (c) acc. to [45]

3.2   Machining Induced Residual Stress Characterization

Down milling with cemented carbide end mills (Kennametal[1] F3AA1200AWL, d = 12 mm, z = 3, for more details see [40]), which represents a typical tool for machining of aerospace aluminum alloys, was carried out on a 5-axis DMG Mori[1] DMU 70 CNC machine to evaluate the effect of the machining on the MIRS. The above mentioned AA7050-T7451 samples with the dimensions $206 \times 102 \times 28.5$ mm^3 were face milled. To neglect the IBRS, the samples in the stress-relieved condition (T7451) with low IBRS were used. The MIRS were explicitly not measured on thin-walled milled but on thick workpieces, since a strong redistribution of the RS, which is associated with the distortion of the thin components, should be avoided to be able to capture the full distortion potential of the MIRS. Three different feeds per tooth f_z and two different cutting speeds v_c were analyzed, resulting in four parameter modes (EM1-4, see Table 1). The width of cut a_e and depth of cut a_p were kept constant at 4 mm and 3 mm respectively, and dry cutting was carried out. To further investigate the influence of the tool type on the MIRS, a second tool, a cutter with indexable inserts (Sandvik[1] R590-110504H-NL H10, d = 50 mm, z = 2, for more details see [42]), which is a typical tool for face milling, was used (Index, see Table 1, $a_e = 40$ mm, $a_p = 1.5$ mm).

Table 1. RS experiment matrix.

acronym	tool	cutting speed v_c in m/min	feed per tooth f_z in mm	IBRS	wp quantity
EM1	end mill	200	0.04	low	3
EM2	end mill	200	0.1	low	3
EM3	end mill	200	0.2	low	3
EM4	end mill	450	0.04	low	3
Index	inserts	730	0.2	low	1

Various RS measuring techniques were investigated:

- incremental hole-drilling (HD) following the ASTM E837-13a standard measuring with strain gauges [ASTM13] (HD-strain)
- HD measuring with an optical laser-based principle, the electronic speckle pattern interferometry (HD-ESPI)
- slotting
- $\cos(\alpha)$ x-ray diffraction (XRD)
- $\sin^2(\psi)$ XRD

The MIRS measured by the different techniques HD, slotting, $\sin^2(\psi)$ XRD were largely consistent [41]. For example, in Fig. 3 the measured MIRS in orthogonal feed direction for machining set EM3 are shown. Root shaped depth profile of compressive RS, which are typical for milling induced RS, were measured. The measurements of the two HD techniques agreed for the normal direction (see Fig. 3b). Similar maximum MIRS (approx. -120 MPa) were found at different depths, which could be attributed to the depth correction applied for the HD-strain measurements (for more information see [41]). However, in shear direction more significant deviations were evident. Furthermore, it was found that MIRS data from HD-strain were most consistent with machining-induced distortion [41].

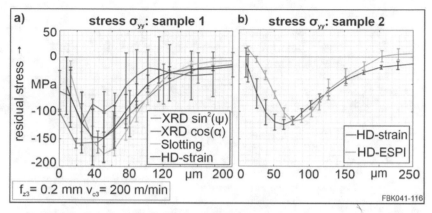

Fig. 3. Comparison of RS measuring techniques for σ_{yy} acc. to [41] (a)

To investigate the repeatability of MIRS for multiple samples and the influence of different machining modes, twelve machining experiments (three for each EM mode) were carried out with three HD-strain measurements each sample. All measured MIRS depth profiles showed compressive MIRS for all three stress components σ_{xx} (feed-direction), σ_{yy} (orthogonal feed-direction), τ_{xy} (shear direction) (see Fig. 4). The normal stresses were similar in their magnitude and lower shear stresses were measured. The maximum of MIRS (MaxRS) for lower feeds (EM1, 2) existed at the shallowest depth (see Fig. 4a). Higher feed per tooth (EM3) led to larger plastically deformed areas and therefore deeper RS and the shift of the MaxRS deeper into the workpiece, due to the increased load on the sample. The repeatability standard deviation (RSTD) indicated that the RS for EM 1, 2 and 3 were repeatable with small variations within one sample and from sample to sample [40]. EM4, where a different cutting speed was chosen, showed more variability compared to the other modes, because machining for EM4 was not stable, which was indicated by the RSTD (see Fig. 4a) and vibrations detected in the force signal of F_z [40].

The different tool geometry and the chosen machining parameters led to smaller and shallower MIRS compared to the ones induced by the regular end mill (see Fig. 4b) due to the decreased mechanical load. Furthermore, a opposite sign of the shear stresses was measured.

For investigating the influence of the IBRS on the MIRS, the solution heat treated and quenched material AA7050-T74 with high IBRS was machined with the machining parameter EM3. From the IBRS measurements (see Sect. 3.1) different regions with variations of IBRS could be identified. HD-strain measurements were applied at locations where IBRS were near zero (Pos. A), tensile (Pos. B) or compressive (Pos. C) (see Fig. 5) [46, 51]. At greater depths (>0.2 mm), the different IBRS are visible. Furthermore, it is evident that the IBRS effected the MaxRS (see Fig. 5).

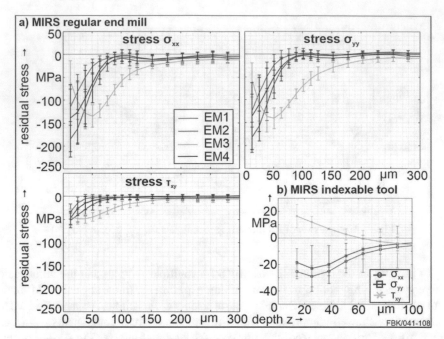

Fig. 4. Measured MIRS: End mill (a) and cutter with indexable inserts (b)

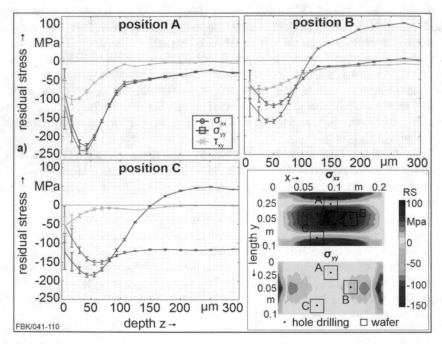

Fig. 5. RS depth profiles on various positions at milled surface acc. to [46, 51]

3.3 Machining Induced Residual Stress as Driver for Distortion

To investigate the distortion potential of the MIRS experiments, where a 1 mm thick wafer was removed at the milled surface via wire electric discharge machining, were developed [40, 41]. The distortion was defined as the out-of plane displacement of the wafer. It was measured with a laser profilometer at points with a 0.2 mm spacing at the backside. The machined surface becomes convex (∩-shaped) due to the compressive MIRS at the milled surface, which induced a bending moment (see Fig. 6a). The maximum distortion was at the top left and bottom right corner because the shear stresses caused a torsional moment in addition to the bending moment induced by the normal RS. Higher or deeper compressive RS resulted in a higher distortion (see Fig. 6c). Variations of MIRS within one machining mode for different samples resulted in a consistent variation of distortion. The highest variation for the wafer distortion was found for EM4 due to its unstable machining [40].

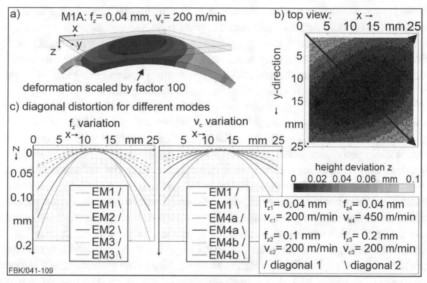

Fig. 6. Qualitative shape of wafer distortion (a), contour plot (b), and diagonal distortions for different machining modes (c) acc. to [40]

3.4 Superposition of IBRS and MIRS and Its Effect on Distortion

The analysis of the effect of the superposition of the MIRS and the IBRS on distortion was done the same way as described in Sect. 3.3: Wafers were cut out at the positions A, B and C according to the HD measurement locations (see Fig. 5). The distortion shape and level changed compared to the low IBRS wafers due to the higher IBRS (see Fig. 8). The high IBRS act as a preload [46]. When the IBRS along the milling direction are tensile, the material flows more in the direction of tension, leading to a rotation of the convex wafer distortion away from the diagonal, closer to the milling direction (see

Fig. 8b) [46]. For wafers in regions of compressive IBRS along the milling direction, the opposite occurs, and the convex distortion shape rotates even further away from the milling direction (see Fig. 8c).

The distortion of more complex workpieces than the flat wafers was analyzed as following: A small thin-walled structural component of the size 200x98x20 mm^3 with one rib in the center surrounded by two pockets was milled from the initial block (206 $\times$ 102 $\times$ 28.5 mm^3). The influence of different IBRS (low vs. high), MIRS, machining strategy (zig vs. spiral out), and wall thickness (3 mm vs. 7 mm) on the distortion was analyzed. For the 3 mm wall thickness (7 mm), about 84% (67%) of the initial material was removed. First the outer walls were milled by side milling with the regular end mill (v_c = 450 m/min; f_z = 0.055 mm; a_p = 22 mm; roughing a_e = 2.5 mm; finishing a_e = 0.5 mm). Second the back and top side were face milled with the parameter set Index (see Table 1). To enhance the clamping, which was realized by side clamps, additional holes for clamping with screws were drilled. Finally, the pockets were milled with EM1 or EM3 to induce different MIRS. To realize high feed rates, the pockets were milled in multiple layers. To further analyze the effect of the milling path on the distortion, two strategies were investigated for the 3 mm samples: First, the milling of the pockets was done in alternating order in zig strategy (pocket milling paths from left to right). Second, a spiral milling path following the contour of the workpiece (inside-out) was used. The distortion was measured on the backside of the sample before (Pre-) and after step 5 (Post-) with a coordinate measuring machine and a spacing of 2 mm.

A general comparison of the distortion of the low and high IBRS samples (independent of their wall thickness and machining mode) showed that their distortion shape and magnitude differ (see Fig. 7). The low IBRS samples, machined with the zig strategy, showed a X-shaped distortion with its maximum distortion near the top left and bottom right corner and its minimum at the other two corners. Like the distortion behavior of the wafers, the shear MIRS were responsible for this distortion shape. In contrast, the high IBRS samples become convex ($\cap$-shaped), and the maximum distortion was found towards the left and right edge. The magnitude of the maximum distortion was approximately 0.6 mm, which was about five times higher than for the low stress samples. This indicates that for high IBRS samples the IBRS are driving the distortion, because their RS are much higher as the MIRS, and they are contained in the entire bulk of the sample. The removal of the material led to a disequilibrium of the IBRS. The distortion was the result of the stresses gaining equilibrium again. Nevertheless, there was a systematic influence of the combined effect of both RS types evident for thin wall thicknesses (3 mm), where the maximum distortion was also found at the two opposite corners. The machining EM3, inducing more MIRS deeper into the material, led to higher distortions than EM1 (see Fig. 7). [42].

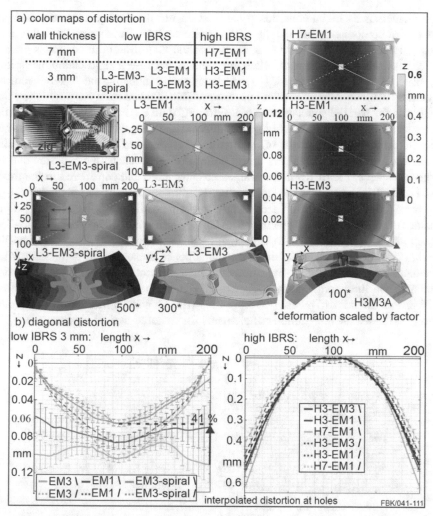

Fig. 7. Color maps of distortion, diagonal distortion of each configuration acc. to [42]

4   Simulation Models

Two different simulation models were developed: First, a distortion prediction model, which uses the RS as input to calculate the distortion. Second, a cutting model, that simulates the tool workpiece interaction due to the given cutting parameters to predict the resulting MIRS. It was investigated whether that model can substitute the MIRS measurements required for the distortion model.

4.1   Distortion Prediction Model

Using a static, linear elastic FEM model, realized in ABAQUS[1], the distortion due to the RS was simulated [42, 43]. The measured MIRS and the measured IBRS were implemented as an initial condition (type = stress) to the final part shape (wafer and thin-walled component) and the distortion was calculated because of the RS gaining re-equilibrium. The assumption was made that the distortion only occurs as soon as the sample is removed from the clamping device, because a rigid clamping strategy was used in the experiments [42]. That means that simulating the material removal process by element deletion was not necessary. This allowed for a fast simulation time (< 30 min for parallel simulation on a desktop PC with 8 cores). The measured MIRS (plane stress: σ_{xx}, σ_{yy}, τ_{xy}) were linearly interpolated over the depth z at the element centroids in the boundary layer of the respective milled surfaces. For the thin-walled structure in addition to the MIRS at the bottom of the pockets, the backside and top of the workpiece, the MIRS in the walls were measured and considered in the model [43]. For depths greater than the last measured depth, the measured in-plane IBRS (σ_{xx}, σ_{yy}) were linearly interpolated accordingly to their position x,y (see Fig. 2). To also consider the true milling path and therefore the exact direction of the MIRS, the G-Code was exported from the CAD/CAM system [43]. In this way, the direction of milling could be identified at each location. The elements of those regions containing MIRS were detected and the MIRS were assigned to each element according to their direction, calculated via the coordinate transformation. The mesh was refined at the machined surfaces in z-direction to precisely resolve the MIRS. A coarser mesh was used in other regions to reduce the total number of elements for calculation time reasons.

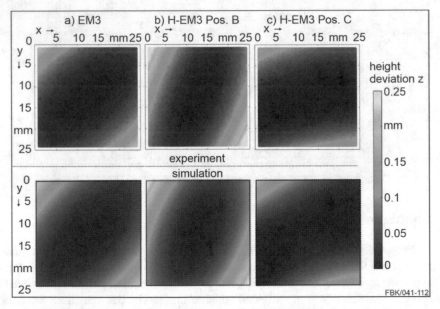

Fig. 8. Comparison of measured and predicted wafer distortion acc. to [51]

Figure 8 shows the comparison between the measured wafer distortion and the simulated one for machining conditions EM3/EM3-HS at different positions (see also

Sect. 3.2.). The results show that the simulation can predict the shape and level of distortion qualitatively and quantitatively for each scenario [40, 46].

The simulation model is also able to predict the shape of the distortion of the thin-walled structure qualitatively for all different configurations (see Fig. 9). All the different effects discussed in Sect. 3.3, such as the X-shape for the zig strategy of low, the U-shape for spiral strategy of low and the ∩-shape of distortion of high IBRS samples, are covered by the simulation. Furthermore, the magnitudes of distortion for simulated and measured distortions were on a similar level [42]: The relative error of the simulation for the maximum distortion found at the samples with a wall thickness of 3 mm and machined with the zig (spiral) strategy was 9% (−19%). The deviation for machining with the spiral milling strategy was higher, because in general lower distortions were measured and therefore deviations in the measured RS, used as input, have a bigger influence on the distortion prediction accuracy. In addition, the application of only MIRS or IBRS, showed that the MIRS in the pockets, especially the shear stresses, are driving the distortion for milling the investigated thin-walled structure with 3 mm wall-thickness and stress relieved (T7451) material when using the zig strategy [42]. In contrast, the low IBRS are driving the distortion when choosing the spiral strategy, because the shear MIRS in the pockets equilibrate each other almost fully. But still, for both strategies, the superposition of the IBRS and the MIRS is evident [43].

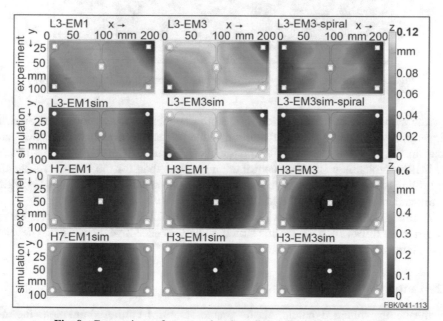

Fig. 9. Comparison of measured and predicted distortion acc. to [43]

Since the model was validated, it could be used for investigating more use cases than experimentally examined: A mutual influence of the part geometry, topology respectively (angle of stiffeners, wall thickness, size, complexity), and the machining strategy (IBRS configuration, milling path) on the part distortion was evident [47]. For example, a change of the angle of stiffener from 90° towards 45° led to a significantly reduced distortion (−66%) of the part machined in zig milling strategy (see Fig. 10), because the sample was stiffened along the principal direction with maximum displacement. Whereas machining in a spiral path increased the distortion for the geometry with 45° stiffener because shear stresses did not fully equilibrate each other anymore.

A decreasing wall thickness was found to increase the distortion with accompanying more dominant MIRS in comparison to low IBRS.

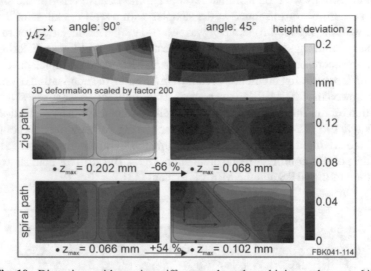

Fig. 10. Distortions with varying stiffener angle and machining path acc. to [47]

The simulation of the distortion of a more complex part, a small-scaled airplane wing-rib, showed similarities in distortion shape as for the investigated smaller parts (see Fig. 11): For milling with zig strategy, the X-shaped distortion was present with the MIRS dominating. But their contribution was less than for the small parts. Again, for milling with a spiral strategy the maximum distortion was reduced by −45% with a minimization of the effect of the MIRS.

Defining a universal crucial wall thickness when MIRS are the main factor for the distortion for thin-walled monolithic structural parts is deceptive, because that depends on the part geometry and the machining strategy [47].

4.2 Cutting Model to Predict the MIRS

By means of explicit, dynamic, elastic-plastic 3D FEM cutting simulations the tool workpiece interaction was modeled in ABAQUS[1] to predict the MIRS [48]. Two different cutting conditions, one for each tool and milling process (pocket milling EM3, face milling Index, see Table 1), were modelled (see Fig. 12). The tools were assumed as rigid bodies, neglecting wear. This assumption was valid, because the elastic modulus of the cemented carbide tools was significantly higher, resulting therefore in a low elastic deflection compared to the large plastic deformation of the aluminum workpiece.

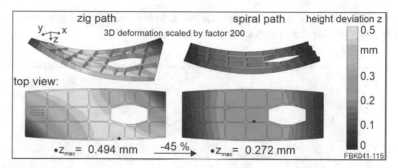

Fig. 11. Predicted distortion of structural part with varying mach. path acc.to [47]

Besides, only one revolution (EM3), two respectively (Index), of the respective tool were simulated. To furthermore save computational time, the workpieces were modelled smaller as in experiments. To account for thermal and mechanical effects, thermo-mechanical elements (C3D8RT) were chosen for the workpiece. The contact between workpiece and tool was modeled using a general contact interaction with Coulomb friction. The elastic-perfectly plastic material behavior was modeled temperature dependent with given values from the literature [49]. Material damage was implemented by the Johnson-Cook damage initiation criterion, which is a special case of the ductile criterion. The JC damage parameters were chosen according to the literature [50]. Although the simulations were run on the high-performance computer "Elwetritsch" at the TU Kaiserslautern the simulation time was 7 to 10 days. The MIRS, resulting from the plastic deformation and temperature gradients during the cutting process, were analyzed in the boundary layer and the machined surface. Therefore, the stresses at nodes in an area with the size that is equal to the measurement area were extracted and averaged for each element depth. Besides, the standard deviation was computed. The predicted MIRS depth profiles showed the typical root-shape of compressive RS (see Fig. 12). In general, for both milling conditions the magnitude of the simulated MIRS were on a similar level as the measured ones and the sign of the shear stresses was predicted correctly. However, there were still significant deviations found in comparison to reality: For EM3 lower penetration depth and lower values of the MaxRS were predicted (see Fig. 12a). For the Index case higher penetration depth and higher values of the MaxRS were simulated (see Fig. 12b). Possible reasons are that the microscopic geometry of the tools deviated from reality (cutting edge radius: CAD model ideal sharp vs. reality ~ 10 μm). Furthermore, the model is only a simplification of reality. When using the predicted MIRS from the cutting model as input for distortion model, the same shape but significant differences in magnitude of distortion were predicted (see [48]). Another disadvantage of this approach is, that modelling machining with lower feeds per tooth than the simulated 0.2 mm would require smaller elements in the cutting area, which would increase the already high simulation time drastically. In general, modelling the MIRS is difficult, because accurate material models are required. Also, the uncertainty is too high, meaning RS measurements are required anyway due to validation purposes. Besides, it was shown that measurements were transferable: Meaning it would be possible to carry out the RS measurement on test specimens that have been machined with the same machining strategy as the component itself. In summary, using measured residual stress data as input for the distortion model is favorable.

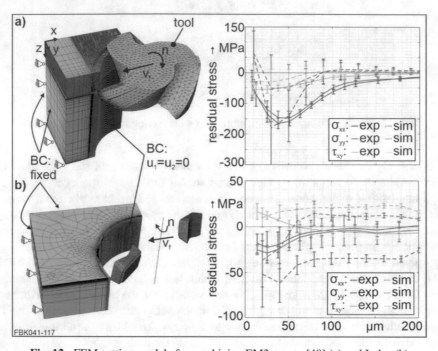

Fig. 12. FEM cutting models for machining EM3 acc. to [48] (a) and Index (b)

5   Development of Compensation Techniques

It was shown both experimentally (see Sect. 3.3) and simulatively (see Sect. 4.1) that the distortion can be reduced by changing the machining strategy from zig to spiral milling: A distortion reduction of 41% was achieved for small parts with 3 mm thick walls and 45% for a more complex structural part. Especially for smaller wall thicknesses the potential of minimizing the distortion by changing the direction of the milling is higher than for parts with thicker walls. [43]. Furthermore, by applying similar MIRS at the backside of the sample than those at the bottom of the pockets, an opposite bending moment could be induced to minimize the distortion. Adjusting the part topology, for example by aligning the stiffener along the principal direction of stresses, decreased the distortion as well. Those methods fall into the category of precontrol compensation techniques [5] or also called offline methods [22], which can be divided into the following categories [43]:

– The **process parameters**, such as the tool properties and cutting parameters effect the MIRS (depth and magnitude) and therefore the part distortion directly.

- The **process strategy** effects both the MIRS and the IBRS. The positioning of the part in the semi-finished product determines the re-distribution of the IBRS. Subsequent cutting steps remove layers of MIRS and introduce others. The milling path effects the direction of MIRS and therefore the induced bending moments.
- The **part topology** determines the location of the removed material and therefore the redistribution of the IBRS effecting the distortion. The stiffness of the structural component itself is also affected.

Due to the higher distortion of high IBRS samples other minimization techniques are required: For example, by using the developed distortion prediction model in advance to machining, the distortion could be compensated by milling the inverse distortion onto the backside of the sample. This led to a distortion reduction by 77% [43].

6 Summary

In our research the effect of the residual stresses, the machining strategy, the part topology and the geometry, including the wall-thickness, on distortion were investigated experimentally, and simulatively by validated virtual models based on the finite-element method. First, the effect of each RS type on the part distortion was analyzed individually before understanding their superposition. A repeatability analysis of the MIRS formed the basis and showed that for stable machining, repeatable MIRS led to repeatable distortions [40]. Furthermore, a set of different machining parameters were identified causing different MIRS, where higher and deeper MIRS resulted in higher distortions. Hereby a simple experiment was developed highlighting the distortion potential of RS in the boundary layer of parts: A 1 mm thick wafer was removed at the milled surface. A static, linear elastic finite element model was developed to simulate the distortion due to the measured RS in a short time [42]. The model considers all RS (IBRS and MIRS) contained in the entire part at different locations as well as the milling path. It was validated by various experiments (different geometries, RS, milling paths). It could be shown that the shear MIRS are crucial and contribute much to distortion (when not compensated for), because they induce a torsional bending moment in addition to the bending moment of the normal MIRS [42].

In general, the part topology, part size and the machining strategy have a mutual impact on the part distortion. By decreasing the wall thickness the distortion increases with a shift of the RS type dominating the distortion towards the MIRS. Defining a universal crucial wall thickness when MIRS are the main factor for the distortion for thin-walled monolithic structural parts is deceptive, because of the dependence on the part geometry and the machining strategy. For the investigated geometry (wall thickness < 3 mm) and machining parameters, it could be shown that for low IBRS samples (stress relieved), the MIRS introduced in the surface layer of the pockets are driving the distortion when a zig milling path strategy is used [42]. In comparison, when milling the pockets in spiral form, the low IBRS dominate the lower distortion [43]. For high IBRS samples the IBRS are driving the distortion. Nevertheless, there is a systematic influence of the combined effect of both RS types found for thin wall thicknesses (<3 mm) and the zig milling strategy [42]. The knowledge gained from investigating smaller and simpler parts (size and complexity) could partially be transferred to bigger parts.

The process parameters, the part topology and the process strategy were identified as precontrol distortion compensation techniques, because they effect either the MIRS, the IBRS or both and therefore the distortion. An appropriate milling strategy, applying opposite bending moments and balancing shear stresses, minimized the distortion.

Furthermore, a 3D FEM cutting simulation was developed [Webe21c]. It was able to predict the MIRS for high feed machining qualitatively. However, they could not be used as an adequate replacement for the measured MIRS for the distortion model.

In future research, the investigations will be expanded to include more cutting parameters, especially the investigation of cooling lubricants or cryogenic machining and its effect on the MIRS and part distortion.

Acknowledgements. The authors would like to thank the German Research Foundation (DFG, Germany) and the National Science Foundation (NSF, USA) for the financial support within the project AU 185/64-1 (351381681) "NSF DFG Collaboration to Understand the Prime Factors Driving Distortion in Milled Aluminum Workpieces" (NSF funding Award No. 1663341) and within the International Research Training Group 2057 – Physical Modeling for Virtual Manufacturing (IRTG2057) 252408385.

Any opinions, findings, conclusions or recommendations expressed in this material are those of the authors and do not necessarily reflect the views of the NSF or DFG.

[1]Naming of specific manufacturers is done solely for the sake of completeness and does not necessarily imply an endorsement of the named companies nor that the products are necessarily the best for the purpose.

References

1. Wanhill, R.J.H.: Chapter 15 - Aerospace applications of aluminum-lithium alloys. In: Prasad, N.E., Gokhale, A.A., Wanhill, R.J.H. (eds.) Aluminum-Lithium Alloys, Processing, Properties and Applications, pp. 503–535. Butterworth-Heinemann, Elsevier Inc., Oxford (2014)
2. Dursun, T., Soutis, C.: Recent developments in advanced aircraft aluminium alloys. Mater. Des. **56**, 862–871 (2014). https://doi.org/10.1016/j.matdes.2013.12.002
3. Berky, E.: Aerostructures made in Augsburg. International ARO Seminar Aluminium-HSC Machining, Reutte (A) (2001)
4. Denkena, B., Dreier, S.: Simulation of residual stress related part distortion. In: Proceedings of the 4th Machining Innovations Conference, pp. 105–113 (2014). https://doi.org/10.1007/978-3-319-01964-2_15
5. Li, J.-G., Wang, S.: Distortion caused by residual stresses in machining aeronautical aluminum alloy parts: recent advances. Int. J. Adv. Manuf. Technol. **89**(1–4), 997–1012 (2016). https://doi.org/10.1007/s00170-016-9066-6
6. Cardarelli, F.: Materials Handbook: A Concise Desktop Reference. Springer, London (2018). https://doi.org/10.1007/978-1-84628-669-8
7. https://de.statista.com/themen/2514/aluminium (2022)
8. Alamdari, H.: Aluminium production process: challenges and opportunities. Metals **7**(4), 133 (2017). https://doi.org/10.3390/met7040133
9. Klocke, F.: Fertigungsverfahren 1 – Zerspanung mit geometrisch bestimmter Schneide. Springer Vieweg, Berlin (2018)

10. Warren, A.S.: Developments and challenges for aluminum – a boeing perspective. Mater. Forum **28**, 24–31 (2004)
11. Zhou, B., Liu, B., Zhang S.: The advancement of 7XXX series aluminum alloys for aircraft structures: a review. Metals **11**(5), 718 (2021). https://doi.org/10.3390/met11050718
12. Santos, M.C., Machado, A.R., Sales, W.F., Barrozo, M.A.S., Ezugwu, E.O.: Machining of aluminum alloys: a review. Int. J. Adv. Manuf. Technol. **86**(9–12), 3067–3080 (2016). https://doi.org/10.1007/s00170-016-8431-9
13. Totten, G.E.: Handbook of Residual Stress and Deformation of Steel. ASM International (2002)
14. Schajer, G.S., Ruud, C.O.: Overview of residual stress and their measurement. In: Schajer, G.S. (ed.) Practical Residual Stress Measurement Methods. Wiley, West Sussex (2013)
15. Prime, M.B., Hill, M.R.: Residual stress, stress relief, and inhomogeneity in aluminum plate. Scr. Mater. **46**(1), 77–82 (2002). https://doi.org/10.1016/S1359-6462(01)01201-5
16. Wyatt, J.E., Berry, J.T.: A new technique for the determination of superficial residual stress associated with machining and other manufacturing processes. J. Mater. Process. Technol. **171**(1), 132–140 (2006). https://doi.org/10.1016/j.jmatprotec.2005.06.067
17. Denkena, B., Boehnke, D., de Leon, L.: Machining induced residual stress in structural aluminum parts. Prod. Eng. Res. Devel. **2**, 247–253 (2008). https://doi.org/10.1007/s11740-008-0097-1
18. Perez I., et al.: Effect of cutting speed on the surface integrity of face milled 7050-T7451 aluminium workpieces. In: Procedia CIRP 71 – Proceedings to the 4th CIRP Conference on Surface Integrity, pp. 460–465 (2018). 10.101 6/j.procir.2018.05.034
19. Denkena, B., de Leon, L.: Milling induced residual stresses in structural parts out of forged aluminium alloys. Int. J. Mach. Mach. Mater. **4**(4), 335–344 (2008). https://doi.org/10.1504/IJMMM.2008.023717
20. Rao, B., Shin, Y.: Analysis on high-speed face-milling of 7075–T6 aluminum using carbide and diamond cutters. Int. J. Mach. Tools Manuf. **41**, 1763–1781 (2001). https://doi.org/10.1016/S0890-6955(01)00033-5
21. Tang, Z.T., Liu, Z.Q., Wan, Y., Ai, X.: Study on residual stresses in milling aluminium alloy 7050-T7451. In: Yan, X.T., Jiang, C., Eynard, B. (eds.) Advanced Design and Manufacture to Gain a Competitive Edge, pp. 169–178, Springer, London (2008). https://doi.org/10.1007/978-1-84800-241-8_18
22. Aurrekoetxea, M., Llanos, I., Zelaieta, O., López de Lacalle, L.N.: Towards advanced prediction and control of machining distortion: a comprehensive review. Int. J. Adv. Manuf. Technol. 1–26, 2823–2848 (2022). https://doi.org/10.1007/s00170-022-10087-5
23. Jayanti, S., et al.: Predictive modeling for tool deflection and part distortion of large machined components. In: Procedia CIRP 12 - Proceedings of the 8th CIRP Conference on Intelligent Computation in Manufacturing Engineering, pp. 37–42 (2013). https://doi.org/10.1016/j.procir.2013.09.008
24. Dreier, S.: Simulation des eigenspannungsbedingten Bauteilverzugs. Dissertation, Hannover (2018)
25. Huang, X., Sun, J., Li, J.: Finite element simulation and experimental investigation on the residual stress-related monolithic component deformation. Int. J. Adv. Manuf. Technol. **77**(5–8), 1035–1041 (2014). https://doi.org/10.1007/s00170-014-6533-9
26. Madariaga, A., Perez, I., Arrazola, P.J., Sanchez, R., Ruiz, J.J., Rubio, F.J.: Reduction of distortions in large aluminium parts by controlling machining-induced residual stresses. Int. J. Adv. Manuf. Technol. **97**(1–4), 967–978 (2018). https://doi.org/10.1007/s00170-018-1965-2
27. Ma, Y., Zhang, J., Yu, D., Feng, P., Xu, C.: Modeling of machining distortion for thin-walled components based on the internal stress field evolution. Int. J. Adv. Manuf. Technol. **103**(9–12), 3597–3612 (2019). https://doi.org/10.1007/s00170-019-03736-9

28. Gulpak, M., Sölter, J., Brinksmeier, E.: Prediction of shape deviations in face milling of steel. In: Procedia CIRP 8 – Proceedings of the 14th Conference on Modeling of Machining Operations, pp. 15–20 (2013). https://doi.org/10.1016/j.procir.2013.06.058

29. Jawahir, I.S., et al.: Surface integrity in material removal processes: recent advances. CIRP Ann. **60**(2), 603–626 (2011). https://doi.org/10.1016/j.cirp.2011.05.002

30. Astakhov, V.P., Outeiro, J.C.: Metal cutting mechanics, finite element modelling. In: Davim, J.P. (eds.) Machining. Springer, London (2008). https://doi.org/10.1007/978-1-84800-213-5_1

31. Fan, L., et al.: Control of machining distortion stability in machining of monolithic aircraft parts. Int. J. Adv. Manuf. Technol. **112**(11–12), 3189–3199 (2021). https://doi.org/10.1007/s00170-021-06605-6

32. Li, B., Jiang, X., Yang, J., Liang, S.Y.: Effects of depth of cut on the redistribution of residual stress and distortion during the milling of thin-walled part. J. Mater. Process. Technol. **216**, 223–233 (2015). https://doi.org/10.1016/j.jmatprotec.2014.09.016

33. Dreier, S., Brüning, J., Denkena, B.: Simulation based reduction of residual stress related part distortion. Materialwiss. Werkstofftech. **47**(8), 710–717 (2016). https://doi.org/10.1002/mawe.201600604

34. Marusich, T.D., Usui, S., Marusich, K.J.: Finite element modeling of part distortion. In: Xiong, C., Liu, H., Huang, Y., Xiong, Y. (eds.) ICIRA 2008. LNCS (LNAI), vol. 5315, pp. 329–338. Springer, Heidelberg (2008). https://doi.org/10.1007/978-3-540-88518-4_36

35. Keleshian, N., Kyser, R., Rodriquez, J.: On the distortion and warpage of 7249 aluminum alloy after quenching and machining. J. Mater. Eng. Perform. **20**, 1230–1234 (2011). https://doi.org/10.1007/s11665-010-9756-4

36. Chantzis, D., Van-der-Veen, S., Zettler, J., Sim, W.M.: An industrial workflow to minimise part distortion for machining of large monolithic components in aerospace industry. In: Procedia CIRP 8 – Proceedings of 14th CIRP Conference on Modeling of Machining Operations, pp. 281–286 (2013). https://doi.org/10.1016/j.procir.2013.06.103

37. Zhang, Z., Li, L., Yang, Y., He, N., Zhao, W.: Machining distortion minimization for the manufacturing of aeronautical structure. Int. J. Adv. Manuf. Technol. **73**(9–12), 1765–1773 (2014). https://doi.org/10.1007/s00170-014-5994-1

38. Cerutti, X., Mocellin, K.: Influence of the machining sequence on the residual stress redistribution and machining quality: analysis and improvement using numerical simulations. Int. J. Adv. Manuf. Technol. **83**(1–4), 489–503 (2015). https://doi.org/10.1007/s00170-015-7521-4

39. Weber, D., Kirsch, B., D'Elia, C.R., Linke, B.S., Hill, M.R., Aurich, J.C.: Concept to analyze residual stresses in milled thin walled monolithic aluminum components and their effect on part distortion. Production at the leading edge of technology. In: Proceedings of the 9th Congress of the German Academic Association for Production Technology, pp. 287–296 (2019). https://doi.org/10.1007/978-3-662-60417-5_29

40. Weber, D., et al.: Analysis of machining-induced residual stresses of milled aluminum workpieces, their repeatability, and their resulting distortion. Int. J. Adv. Manuf. Technol. **115**(4), 1089–1110 (2021). https://doi.org/10.1007/s00170-021-07171-7

41. Chighizola, C.R., et al.: Intermethod comparison and evaluation of measured near surface residual stress in milled aluminum. Exp. Mech. **61**(8), 1309–1322 (2021). https://doi.org/10.1007/s11340-021-00734-5

42. Weber, D., et al.: Investigation on the scale effects of initial bulk and machining induced residual stresses of thin walled milled monolithic aluminum workpieces on part distortions: experiments and finite element prediction model. In: Procedia CIRP 102 – Proceedings of the 18th CIRP Conf. on Modeling of Machining Operations, pp. 337–342 (2021). https://doi.org/10.1016/j.procir.2021.09.058

43. Weber, D., et al.: Simulation based compensation techniques to minimize distortion of thin-walled monolithic aluminum parts due to residual stresses. CIRP J. Manuf. Sci. Technol. **38**, 427–441 (2022). https://doi.org/10.1016/j.cirpj.2022.05.016

44. ASTM International, Standard Test Method for Determining Residual Stresses by the Hole-Drilling Strain-Gage Method E837-13a. West Conshohocken (2013)

45. Chighizola, C.R., Hill, M.R.: Two-dimensional mapping of bulk residual stress using cut mouth opening displacement. Exp. Mech. **62**(1), 75–86 (2021). https://doi.org/10.1007/s11340-021-00745-2

46. Jonsson, J.E., et al.: Milling-induced residual stress and distortion under variations of bulk residual stress. In: Proceedings of the 31st ASM Heat Treating Society Conference, pp. 96–99 (2021). https://doi.org/10.31399/asm.cp.ht2021exabp0096

47. Weber, D., et al.: Simulation based investigation on the effect of the topology and size of milled thin-walled monolithic aluminum parts on the distortion due to residual stresses. In: 11th International Conference on Residual stresses (2022). hal-03869964

48. Weber, D., et al.: Finite element simulation combination to predict the distortion of thin walled milled aluminum workpieces as a result of machining induced residual stresses. In: Open Access Series in Informatics (OASIcs) 89 - 2nd International Conference of the DFG International Research Training Group 2057 – Physical Modeling for Virtual Manufacturing (iPMVM 2020), pp. 11:1–11:21 (2021). https://doi.org/10.4230/OASIcs.iPMVM.2020.11

49. Koc, M., Culp, J., Altan, T.: Prediction of residual stresses in quenched aluminum blocks and their reduction through cold working processes. J. Mater. Process. Technol. **174**(1–3), 342–354 (2006). https://doi.org/10.1016/j.jmatprotec.2006.02.007

50. Lesuer, D.R., Kay, G.J., LeBlanc, M.M.: Modeling large-strain, high-rate deformation in metals. In: Third Biennial Tri-Laboratory Engineering Conference Modeling and Simulation, Pleasanton, CA, November 3–5, 1999

51. Chighizola, C.R., et al.: The effect of bulk residual stress on milling-induced residual stress and distortion. Exp. Mech. (2022). https://doi.org/10.1007/s11340-022-00843-9

Prediction of Thermodynamic Properties of Fluids at Extreme Conditions: Assessment of the Consistency of Molecular-Based Models

J. Staubach and S. Stephan[(✉)]

Laboratory of Engineering Thermodynamics (LTD), RPTU Kaiserslautern, Kaiserslautern, Germany
{jens.staubach,simon.stephan}@rptu.de

Abstract. For machining processes, such as drilling, grinding, and cutting, fluids play a crucial role for lubrication and cooling. For adequately describing such processes, robust models for the thermophysical properties of the fluids are a prerequisite. In the contact zone, extreme conditions prevail, e.g. regarding temperature and pressure. As thermophysical property data at such conditions are presently often not available, predictive and physical models are required. Molecular-based equations of state (EOS) are attractive candidates as they provide a favorable trade-off between computational speed and predictive capabilities. Yet, without experimental data, it is not trivial to assess the physical reliability of a given EOS model. In this work, Brown's characteristic curves are used to assess molecular-based fluid models. Brown's characteristic curves provide general limits that are to be satisfied such that a given model is thermodynamically consistent. Moreover, a novel approach was developed, which uses pseudo-experimental data obtained from molecular simulations using high-accurate force fields. The method is generalized in a way that it can be applied to different force field types, e.g. model potentials and complex real substances. The method was validated based on the (scarcely) available data in the literature. Based on this pseudo-experimental data, different thermodynamic EOS models were assessed. Only the SAFT-VR Mie EOS is found to yield thermodynamically consistent results in all cases. Thereby, robust EOS models were identified that can be used for reliably modeling cutting fluids at extreme conditions, e.g. in machining processes.

1 Introduction

Cutting fluids experience extreme conditions in a contact zone of machining processes (cf. Fig. 1), i.e. large temperature and pressure as well as extreme gradients in these properties [8, 47, 49]. The pressure in a tribological contact zone can be up to several GPa and the temperature up to 1 000 K [6, 8]. For modeling such processes, reliable and robust models for describing the thermophysical properties of the fluids, e.g. the heat capacity, compressibility, and density at a given temperature and pressure are required. Yet, classical laboratory experiments for determining thermophysical properties at such conditions are practically not feasible. Also, empirical models fitted to the available data

J. C. Aurich et al. (Eds.): IRTG 2023, *Proceedings of the 3rd Conference on Physical Modeling for Virtual Manufacturing Systems and Processes*, pp. 170–188, 2023.
https://doi.org/10.1007/978-3-031-35779-4_10

at moderate conditions often exhibit an unrealistic extrapolation behavior to extreme temperature and pressure. Hence, physical predictive and reliable models are required.

Models based on molecular thermodynamics can be favorably used for predicting thermophysical properties due to their physical kernel, i.e. mathematical structure. In particular, the strong physical background of these models often allows successful extrapolations to conditions that were not considered for the model development such as extreme pressure and temperature. Both molecular dynamics (MD) simulations and molecular-based equation of state (EOS) models are attractive candidates for such applications.

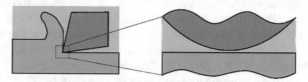

Fig. 1. Lubricated contact zone of a machining process between a workpiece (light gray) and a tool (dark gray). In the contact zone, extreme temperature and pressure prevail in the cutting fluid (blue).

The most popular modeling approach based on molecular thermodynamics is molecular simulation, i.e. molecular dynamics (MD) or Monte Carlo (MC) simulations based on classical molecular force fields. In molecular simulations, matter is modeled on the atomistic scale using a particle-based scheme. These particles interact with intermolecular potentials, i.e. the force fields. The accuracy of a molecular simulation primarily depends on the quality of the force field, which are available today for a large number of substances [46]. Molecular simulation has been extensively used for predicting thermophysical properties at extreme conditions, e.g. Refs. [32, 34, 35]. Due to the strong physical basis and reliability of the predictions, molecular simulations data is at times used as pseudo-experimental data where no 'real' experimental data is available [32, 34]. Yet, molecular simulations (MD or MC) are computationally expensive. For predicting thermophysical properties at a single state point, on the order of magnitude of 10^2 CPUh are required. An alternative modeling approach are molecular-based equation of state (EOS) models. They are algebraic models formulated in the Helmholtz energy as a function of the temperature and density, i.e. $a = a(T, \rho)$. Thereby, fluid properties at a given state point can be evaluated in milliseconds. Also, since the Helmholtz energy ansatz is a thermodynamic fundamental expression, all other thermodynamic properties can be derived from such a model – including phase equilibria, which are for example important for describing cavitation in a contact zone. Moreover, equation of state models can be favorably combined with further physical theories for modeling for example transport and interfacial properties, e.g. entropy scaling [20] and density gradient theory [40, 44]. In particular, molecular-based EOS models can be directly integrated for scale bridging in macroscopic models, e.g. in phase field or CFD simulations [15]. Molecular-based EOS come along with the drawback that approximations and assumptions are made within the model formulation. Hence, compared to molecular simulation, molecular-based EOS have a less strong physical backbone. A good extrapolation behavior can therefore not be presumed a priori.

Different strategies have been proposed for the assessment of EOS models [1, 2, 17, 39, 48]. The thermodynamic consistency of a pure component model can for example be assessed using ab initio virial coefficients in the low-density limit and the Nezbeda compressibility or Clausius-Clapeyron test for the vapor-liquid equilibrium [30, 48]. An interesting strategy for testing the extrapolation behavior at extreme pressure and temperature liquid states is based on Brown's characteristic curves. Brown's characteristic curves define lines on the thermodynamic equilibrium surface and are named Zeno, Amagat, Boyle, and Charles curve. Each of the four characteristic curves has – for a given thermophysical property – the same behavior as the ideal gas [9]. These curves are located within a large pressure and temperature range. For a given molecular fluid, Brown's characteristic curves are known to exhibit certain thermodynamic features [9]. Therefore, Brown's characteristic curves have become an important tool for the assessment of the extrapolation behavior of new EOS [1, 39, 43]. Nevertheless, due to the extreme conditions, there is practically no experimental data available on the characteristic curves. Therefore, Brown's characteristic curves are usually only used for a qualitative assessment of the behavior of the model, i.e. evaluating the limits and the general shape. It should moreover be noted that providing thermodynamically consistent Brown's characteristic curves is a necessary, but not sufficient condition for a thermodynamic model being physically reasonable, which is also shown in an example in this work.

In this work, a novel approach was developed for testing the extrapolation behavior of EOS models based on Brown's characteristic curves using pseudo-experimental data obtained from molecular simulation. This approach conveniently combines the strong physical backbone and predictive capabilities of molecular force field models [32, 34, 35] with the computational advantages of molecular-based EOS. By using molecular simulation pseudo-experimental data, not only the thermodynamic consistency of the EOS model regarding Brown's characteristic curves can be assessed, but also the accuracy of the model can be simultaneously evaluated. The computational procedure for determining Brown's characteristic curves for a given force field model is adopted from [51]. The new approach for assessing EOS models with pseudo-experimental characteristic curve data was tested on both model fluids and real substances. This paper is outlined as follows: First, the methods used for determining the characteristic curves are described. Then, the results for the different substances are presented and discussed.

2 Methods

2.1 Brown's Characteristic Curves

The characteristic curves of a molecular fluid were postulated by E.H. Brown [9] as curves on the thermodynamic pvT surface, along which the compressibility factor Z or its derivatives are identical to the values of an ideal gas. Brown's characteristic curves include state points at extreme conditions regarding temperature and pressure and can therefore be used as a tool for testing the extrapolation behavior of EOS. Moreover, on the thermodynamically consistent surface, Brown deduced from physical arguments that the curves exhibit certain features and limits (details are given below) that can be favorably used for the assessment of models at extreme conditions. The four characteristic curves are:

- Zeno curve Z (also called ideal curve)
- Amagat curve A (also called Joule inversion curve)
- Boyle curve B
- Charles curve C (also called Joule-Thomson inversion curve)

Figure 2 shows a schematic representation of Brown's characteristic curves. The characteristic curves are shown in a double-logarithmic temperature-pressure (pT) diagram. Based on rational thermodynamic arguments, Brown derived several requirements for the characteristic curves to be thermodynamically consistent: Each characteristic curve exhibits a single pressure maximum. Furthermore, the Zeno curve crosses each of the other three curves in one point (cf. Fig. 2), whereas the Boyle, Charles, and Amagat curve do not intersect each other. Moreover, the Charles curve terminates on the vapor-liquid binodal and the Boyle curve terminates on the spinodal. The characteristic curves end at low pressure (corresponds to low densities) at specific temperatures, which are directly linked to the second virial coefficient B [43], which is often known with very high accuracy. In the virial expansion up to second order, the compressibility factor Z can be written as

$$Z = 1 + B\rho. \tag{1}$$

Hence, there is a direct link between the zero-pressure limit of the characteristic curves and the second virial coefficient [29, 43]. In the following, the definition and general features of the four characteristic curves are introduced. A comprehensive introduction to Brown's characteristic curves is given in Refs. [4, 5, 29, 43].

State points on the Zeno curve satisfy the relation

$$Z = \frac{vp}{RT} = 1, \tag{2}$$

where Z is the compressibility factor, v the molar volume, p the pressure, R the molar gas constant, and T the temperature. The definition can be rewritten in terms of the Helmholtz energy as

$$\rho \left(\frac{\partial \tilde{a}}{\partial \rho} \right)_T = 0, \tag{3}$$

where ρ is the molar density and $\tilde{a}$ is the molar Helmholtz energy defined as $\tilde{a} = \frac{A}{N_A k_B T}$ with the Boltzmann constant k_B and the Avogadro constant N_A.

The Zeno curve ends at the so-called Boyle temperature in the zero-pressure (and therefore zero-density) limit, cf. Fig. 2. At the Boyle temperature, the second virial coefficient is zero $B(T_{Boyle}) = 0$, which directly follows from comparing the definition of the Zeno curve Eq. (2) with the virial expansion Eq. (1). The Zeno curve intersects all other characteristic curves. The intersection of the Zeno and the Boyle curve is located in the zero-pressure limit at the Boyle temperature. The temperature of the

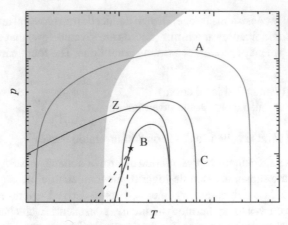

Fig. 2. Schematic representation of Brown's characteristic curves in the temperature-pressure (pT) projection. The Zeno curve Z (red), Amagat curve A (orange), Boyle curve B (blue), and Charles curve C (pink) are shown. The gray-shaded area indicates the solid phase region. The VLE, spinodals, and critical point of the fluid are indicated by the solid black line, the dashed black line, and the star, respectively.

intersection with the Charles curve is approximately the critical temperature of the fluid, cf. Fig. 2. The intersection of the Zeno curve with the Amagat curve is located at very low temperatures – for many substances in the solid phase region.

The Boyle curve is defined by the following relations:

$$\left(\frac{\partial Z}{\partial 1/\rho}\right)_T = 0, \qquad \left(\frac{\partial Z}{\partial p}\right)_T = 0. \tag{4}$$

These definitions can be rewritten in terms of the Helmholtz energy as

$$\rho\left(\frac{\partial \tilde{a}}{\partial \rho}\right)_T + \rho^2\left(\frac{\partial^2 \tilde{a}}{\partial \rho^2}\right)_T = 0. \tag{5}$$

The Boyle curve starts at the Boyle temperature in the zero-pressure limit, cf. Fig. 2. The Boyle curve has a bell shape and ends at the spinodal in the metastable region of the fluid, where the first and second density derivative of the Helmholtz energy are zero.

The Charles curve is defined by one of the following relations:

$$\left(\frac{\partial Z}{\partial 1/\rho}\right)_p = 0, \quad \text{or} \quad \left(\frac{\partial Z}{\partial T}\right)_p = 0, \quad \text{or} \quad \left(\frac{\partial H}{\partial p}\right)_T = 0, \tag{6}$$

where H is the enthalpy. These relations can be rewritten in terms of the Helmholtz energy as

$$\rho\left(\frac{\partial \tilde{a}}{\partial \rho}\right)_T + \rho^2\left(\frac{\partial^2 \tilde{a}}{\partial \rho^2}\right)_T + \frac{\rho}{T}\frac{\partial \tilde{a}}{\partial \rho \, \partial 1/T} = 0. \tag{7}$$

The Charles curve starts in the zero-pressure limit at high temperatures where the tangent to the second virial coefficient curve passes through the origin, i.e. the equation

$dB/dT = B/T$ holds. At low temperatures, the Charles curve ends at the vapor-pressure curve. The Charles curve has a bell shape and encloses the Boyle curve. The Charles curve is also known as the Joule-Thomson inversion curve, which is of fundamental importance for refrigeration technologies. There are several studies available in the literature investigating the Joule-Thomson inversion curve for refrigerants [10, 12, 13, 22, 52, 53].

The Amagat curve is defined by one of the following relations:

$$\left(\frac{\partial Z}{\partial T}\right)_{\rho} = 0, \quad \text{or} \quad \left(\frac{\partial U}{\partial 1/\rho}\right)_{T} = 0, \tag{8}$$

where U indicates the internal energy. These relations can be rewritten in terms of the Helmholtz energy as

$$\frac{\rho}{T}\frac{\partial \tilde{a}}{\partial \rho \, \partial 1/T} = 0. \tag{9}$$

The Amagat curve starts in the zero-pressure limit at high temperatures, where the second virial coefficient exhibits a maximum. At low temperatures, the Amagat curve ends at the vapor-pressure curve. The Amagat curve has a bell shape and encloses the Boyle curve and the Charles curve, cf. Fig. 2.

To be thermodynamically consistent, all four curves are required to exhibit a concave shape throughout in the double logarithmic pT projection. Moreover, for model fluids, the zero-pressure limit state point can be computed exactly using the virial route [51]. Hence, the zero-pressure limit is known exactly for model fluids.

2.2 Substances

In this work, both model fluids and real substances were studied to demonstrate and test the novel approach. As model fluids, the Lennard-Jones (LJ) fluid, the Lennard-Jones truncated and shifted (LJTS) fluid, and five Mie fluids were studied. For the studied model fluids, spherical molecules were considered that interact with the respective interaction potential. The model potentials used here can be employed for describing a large number of real substances, for example available in the MolMod force field database [46]. The underlying interaction potential are moreover important, as they are often used as building block in complex molecular force fields of real substances [28, 46]. Three real substances were studied in this work: toluene, ethanol, and dimethyl ether. The three substances strongly differ regarding the molecular structure and intermolecular interactions such that the novel approach is tested here for different situations.

The (full) Lennard-Jones potential u_{LJ} is defined as

$$u_{LJ}(r) = 4\varepsilon\left[\left(\frac{\sigma}{r}\right)^{12} - \left(\frac{\sigma}{r}\right)^{6}\right], \tag{10}$$

with the energy parameter ε, the size parameter σ, and the distance between two particles r. The LJ potential models repulsive interactions between particles at very small distances and attractive (dispersive) interactions at intermediate distances. The exponent

12 characterizes the repulsive interactions and the exponent 6 the attractive interactions between particles. The LJ potential is relatively simple, but a good approximation for intermolecular interactions. It has been extensively used since the early days of computer simulations and, accordingly, high quality benchmark data is available for the LJ fluid [48]. The Lennard-Jones truncated and shifted (LJTS) potential u_{LJTS} is computationally much cheaper, but simplifies the dispersive long-range interactions. The LJTS potential is defined as

$$u_{LJTS}(r) = \begin{cases} u_{LJ}(r) - u_{LJ}(r_c) & r \leq r_c \\ 0 & r > r_c, \end{cases} \tag{11}$$

with the cutoff radius $r_c = 2.5\sigma$. The Mie potential is a generalization of the LJ potential defined as

$$u_{Mie}(r) = C\varepsilon\left[\left(\frac{\sigma}{r}\right)^{\lambda_n} - \left(\frac{\sigma}{r}\right)^{\lambda_m}\right], \tag{12}$$

with

$$C = \frac{\lambda_n}{\lambda_n - \lambda_m}\left(\frac{\lambda_n}{\lambda_m}\right)^{\frac{\lambda_m}{\lambda_n - \lambda_m}}. \tag{13}$$

Hence, the Mie λ_n, λ_m potential has two additional parameters: the exponent for the repulsive interactions λ_n and the exponent for the attractive interactions λ_m. Often, $\lambda_m = 6$ is chosen, while λ_n is often used as an additional adjustable parameter. In this work, five Mie fluids were considered with $(\lambda_n, \lambda_m) = (8,6)$ $(12,6)$ $(20,6)$ $(12,4)$ $(12,8)$.

For the real substances, toluene, ethanol, and dimethyl ether were chosen as high-accurate force fields for these substances are available in the literature [18, 25, 38, 46]. The force fields for toluene, ethanol, and dimethyl ether are based on a united-atom approach and modeled as rigid bodies, where the hydrogen atoms are not explicitly modeled. The toluene force field [25] has seven Lennard-Jones interactions sites and six point quadrupoles modeling the π orbital. The ethanol force field [38] has three Lennard-Jones interactions sites and three point charges modelling the alcohol group and enabling h-bonding. The dimethyl ether force field [18] has three Lennard-Jones interactions sites and a single point dipole modeling the overall polarity of the molecule.

Also, equation of state models for these substances are available in the literature. All three substances are base components in the chemical industry. Moreover, the three substances have strongly different molecular architecture and interactions: ethanol strongly forms hydrogen bonds, toluene is an aromatic compound with strong quadrupolar interactions caused by π-orbitals, and dimethyl ether has strong dipolar interactions. Hence, the three substances challenge the predictive capabilities of the molecular-based EOS models regarding different aspects.

2.3 Molecular Simulation

In this work, reference data for the assessment of the EOS models was generated using molecular dynamics simulations. Hence, the MD data is considered as pseudo-experimental data for the characteristic curves. In particular, high accurate molecular force field models were used that are known to provide excellent agreement with experimental data [18, 25, 38].

For determining the characteristic curve state points from a given force field, a new simulation method was developed by our group [51]. For the development, a large number of sampling routes were compared. Based on this systematic approach, for each characteristic curve, the most favorable simulation route was determined [51]. The new method has several advantages, e.g. a rigorous method for determining the statistical uncertainties and evaluating the thermodynamic self-consistency of the results [51]. Nevertheless, it should be noted that also systematical uncertainties – as well-known from laboratory experiments – may affect the reproducibility of the simulation results [36, 42], e.g. the simulation engines and implemented algorithms may have some minor influence on the results.

Molecular dynamics simulations were performed with the molecular simulation tool $ms2$ [21, 33] using at least $N = 2,000$ particles. For each characteristic curve, at least 14 state points were determined. For each characteristic curve state point, five state points in the vicinity of the initial guess characteristic curve were considered as a set. Each simulation of a simulation set was carried out at the same temperature. The simulations were carried out in the NVT ensemble for the Zeno, Boyle, and Charles curve. The NPT ensemble was used for the simulations of the Amagat curve. From the simulation results of a set, the characteristic curve state point was determined using the respective relations. In Table 1, the relations used for determining the characteristic curve state points are listed. For the zero-pressure limit, cluster integrals were evaluated for determining the characteristic curve state point via the virial route with high accuracy [51]. For the studied model fluids, these integrals can be evaluated exactly. For the real substance fluids, the integrals were evaluated using Monte Carlo simulations for sampling different orientations and distances between two molecules. For both, the model fluids and the real substances, the second virial coefficient was computed in a wide temperature range. From these results, the zero-pressure limit state points of the characteristic curves were obtained. Determining Brown's characteristic curves from molecular simulation for a given substance requires approximately 10^4 CPU hours. However, the force field type has a significant influence on that. Details on the molecular simulation methodology are given in Ref. [51].

Table 1. Conditions used for determining characteristic curve state points using MD simulations based on the method proposed in Ref. [51].

Zeno	Amagat	Boyle	Charles
$Z - 1 = 0$	$\left(\frac{\partial U}{\partial 1/\rho}\right)_T = 0$	$\left(\frac{\partial Z}{\partial 1/\rho}\right)_T = 0$	$\left(\frac{\partial H}{\partial p}\right)_T = 0$

2.4 Molecular-Based Equation of States

Molecular-based EOS have been primarily developed within the chemical engineering community for modelling in particular phase equilibria of mixtures [11, 14]. Yet, due to their physical general mathematical backbone, they can be favorably applied also in other fields for modeling properties of fluids. The most popular molecular-based EOS is PC-SAFT [24]. Molecular-based EOS are formulated as

$$\tilde{a} = \tilde{a}_{rep} + \tilde{a}_{att} + \tilde{a}_{chain} + \tilde{a}_{assoc} + \tilde{a}_D + \tilde{a}_Q, \tag{14}$$

where $\tilde{a}$ indicates the configurational Helmholtz energy, $\tilde{a}_{rep}$ the contribution due to repulsive interactions between particles, $\tilde{a}_{att}$ the contribution due to dispersive attractive interactions, $\tilde{a}_{chain}$ the contribution due to the chain length of molecules, $\tilde{a}_{assoc}$ the contribution due to h-bonding interactions [19], $\tilde{a}_D$ the contribution due to dipole interactions [23], and $\tilde{a}_Q$ the contribution due to quadrupole interactions. Each contribution term has one or two substance-specific parameters that characterize features of the molecule or intermolecular interactions, e.g. the chain term $\tilde{a}_{chain}$ comprises chain length parameter m and the attractive contribution $\tilde{a}_{att}$ the dispersion energy ε. Details on the molecular parameters of this model class are given in the review of Economou [19]. Moreover, $\tilde{a}$ is a function of the temperature T and the density ρ (and for mixtures the composition vector $\underline{x}$), which also holds for the individual terms on the RHS of Eq. (14). According to Eq. (14), the Helmholtz energy contributions from the different molecular interactions and features are independent. This is an approximation made in the model, which may seem crude, but has proven reliable and flexible. Thermodynamic properties can be straightforwardly computed from Eq. (14) using well-known relations between the derivatives of the Helmholtz energy with respect to the density and inverse temperature, e.g. the pressure can be computed as $p = \rho T R(1 + \partial \tilde{a}/\partial \rho)$ and the heat capacity can be computed as $c_v = -(\partial^2 \tilde{a}/\partial(1/T)^2)$. Only derivatives of the Helmholtz energy with respect to the density and inverse temperature (and eventually composition) are required for the calculation of thermodynamic properties of interest from a given EOS. In this work, the characteristic curves were computed from different EOS models using the thermodynamic definitions given by Eqs. (3), (5), (7), and (9).

Two different molecular-based EOS frameworks were considered in this work: PC-SAFT [24] and SAFT-VR Mie [27]. The latter was directly used for describing the LJ and Mie model fluid as well as for the real substances toluene, ethanol, and dimethyl ether. For describing the LJ and LJTS model fluid within the PC-SAFT framework, the models from Refs. [26, 45] were used, which are re-parametrized PC-SAFT monomer models. For comparison, also empirical EOS models were used in some cases [37, 50, 54]. For the LJ fluid, the molecular-based EOS of Stephan et al. [45] and that of Lafitte et al. [27] were used. For the LJTS fluid, the molecular-based EOS of Heier et al. [26] and the empirical EOS of Thol et al. [50] were used. For the Mie fluid, the EOS of Lafitte et al. [27] was used. For modelling the three real substances, toluene, ethanol, and dimethyl ether, both the SAFT-VR Mie EOS and the PC-SAFT EOS were used. Moreover, for ethanol, the empirical EOS from Schroeder et al. [37] and, for dimethyl ether, the empirical EOS from Wu et al. [54] were used. The PC-SAFT and SAFT-VR Mie model parameters for all three real substances are summarized in Table 2. For dimethyl ether, a new SAFT-VR Mie model was parametrized within this work. The other model parameters were taken from the literature [3, 23, 24, 27].

Table 2. Model parameters used for the SAFT-VR Mie and PC-SAFT EOS. The model parameters were taken from Refs. [3, 23, 24, 27]. The model parameters for dimethyl ether for the SAFT-VR Mie EOS were obtained within this work. Columns indicate the dispersion energy ε, particle size parameter σ, attractive λ_n and repulsive λ_m exponent, chain length m, association volume κ, association energy ϵ, and dipole moment μ.

Substance	$\varepsilon/k_B/K$	$\sigma/\text{Å}$	λ_n	λ_m	m	κ	$\epsilon/k_B/K$	μ/D
PC-SAFT								
Toluene	285.69	3.7169	-	-	2.8149	-	-	-
Ethanol	191.31	3.1477	-	-	2.4382	0.03481	2599.8	1.7
Dimethyl ether	210.29	3.2723	-	-	2.2634	-	-	1.3
SAFT-VR Mie								
Toluene	409.73	4.2770	16.334	6	1.9977	-	-	-
Ethanol	168.15	3.4914	7.6134	6	1.9600	0.34558	2833.7	-
Dimethyl ether	216.74	3.306	12	6	2.2340	-	-	-

3 Results

For the real substance fluids, classical SI units are used for presenting the results. For the model fluids, the Lennard-Jones units are used, i.e. using the potential well depth ε, the particle size parameter σ (cf. Eq. (10)), and the mass of the particle M as well as the Boltzmann constant k_B constitute the base unit system. Details on the LJ unit system are given in Ref. [48].

3.1 Lennard-Jones Fluids

Figures 3 and 4 show the results for the studied Lennard-Jones fluids: Fig. 3 for the (full) LJ fluid and Fig. 4 for the LJTS fluid. In both cases, results from different EOS and molecular simulation are shown. The latter is taken as reference.

For the LJ fluid (cf. Fig. 3), two molecular simulation data sets are compared. Overall, the data from Deiters and Neumaier [16] and the results using the new simulation method from our group [51] are in excellent agreement. In particular, both data sets are in excellent agreement with the zero-density limit data point obtained from the exact virial route. Moreover, the simulation results for all four characteristic curves exhibit a smooth trend. For the LJ fluid, two molecular-based EOS were used for predicting the characteristic curves. The EOS from Stephan et al. [45] (re-parametrized PC-SAFT monomer) yields a realistic Zeno, Boyle, and Charles curve. For the Amagat curve, however, the re-parametrized PC-SAFT monomer yields an erratic behavior, cf. Fig. 3. The EOS from Lafitte et al. [27], on the other hand, is in excellent agreement with the reference data. Moreover, this EOS yields physically realistic predictions for all four characteristic curves. Interestingly, the reference data and the EOS from Lafitte et al. agree very well even in the VLE metastable region.

For the LJTS fluid (cf. Fig. 4), only the data set determined by our group is available [51]. The simulation results show a very smooth trend and are also consistent with the

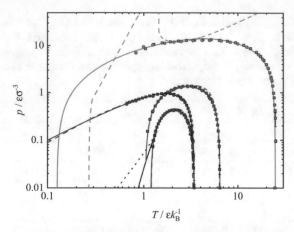

Fig. 3. Characteristic curves of the Lennard-Jones 12,6 fluid. Symbols indicate MD simulation results from Deiters and Neumaier (o) [16] and from this work (□). Lines correspond to predictions from the SAFT-VR Mie EOS (solid lines) and the EOS from Stephan et al. (dashed lines). Results for the Boyle curve (blue), Zeno curve (red), Charles curve (pink), and Amagat curve (orange). The vapor pressure curve (—) and the spinodal curves (---) as well as the critical point (★) are shown – computed from the EOS from Stephan et al.

exact zero-pressure limit results obtained from the virial route. Based on the molecular simulation reference data, two EOS models are assessed: the EOS of Thol et al. [50] and the EOS of Heier et al. [26]. The latter is also a re-parametrized PC-SAFT monomer, which is the reason that it yields a spurious Amagat curve. It was shown by Boshkhova and Deiters [7] that these defects are caused by a problematic formulation of the repulsive term of the underlying EOS framework. In the range of the Zeno, Boyle, and Charles curve, the EOS of Heier et al. [26] yields a realistic behavior. The EOS of Thol et al. [50] also exhibits an erroneous Amagat curve, but in the low-temperature regime. Brown showed that the Amagat curve is to intersect the Zeno curve in that regime, which is not the case for the EOS of Thol et al. Nevertheless, the EOS of Thol et al. is in reasonable agreement with the molecular simulation reference data in most cases.

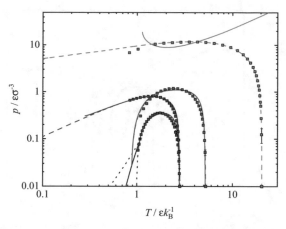

Fig. 4. Characteristic curves of the Lennard-Jones truncated and shifted (LJTS) fluid. Symbols □indicate MD simulation results [51]. Lines correspond to predictions from the LJTS EOS from Heier et al. (solid lines) and the EOS from Thol et al. (dashed lines). Results for the Boyle curve (blue), Zeno curve (red), Charles curve (pink), and Amagat curve (orange). The vapor pressure curve (—) and the spinodal curves (---) as well as the critical point (★) are shown.

3.2 Mie Fluids

Figure 5 and 6 show the results for the studied Mie fluids. Figure 5 shows the results for Mie fluids with different repulsive exponent λ_n; Fig. 6 the results for Mie fluids with different attractive exponent λ_m.

While a large number of equations of state are available in the literature for the classical (full) LJ fluid, only some Mie EOS are presently available. Here, the Mie EOS from Lafitte et al. [27] was considered. Overall, the predictions from the Mie EOS are in excellent agreement with the molecular simulation reference data – for all studied λ_n, λ_m combinations. This is impressive considering the fact that only data for moderate conditions was used for the model development [27]. This highlights the advantage of the physically-based model to reliable extrapolate to extreme temperatures and pressures. Only, some small deviations to the reference data are observed for the Amagat curve, cf. Fig. 5 and 6.

Overall, the EOS of Lafitte et al. captures the effect of both the repulsive and the attractive exponent well and is in excellent agreement with the pseudo-experimental data – in a very wide temperature and pressure range. Hence, the EOS of Lafitte et al. is an excellent candidate for modeling real substance components and testing their extrapolation behavior.

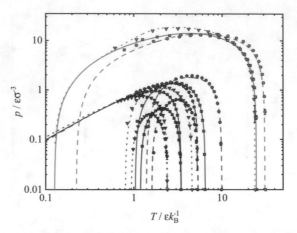

Fig. 5. Characteristic curves of the Mie $\lambda_n,6$ fluid. Symbols indicate MD simulation results [41]. Lines correspond to predictions from the SAFT-VR Mie EOS. Results for three $\lambda_n,6$ Mie fluids: 8,6 (o and dashed lines), 12,6 (□ and solid lines), and 20,6 (▼ and dotted lines). Results for the Boyle curve (blue), Zeno curve (red), Charles curve (pink), and Amagat curve (orange).

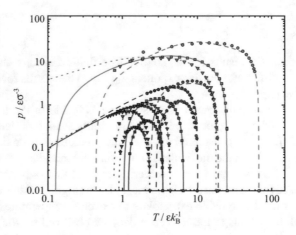

Fig. 6. Characteristic curves of the Mie $12,\lambda_m$ fluid. Symbols indicate MD simulation results [41]. Lines correspond to predictions from the SAFT-VR Mie EOS. Results for three 12, λ_m Mie fluids: 12,4 (o and dashed lines), 12,6 (□ and solid lines), and 12,8 (▼ and dotted lines). Results for the Boyle curve (blue), Zeno curve (red), Charles curve (pink), and Amagat curve (orange).

3.3 Toluene, Ethanol, and Dimethyl Ether

For the model fluids (cf. Sect. 3.2), the molecular-based EOS only used a repulsive and an attractive term for modeling the simple spherical particles. For the real substances discussed here, also the chain contribution and the association contribution were used. This significantly increases the complexity of the models.

Figure 7 shows the results for toluene. Again, molecular simulation data is taken as reference. These data are used for the assessment of the two EOS models: the PC-SAFT

EOS and the SAFT-VR Mie EOS. For the PC-SAFT EOS, a spurious Amagat curve is obtained. This is not surprising as this defect is inherited from the repulsive term (cf. Sect. 3.2). The SAFT-VR Mie EOS predictions, on the other hand, are physically reasonable and in good agreement with the pseudo-experimental data. Only, the temperature of the characteristic curves in the vicinity of the zero-pressure limit is overestimated by the SAFT-VR Mie EOS.

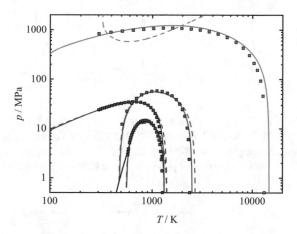

Fig. 7. Characteristic curves of toluene. Symbols (□) indicate MD simulation results [51]. Lines are predictions from the SAFT-VR Mie EOS (solid lines) and the PC-SAFT EOS (dashed lines). Results for the Boyle curve (blue), Zeno curve (red), Charles curve (pink), and Amagat curve (orange). The vapor pressure curve (—) predicted from the SAFT-VR Mie EOS is shown.

For ethanol and dimethyl ether, only the Charles curve was studied, cf. Figures 8 and 9, respectively.

The molecular simulation reference data is compared with the predictions from three EOS. Both the PC-SAFT and SAFT-VR Mie EOS were used for both substances. Moreover, the empirical EOS from Schroeder et al. was used for modeling ethanol and the empirical EOS from Wu et al. was used for modeling dimethyl ether. Both empirical EOS yield an unphysical behavior, i.e. a convex shape at high pressure. The PC-SAFT EOS yields physically reasonable Charles curve for both components. For both the PC-SAFT and SAFT-VR Mie EOS, the results are qualitatively in accordance with Brown's postulates. For ethanol, the PC-SAFT results show best quantitative agreement with the pseudo-reference data. For dimethyl ether, the SAFT-VR Mie results show the best agreement with the reference data. This is probably due to the underlying substance model parameters, cf. Table 2.

4 Conclusions

In this work, the extrapolation behavior of physically-based thermodynamic models was studied. Therefore, a novel approach was developed that uses molecular dynamics simulations for generating pseudo-experimental data. Using that data as a reference, different

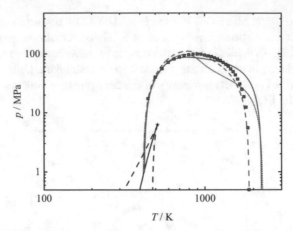

Fig. 8. Charles curve of ethanol. Symbols ▪indicate MD results [31]. Lines indicate predictions from different EOS: SAFT-VR Mie (solid line), PC-SAFT (dashed line), and the empirical multi-parameter EOS from Schroeder et al. [37] (dotted line). The vapor pressure curve (—) and the spinodal curves (---) as well as the critical point (★) were computed from the EOS of Schroeder et al.

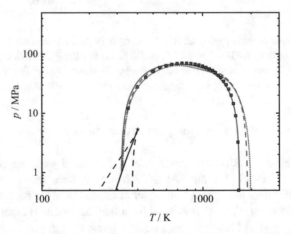

Fig. 9. Charles curve of dimethyl ether. Symbols ▪indicate MD results [31]. Lines indicate predictions from different EOS: SAFT-VR Mie (solid line), PC-SAFT (dashed line), and the empirical multi-parameter EOS from Wu et al. [54] (dotted line). The vapor pressure curve (—) and the spinodal curves (---) as well as the critical point (★) were computed from the EOS of Wu et al.

molecular-based equation of state models were assessed regarding their applicability for modeling fluid properties at extreme temperature and pressure.

In a first step, a new simulation method was developed for determining Brown's characteristic curves from a given molecular force field model [51]. This method combines statistical mechanics for (exactly) computing the zero-pressure limit state point of the characteristic curve via the virial route with classical MD simulations in the high-pressure regime. The method can be applied to both simple model fluids as well as complex real molecular substances. It has not yet been tested on electrolyte and polymer systems, which would be interesting for future work. Moreover, we have only addressed pure substances here. In technical applications, however, fluid mixtures are in practically all cases present. An extension of the computational approach to mixtures would therefore also be interesting for a future work.

The performance of two molecular-based EOS frameworks was compared, i.e. the PC-SAFT EOS and the SAFT-VR Mie EOS. The PC-SAFT model comprises more strongly simplifying approximations compared to the SAFT-VR Mie EOS. This is probably the reason that the SAFT-VR Mie EOS is found to extrapolate significantly better to extreme conditions regarding temperature and pressure. The PC-SAFT EOS, which is very frequently used in chemical engineering, on the other hand exhibits an artificial Amagat curve and should therefore not be applied at extreme temperatures and pressures. Hence, the SAFT-VR Mie EOS is found to be an excellent candidate for modeling fluid properties at extreme conditions, e.g. in tribological contact processes. Impressively, the SAFT-VR Mie EOS does not only satisfy the requirements of Brown's characteristic curves, the predictions of the EOS are in most cases in excellent agreement with the pseudo-experimental reference data. This is surprising considering the fact that the SAFT-VR Mie EOS was parametrized using data at moderate conditions alone. This supports the fact that models with a strong physical backbone can provide an excellent extrapolation behavior. In some cases, also empirical EOS were used for comparison and found to exhibit an unphysical behavior in some state region.

Based on the novel approach using MD pseudo-experimental reference data, for the first time, Brown's characteristic curves were used for a quantitative assessment of thermodynamic equation of state models. For future work, it would be interesting to study long chain alkane molecules, which are important substances used as lubricants. Moreover, it would be interesting to study the reproducibility of Brown's characteristic curves across different MD codes as well as the influence of the force field used for modeling a given substance.

Acknowledgement. The authors gratefully acknowledge access to the ELWE supercomputer at Regional University Computing Center Kaiserslautern (RHRK) under the grant TUKL-MTD.

References

1. Al-Saifi, N.M., Elliott, J.R.: Avoiding artifacts in noncubic equations of state. Ind. Eng. Chem. Res. **61**(42), 15661–15677 (2022)
2. Al-Saifi, N.M.: Simulation-based equations of state for the Lennard-Jones fluid: apparent success and hidden failure. AIChE J. **66**(7), e16244 (2020)
3. Al-Saifi, N.M., Hamad, E.Z., Englezos, P.: Prediction of vapor–liquid equilibrium in water–alcohol–hydrocarbon systems with the dipolar perturbed-chain SAFT equation of state. Fluid Phase Equilib. **271**, 82–93 (2008)
4. Apfelbaum, E.M., Vorob'ev, V.S., Martynov, G.A.: Virial expansion providing of the linearity for a unit compressibility factor. J. Phys. Chem. A **108**, 10381–10385 (2004)
5. Apfelbaum, E.M., Vorob'ev, V.S., Martynov, G.A.: Triangle of liquid-gas states. J. Phys. Chem. B 110, 8474–8480 (2006)
6. Bair, S.: High-Pressure Rheology for Quantitative Elastohydrodynamics. Elsevier, Amsterdam (2007)
7. Boshkova, O.L., Deiters, U.K.: Soft repulsion and the behavior of equations of state at high pressures. Int. J. Thermophys. **31**(2), 227–252 (2010)
8. Brinksmeier, E., Meyer, D., Huesmann-Cordes, A.G., Herrmann, C.: Metalworking fluids – mechanisms and performance. CIRP Ann. Manuf. Technol. **64**, 605–628 (2015)
9. Brown, E.H.: On the thermodynamic properties of fluids. Bulletin de l'Institut International du Froid Annexe **1**, 169–178 (1960)
10. Chacin, A., Vazquez, J., Müller, E.: Molecular simulation of the Joule-Thomson inversion curve of carbon dioxide. Fluid Phase Equilib. **165**, 147–155 (1999)
11. Chapman, W.G., Gubbins, K.E., Jackson, G., Radosz, M.: New reference equation of state for associating liquids. Ind. Eng. Chem. Res. **29**, 1709 (1990)
12. Colina, C.M., Lisal, M., Siperstein, F.R., Gubbins, K.E.: Accurate CO2 Joule-Thomson inversion curve by molecular simulations. Fluid Phase Equilib. **202**, 253–262 (2002)
13. Colina, C.M., Müller, E.A.: Molecular simulation of Joule-Thomson inversion curves. Int. J. Thermophys. **20**, 229–235 (1999)
14. Cotterman, R.L., Schwarz, B.J., Prausnitz, J.M.: Molecular thermodynamics for fluids at low and high densities. Part I: pure fluids containing small or large molecules. AIChE J. **32**(11), 1787–1798 (1986)
15. Diewald, F., et al.: Molecular dynamics and phase field simulations of droplets on surfaces with wettability gradient. Comput. Methods Appl. Mech. Eng. **361**, 112773 (2020)
16. Deiters, U.K., Neumaier, A.: Computer simulation of the characteristic curves of pure fluids. J. Chem. Eng. Data **61**(8), 2720–2728 (2016)
17. Deiters, U.K., De Reuck, K.M.: Guidelines for publication of equations of state I. Pure fluids. Pure Appl. Chem. **69**(6), 1237–1250 (1997)
18. Eckl, B., Vrabec, J., Hasse, H.: Set of molecular models based on quantum mechanical ab initio calculations and thermodynamic data. J. Phys. Chem. B **112**, 12710–12721 (2008)
19. Economou, I.: Statistical associating fluid theory: a successful model for the calculation of thermodynamic and phase equilibrium properties of complex fluid mixtures. Ind. Eng. Chem. Res. **41**, 953–962 (2002)
20. Fertig, D., Hasse, H., Stephan, S.: Transport properties of binary Lennard-Jones mixtures: Insights from entropy scaling and conformal solution theory. J. Mol. Liq. **367**, 120401 (2022)
21. Fingerhut, R., et al.: ms2: a molecular simulation tool for thermodynamic properties, release 4.0. Comput. Phys. Commun. **262**, 107860 (2021)
22. Figueroa-Gerstenmaier, S., Lisal, M., Nezbeda, I., Smith, W.R., Trejos, V.M.: Prediction of isoenthalps, Joule-Thomson coefficients and Joule-Thomson inversion curves of refrigerants by molecular simulation. Fluid Phase Equilib. **375**, 143–151 (2014)

23. Gross, J., Vrabec, J.: An equation-of-state contribution for polar components: dipolar molecules. AIChE J. **52**, 1194–1204 (2006)
24. Gross, J., Sadowski, G.: Perturbed-chain SAFT: an equation of state based on a perturbation theory for chain molecules. Ind. Eng. Chem. Res. **40**, 1244–1260 (2001)
25. Guevara-Carrion, G., Janzen, T., Munoz-Munoz, Y.M., Vrabec, J.: Mutual diffusion of binary liquid mixtures containing methanol, ethanol, acetone, benzene, cyclohexane, toluene, and carbon tetrachloride. J. Chem. Phys. **144**, 124501 (2016)
26. Heier, M., Stephan, S., Liu, J., Chapman, W.G., Hasse, H., Langenbach, K.: Equation of state for the Lennard-Jones truncated and shifted fluid with a cut-off radius of 2.5 based on perturbation theory and its applications to interfacial thermodynamics. Mol. Phys. **116**(15), 2083–2094 (2018)
27. Lafitte, T., et al.: Accurate statistical associating fluid theory for chain molecules formed from Mie segments. J. Chem. Phys. **139**, 154504 (2013)
28. Mick, J.R., Barhaghi, M.S., Jackman, B., Rushaidat, K., Schwiebert, L., Potoff, J.: Optimized Mie potentials for phase equilibria: Application to noble gases and their mixtures with n-alkanes. J. Chem. Phys. **143**, 114504 (2015)
29. Neumaier, A., Deiters, U.K.: The characteristic curves of water. Int. J. Thermophys. **37**(9), 96 (2016)
30. Nezbeda, I.: Vapour–liquid equilibria from molecular simulations: some issues affecting reliability and reproducibility. Mol. Phys. **117**, 2814 (2019)
31. Rößler, J., Antolović, I., Stephan, S., Vrabec, J.: Assessment of thermodynamic models via Joule-Thomson inversion. Fluid Phase Equilib. **556**, 113401 (2022)
32. Rutkai, G., Thol, M., Lustig, R., Span, R., Vrabec, J.: Communication: Fundamental equation of state correlation with hybrid data sets. J. Chem. Phys. **139**, 041102 (2013)
33. Rutkai, G., et al.: ms2: a molecular simulation tool for thermodynamic properties, release 3.0. Comput. Phys. Commun. **221**, 343 (2017)
34. Saager, B., Fischer, J.: Predictive power of effective intermolecular pair potentials: MD simulation results for methane up to 1000 MPa. Fluid Phase Equilib. **57**(1), 35–46 (1990)
35. Schmitt, S., Fleckenstein, F., Hasse, H., Stephan, S.: Comparison of force fields for the prediction of thermophysical properties of long linear and branched alkanes. J. Phys. Chem. B (2023, in press)
36. Schappals, M., et al.: Round robin study: Molecular simulation of thermodynamic properties from models with internal degrees of freedom. J. Chem. Theory Comput. **13**, 4270 (2017)
37. Schroeder, J.A., Penoncello, S.G., Schroeder, J.S.: A fundamental equation of state for ethanol. J. Phys. Chem. Ref. Data **43**, 043102 (2014)
38. Schnabel, T., Vrabec, J., Hasse, H.: Henry's law constants of methane, nitrogen, oxygen and carbon dioxide in ethanol from 273 to 498 K: prediction from molecular simulation. Fluid Phase Equilib. **233**, 134–143 (2005)
39. Span, R., Wagner, W.: On the extrapolation behavior of empirical equations of state. Int. J. Thermophys. **18**(6), 1415–1443 (1997)
40. Staubach, J., Stephan, S.: Interfacial properties of binary azeotropic mixtures of simple fluids: Molecular dynamics simulation and density gradient theory. J. Chem. Phys. **157**, 124702 (2022)
41. Stephan, S., Urschel, M.: Characteristic curves of the Mie fluid. J. Mol. Liq. **383**, 122088 (2023)
42. Stephan, S., Dyga, M., Alhafez, I.A., Lenard, J., Urbassek, H., Hasse, H.: Reproducibility of atomistic friction computer experiments: a molecular dynamics simulation study. Mol. Simul. **47**(18), 1509–1521 (2021)
43. Stephan, S., Deiters, U.: Characteristic curves of the Lennard-Jones fluid. Int. J. Thermophys. **41**, 147 (2020)

44. Stephan, S., Hasse, H.: Molecular interactions at vapor-liquid interfaces: Binary mixtures of simple fluids. Phys. Rev. E **101**, 012802 (2020)
45. Stephan, S., Staubach, J., Hasse, H.: Review and comparison of equations of state for the Lennard-Jones fluid. Fluid Phase Equilib. **523**, 112772 (2020)
46. Stephan, S., Horsch, M., Vrabec, J., Hasse, H.: MolMod - an open access database of force fields for molecular simulations of fluids. Mol. Simul. **45**, 806–814 (2019)
47. Stephan, S., Dyga, M., Urbassek, H., Hasse, H.: The influence of lubrication and the solid-fluid interaction on thermodynamic properties in a nanoscopic scratching process. Langmuir **35**, 16948 (2019)
48. Stephan, S., Thol, M., Vrabec, J., Hasse, H.: Thermophysical properties of the Lennard-Jones fluid: database and data assessment. J. Chem. Inf. Model. **59**, 4248–4265 (2019)
49. Stephan, S., Lautenschlaeger, M.P., Alhafez, I.A., Horsch, M.T., Urbassek, H.M., Hasse, H.: Molecular dynamics simulation study of mechanical effects of lubrication on a nanoscale contact process. Tribol. Lett. **66**(4), 1–13 (2018)
50. Thol, M., Rutkai, G., Span, R., Vrabec, J., Lustig, R.: Equation of State for the Lennard-Jones Truncated and Shifted Model Fluid. Int. J. Thermophys. **36**(1), 25–43 (2014)
51. Urschel, M., Stephan, S.: Determining Brown's characteristic curves using molecular simulation. J. Chem. Theory Comput. **19**, 1537–1552 (2023)
52. Vrabec, J., Kumar, A., Hasse, H.: Joule-Thomson inversion curves of mixtures by molecular simulation in comparison to advanced equations of state: Natural gas as an example. Fluid Phase Equilib. **258**, 34–40 (2007)
53. Vrabec, J., Kedia, G.K., Hasse, H.: Prediction of Joule-Thomson inversion curves for pure fluids and one mixture by molecular simulation. Cryogenics **45**, 253–258 (2005)
54. Wu, J., Zhou, Y., Lemmon, E.W.: An equation of state for the thermodynamic properties of dimethyl ether. J. Phys. Chem. Reference Data **40**, 023104 (2011)

A Methodology for Developing a Model for Energy Prediction in Additive Manufacturing Exemplified by High-Speed Laser Directed Energy Deposition

S. Ehmsen[1(✉)], M. Glatt[1], B. S. Linke[2], and J. C. Aurich[1]

[1] Institute for Manufacturing Technology and Production System, RPTU
Kaiserslautern-Landau, Kaiserslautern, Germany
svenja.ehmsen@rptu.de

[2] Department for Mechanical and Aerospace Engineering, University of California Davis,
Davis, USA

Abstract. The need for energy-efficient manufacturing technologies is growing due to the increasing pressure from climate change, consumers, and regulations. Additive manufacturing is claimed to be a sustainable manufacturing technology, especially for individualized products and small batches. To include the energy demand in the decision-making process on whether a part should be manufactured by additive or rather by subtractive or formative manufacturing, the energy demand which arises during manufacturing of a part must be predicted before the manufacturing process. For this, individual energy prognosis models are needed for each individual AM system. This paper, therefore, presents a methodology that enables users to develop a customized model to predict the energy demand of their AM System.

Four steps are necessary to create a model for energy prediction. First, the structure of the investigated system has to be captured. Here the subsystems and their corresponding process parameters are identified. Then the build cycle is analyzed and divided into several process steps in which the power consumption of the subsystems repeatedly follows the same pattern. Afterwards, those process parameters, that have a significant influence on the energy demand of each subsystem are identified within full factorial design of experiments and subsequently analyzed in detail. In the final step, individual models are developed for the energy demand of each subsystem for each process step. These individual models are then aggregated to create an overall model. The application of the methodology is also demonstrated and validated by the example of high-speed laser directed energy deposition.

E_i	Energy demand E of subsystem i
$E_{i,pre\text{-}step}$	Energy demand E of subsystem i during pre-step
$E_{i,in\ step}$	Energy demand E of subsystem i during in-step
$E_{i,post\text{-}step}$	Energy demand E of subsystem i during post-step
$E_{i,ps}$	Energy demand E of subsystem i during process step ps
$E_{i,ps,s}$	Energy demand E of subsystem i during process step ps in section s

J. C. Aurich et al. (Eds.): IRTG 2023, *Proceedings of the 3rd Conference on Physical Modeling for Virtual Manufacturing Systems and Processes*, pp. 189–212, 2023.
https://doi.org/10.1007/978-3-031-35779-4_11

E_{laser}	Energy demand E of the laser
E_{pf}	Energy demand E of the powder feeder
$E_{suction}$	Energy demand E of the suction system
E_{total}	Total energy demand E of the system
E_{ts}	Energy demand E of the trajectory system and the peripherical subsystems
$P_{i,ps}$	Power consumption P of subsystem i during process step ps
$P_{i,ps,basic}$	Basic power consumption P of subsystem i during process step ps
$P_{i,ps,pp}$	Power consumption P of subsystem i during process step ps dependent on process parameter pp
$P_{i,ps,pp,case1}$	Power consumption P of subsystem i during process step ps dependent on process parameter pp in *case 1* of a case distinction
$P_{i,ps,pp,caseC}$	Power consumption P of subsystem i during process step ps dependent on process parameter pp in *case C* of a case distinction
$P_{i,ps,pp,max}$	Power consumption P of subsystem i during process step ps dependent on process parameter pp which is set to the maximal possible setting
$P_{i,ps,pp,min}$	Power consumption P of subsystem i during process step ps dependent on process parameter pp which is set to the minimal possible setting
$P_{i,ps,pp,n}$	Power consumption P of subsystem i during process step ps dependent on process parameter pp which has setting n
$P_{i,ps,pp,50\%}$	Power consumption P of subsystem i during process step ps dependent on process parameter pp which is set to 50%
$P_{i,ps,s}$	Power consumption P of subsystem i during process step ps in section s
$S_{i,ps,pp,set}$	Setting of process parameter PP in process step PS of subsystem i
$S_{i,ps,pp,set,max}$	Maximal possible Setting of process parameter PP in process step PS of subsystem i
t_{ps}	Time t of process step ps
$t_{ps,s}$	Time t of the section s within process step ps

1 Introduction

Energy prices have risen in recent decades, making the purchase of energy a crucial input resource and cost factor for companies [1, 2]. In addition, energy supply and consumption generate greenhouse gas emissions that cause and intensify the climate crisis [1]. For this reason, some greenhouse gas emissions are regulated, e.g., in the EU emission trading system, where the cap, i.e. the EU's annual emissions ceiling, is reduced by 2.2% per year. This increases the pressure on companies to reduce emissions [3]. In addition, consumers are increasingly demanding more sustainable products [4]. Therefore, more and more companies are taking action to increase their energy efficiency to remain competitive [5]. One possible approach to this is the use of energy-efficient manufacturing technologies. In general, there are three categories of manufacturing processes, which are shown in Fig. 1:

- In subtractive manufacturing processes, the geometry of an object is created by removing defined volumes. Examples of subtractive processes are turning and milling.

- In formative manufacturing processes, the volume of the object remains constant but is transformed from one initial shape into another shape. Examples of formative manufacturing processes are forging and casting.
- In additive manufacturing processes, a defined shape is created from scratch by joining material together [6].

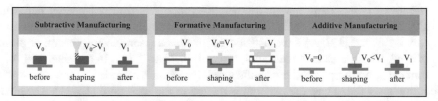

Fig. 1. Categories of manufacturing processes [7]

Each manufacturing technology has different advantages and disadvantages regarding its energy demand and resource consumption, depending on the specific application and the manufactured part. For example, additive manufacturing is claimed to be highly resource-efficient during manufacturing of the part. In addition, the parts are claimed to result in higher energy efficiency during their use due to the possibility of lightweight design [8]. Additive manufacturing thus offers companies the potential to enable energy- and resource-efficient manufacturing [9]. However, this potential is individual for each part and each application. Therefore, it must always be decided individually whether a part is manufactured with additive manufacturing or with another manufacturing process. For this decision-making process, in addition to technical requirements, energy considerations should also be taken into account. But how can the energy demand that arises during additive manufacturing of a part be predicted individually without actually manufacturing the part? To address this issue, this paper aims to develop a procedure for developing a customized energy prediction model for an arbitrary AM machine. Afterwards, this methodology is applied and validated on a high-speed laser directed energy deposition (HS DED-LB) system.

2   State of the Art

2.1   High-Speed Laser Directed Energy Deposition as an Additive Manufacturing Process

2.1.1   Additive Manufacturing

The general term of additive manufacturing (AM) comprises technologies that generate objects based on a geometric representation through the successive application of material. The material is usually added in layers and each new layer is bonded to the previous one, e.g. by fusing using a heat source [10]. Since the first commercial additive manufacturing system in 1987, various process principles have been developed [7, 11]. According to DIN EN ISO 17296, seven categories of additive manufacturing processes can be distinguished:

- Vat Photopolymerization

- Sheet Lamination
- Material Extrusion
- Material Jetting
- Binder Jetting
- Powder Bed Fusion
- Direct Energy Deposition

Hereby, a comparable new and emerging process is directed energy deposition (DED). This process uses focused thermal energy to melt and thus fuse materials as they are deposited [12]. As a thermal source for this, a laser, electron beam, or plasma arc is used, which creates a melt pool, typically 0.25 to 1 mm in diameter and 0.1 to 0.5 mm deep, on the part to be manufactured, into which material, as powder or wire, is continuously fed. Thus, the material is melted in the melt pool [12–14]. The nozzle for the material is located together with the output of the thermal source in the deposition head [14]. The deposition head and the part are continuously moved relative to each other [15]. Thus, the melt pool migrates along the desired contour, the scan path, and solidifies together with the deposited material as the thermal source moves along. This progressively creates a new layer. Depending on the AM machine, either the build platform and thus also the part move, the deposition head moves, or both move. After the layer is completed, the build platform and the application head move the thickness of one layer away from each other, to keep the distance between the deposition head and the build surface constant [14].

The manufacturing principle of DED leads to the following advantages compared to other AM processes which offers great potential for industrial applications.

- Larger material deposition rates lead to shorter process times [16],
- DED has larger build areas compared to other AM processes, which allows the manufacturing of larger parts [17],
- Multiple materials can be used within a build process, creating in-situ generated composites and heterogeneous parts [11, 14],
- DED can be used to remanufacture defective parts or components such as turbine blades [14],
- DED can be used to apply thin layers of corrosion-resistant and wear-resistant materials onto parts to improve their performance and durability [14].

Based on the last advantage, the surface coating with DED, the HS DED-LB found its origin.

2.1.2 High-Speed Laser Directed Energy Deposition

The process of HS DED-LB differs from other DED processes mainly by the powder focus which has been shifted upwards and is about one centimeter above the build surface, as shown in Fig. 2 [18, 19]. Thus, the powder is melted above the melt pool by the laser beam and is therefore applied in a liquid state [20, 21]. The melting of the material by the laser is much faster than the melting of the material in the melt pool. Therefore, with HS DED-LB, significantly faster feed rates of up to 200 m/min can be achieved and the processing time is five to ten times shorter compared to other DED processes [21, 22]. Since the laser mainly melts the powder, the melt pool on the part's surface is smaller, resulting in a lower specific heat input [19].

The material is stored as powder in a container on the powder feeder. Inside the container is a stirrer to avoid agglomeration of the powder. Through an opening at the bottom of the container, the powder falls onto the conveyor disk. The conveyor disc rotates at a defined speed and transports the powder to the inert gas stream, the carrier gas flow, which carries the powder to the nozzle [23]. The nozzle shapes the powder-gas flow into a cone and its tip is the powder focus. The powder focus is ideally at the same point as the focus of the laser beam [21]. The laser beam is generated in a separate subsystem, the laser generator, and guided through the laser optics to the deposition head. To protect the laser optics from contamination from smoke and powder buildup, another inert gas flow, the shielding gas flow, is fed through the inner center of the nozzle. Both inert gas flows also create an inert gas atmosphere at the process point and thus avoid oxidation of the material [23].

Powder that has not been applied is removed from the build chamber by a suction system. The suction system is connected through a connection at the bottom of the AM machine [24].

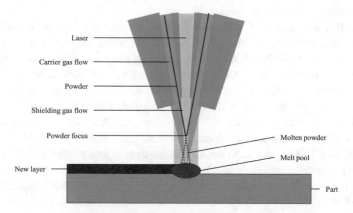

Fig. 2. Principle of HS DED-LB process

The laser optic cannot be moved at high speeds due to its fragility. Therefore, for HS DED-LB systems, the build platform is moved. For this, three linear motors are connected to the building platform via a tripod system. This allows the build platform to be moved in the three Cartesian dimensions [19]. To manufacture a part, the individual subsystems are used simultaneously. Therefore, HS DED-LB is a single-stage process.

2.2 Current Discussion of the Environmental Impact of DED

AM is claimed to have great potential to save resources and is therefore often referred as a sustainable manufacturing technology. However, in the literature, the environmental impact of AM is widely discussed. Through lightweight design, the consolidation of components, and functional integration, a reduction of the environmental impact during the use of additive manufactured parts can be achieved, e.g. through weight reduction [8]. This also leads to high material efficiency, since only the material necessary for the final geometry is required, including a comparatively small addition for post-processing [25, 26]. Therefore, on the one hand, there is less environmental impact from the extraction and processing of the raw material [27] and, on the other hand, less technical scrap and waste are produced [25, 28].

The material efficiency is strongly dependent on the AM process. In powder bed-based processes, for example, only less than 30% of the used metal powder contributes to the part geometry. The remaining 70% must be removed at the end of the process [29]. This is similar to HS DED-LB, where only around 27% of the conveyed powder is deposited [19]. Some of the powder can be reused for reprocessing after a sieving process [30]. However, the reuse is limited due to the decreasing powder quality, which influences the part quality [31]. Moreover, the energy demand is individual for each AM process and each manufactured part [32]. However, the energy demand to manufacture a part with AM is usually higher than manufacturing using formative or subtractive processes [33, 34]. Therefore, from an ecological point of view, AM is only advantageous for individual parts or small batches, since the environmental impact of specific process tools is omitted here [35]. Besides, the specific energy demand (SEC) does not adapt to different process parameter settings.

To take full advantages of AM, eco-design for AM must be considered. This can potentially reduce material consumption during manufacturing as well as energy consumption during use. However, scan paths based on the geometry of the part affect the energy demand. In addition, suitable process parameters must be defined. To compare the energy demand of different parts and process designs, the effects of different scan paths and process parameters must also be considered, without actually manufacturing the part. Thus, an energy prediction model is needed that allows the prediction and comparison of the energy demand during the design stage.

For this, first, it is important to investigate the energy and resource requirements and thus the environmental impact of AM in detail. The energy demand of different AM processes was analyzed for example by BAUMERS ET AL, FREDRIKSON, KELLENS ET AL. and FALUDI ET AL. [36–39]. Regarding DED, several studies have been performed, which can be divided into the following categories.

In the first category, the environmental impact of the DED process or the environmental impact caused during manufacturing of a specific part with DED is determined. A common method used for this purpose is life cycle assessment (LCA) [40]. LIU ET AL. [25] and JIANG ET AL. [41] compare the additive manufacturing of a gear with the combined subtractive and formative manufacturing within the framework of an LCA. An analysis of the ECO-indicator 99 based on the LCA of DED was carried out by LE BOURHIS ET AL. [26, 42] KERBRAT ET AL. [43], and SERRES ET AL. [44]. XIONG ET AL. [45] and MORROW ET AL. [28] also compared the environmental impact of the AM process with subtractive processes. The results of these studies are part-specific and can therefore hardly be transferred to other parts or adapted to other DED processes.

In the second category, a basis for comparison is provided by determining the SEC. WIPPERMAN ET AL. compare DED processes with powder bed fusion and subtractive processes. The study concludes that the SEC of DED is lower than that of powder bed fusion processes [46]. HUANG ET AL. compared DED with subtractive and formative processes and found that DED has the highest SEC [47]. JACKSON ET AL. calculated the SEC which arises during the process chain of manufacturing a part with DED based on literature [48]. The SECs calculated in the presented approaches give a first indication of the energy demand during manufacturing with DED. However, the transferability to other parts is limited here, since these values are individual for the process under consideration and the manufacturing process of the part.

The third category thus includes reusable and customizable energy calculation models. WATSON AND TAMINGER created a calculation model that can be used to determine whether additive or subtractive manufacturing is more energy efficient for manufacturing a metal part [49]. Within the model, the energy requirements for material and powder production, DED, post-processing, and transportation are considered. However, the model is based on average values which do not reflect the influence of different process parameter settings. In addition, this does not allow an analysis of the composition of the energy demand. WEGENER developed a comprehensive model to calculate the energy and resource requirements of DED. Within the energy model, the energy demand of the individual process steps is calculated and the individual power consumption of the individual subsystems is already taken into account [50]. However, the model can only be used to predict energy demand to a limited extent. In addition, it is not shown how the power consumption can be determined as a function of the selected process parameters.

Methodologies to develop energy prediction models already exist for unspecified manufacturing processes. DIETMAIR AND VERL developed a generic method for modeling the energy demand of machines and plants based on a statistical discrete event formulation. The procedure is exemplified by a milling process [51]. SCHMIDT ET AL. developed a methodology for predicting energy demand that can be applied to any manufacturing process and system. The approach aims to achieve a prediction quality of 80% with as little measurement effort as possible. For this purpose, the processes and systems are classified in terms of their complexity with the aid of a decision tree. Based on this, instructions are given for the creation of parametric or empirically based energy prediction models [5]. However, the model is also based on the SEC and can therefore only reflect the various setting of process parameters to a very limited extent.

2.3 Requirements

The studies listed in the previous chapter lack in particular a sufficient level of detail; both the effects of different process parameter settings and different process steps, and subsystems are usually not or only insufficiently considered. As a result, sufficient forecast accuracy for previously unregarded parts or sets of process parameters is not achieved. To achieve this and additionally enable in-depth analysis of the system to identify optimization potentials, the knowledge of when, how much, and which subsystem consumes power is necessary. Based on this, the following requirements for a methodology to develop an individual energy prediction model for an AM process arise:

- The model resulting from the methodology must have a very high forecast quality. The modeled energy demand must therefore not deviate from the actual energy demand by more than 5%.

- The resulting model allows a detailed analysis of the composition of the energy demand within the different process steps and subsystems.
- Within the model developed in the methodology, the influence of the process parameter setting on the energy demand is included.
- The methodology can be used with no or little prior knowledge.
- The model enables the reduction of the experimental effort compared to other prediction models while maintaining the prediction accuracy.

3 Approach for Creating an Energy Prediction Model

To create a model to predict the energy demand for additive manufacturing systems, especially for DED, four steps are necessary. First, the structure of the whole system must be captured. Then, the process and its individual process steps have to be analyzed. Subsequently, the process parameters and their effect on power consumption are investigated by experiments. Based on these results, the model is then developed, validated based on real parts, and, if necessary, further improved.

3.1 Capturing the Structure

The aim of capturing the structure is to obtain basic knowledge of the investigated system and to establish the technical requirements for the following steps. First, all energy-related subsystems of the investigated system are identified and then classified. On the one hand, there are systems whose power consumption remains constant during the entire process and is not changed by process parameters or other possible settings. These can be grouped as peripherical subsystems. On the other hand, there are subsystems whose power consumption changes during the process or depends on process parameters or other settings. For these subsystems, the power consumption must be analyzed individually and later a specified model must be created.

Based on this analysis, energy measurement sensors have to be implemented. Hereby, one energy measurement sensor must measure the power of the entire system. Furthermore, for each subsystem whose power consumption was found to be variable, additional energy measurement sensors must be implemented. Depending on availability and possibility, an energy measurement sensor can be omitted for one of these subsystems. The power consumption of this subsystem can be calculated based on the power consumption of the entire system by subtracting the power consumption of the other subsystems. The sampling rate of the sensors must be matched to the process speed and variability of the machine and the process.

After the initial examination of the subsystems, the adjustable process parameters are identified. These are then structured by assigning them to the subsystems whose settings they change. The identified structure is shown in Fig. 3.

Now, the basic knowledge of the investigated system has been obtained and the energy measurement system provides the opportunity to analyze energy demand.

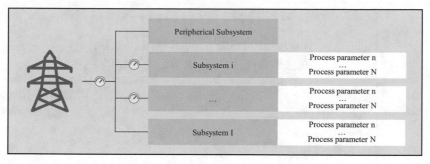

Fig. 3. Exemplary structuring of an investigated system

3.2 Process Analysis

In the next step, the process analysis, the build cycle is examined regarding its individual process steps and the behavior during power consumption of the individual subsystems. For this purpose, in initial experiments, first parts are manufactured and the power consumption is measured. For this purpose, typical or frequently manufactured parts and their corresponding process parameters are selected. The choice of parts, however, is not crucial and can be chosen almost freely. The power data is then examined in two steps.

Firstly, different process steps are defined. If they are not known before, they must be detected during those experiments. Usually, the different process steps can be clearly distinguished by the power consumption of the subsystems. Similar patterns occur for all of the investigated parts. Some subsystems are only active during the actual manufacturing process, others only during preceding or subsequent process steps. For example, many systems require a warm-up before the start and a cool-down after finishing the actual manufacturing process. This knowledge about the different process steps may also exist before those initial experiments.

Subsequently, the behavior of the individual subsystems during the individual process steps is investigated. Here it is examined when power is consumed by the subsystems and how the power consumption behaves. The following characteristics should be identified:

- When or at which process step which subsystem becomes active from standby or off?
- How does the power consumption of the subsystems behave within each process step?
- How does the power consumption behave during start-up, running, and shutdown?

For example, the power consumption of a subsystem can be constant, fluctuate randomly, or follow certain rules. If the power consumption is constant, the subsystem is either on standby or is currently at its set level and consumes power almost constantly, such as the peripherical subsystems and subsystem 4 in Fig. 4. Random fluctuations mostly occur during constant operation, e.g. subsystem 3, and periodic fluctuations by repetitive switching on and off, e.g. subsystem 1 and subsystem 2 in Fig. 4.

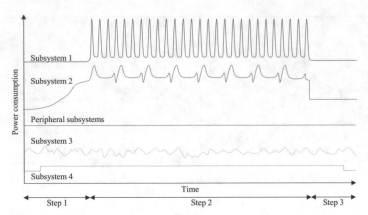

Fig. 4. Example of subdividing a manufacturing process into process steps based on power consumption curves

3.3 Analysis of the Process Parameters

The analysis of the process parameters is conducted in two steps. First, in full factorial design of experiments (DOE), those process parameters are identified, which significantly affect the power consumption of the respective subsystem and if there are interaction effects between the process parameters. Subsequently, in a second step, those process parameters, which were identified as relevant for the power consumption for the subsystem, are analyzed more closely.

During the capturing of the structure of the investigated system, the adjustable process parameters and their respective subsystem are identified. Now, for each subsystem, a full factorial DOE is developed. Here, the principles of random order, repetition for statistical significance, and blocking must be considered. Each test must be performed at least three times. If the results vary greatly, increasing the number of trials may be necessary. Also, if there are many adjustable process parameters for a subsystem, the number of experiments required for a full factorial DOE can become very large. For such cases, a partial factorial DOE may also be appropriate. While performing the experiments, the power consumption is measured, and the collected data are evaluated. Statistical values such as the p-value can then be used to determine, which process parameters affect the power consumption, and thus the energy demand of the subsystems. In addition, interaction effects between process parameters are also identified, as shown in Fig. 5a. Verification of the influence must be carried out for all previously identified process steps. For example, power is often consumed by a subsystem during standby, but the quantity is usually independent of process parameters. In the further steps of creating the model, only the significant process parameters are considered.

The relevant process parameters are analyzed in more detail in the second step. For this purpose, in further experiments, only the investigated process parameter is varied from the minimum setting to the maximum setting in several steps, while all other process parameters are kept constant. Here, the trials are also performed at least three times in random order and grouped into blocks. Based on these trials, it is investigated how the power consumption behaves with increasing process parameter settings. For

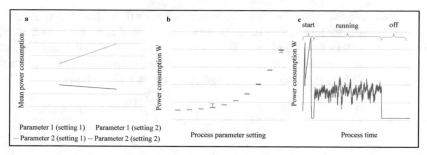

Fig. 5. Examples for evaluating the trials, a) for identifying relevant process parameters and their relation, b) for evaluating the changing power consumption with an increasing setting of the process parameter, c) for identifying the behavior during the process

example, the averaged power consumption can increase, with a gradual increase in the process parameter, either approximately linearly or over proportionally, as shown in Fig. 5b. Each measured power curve is analyzed in detail. The individual phases of standby, start-up, constant operation, and shutdown, as well as other possible phases, are evaluated separately, as shown in Fig. 5c.

3.4 Creating the Model

The procedure for creating the model for energy prediction follows a structured approach, which is shown in Fig. 6.

In general, the total energy demand of a system E_{total} is composed of the energy demand of its subsystems E_i. Therefore Eq. 1 applies.

$$E_{total} = \sum_{i=1}^{I} E_i \qquad (1)$$

In the previous steps, the power consumption of each subsystem and its corresponding process parameters was analyzed in detail. These findings are now being transferred for model creation. To develop a model that can predict energy demand as accurately as possible and, at the same time, analyze the composition of the energy demand of the entire system, a separate model is created for each subsystem.

During process analysis, the process was divided into several process steps within which the power consumptions of the individual subsystems are structurally similar. Thus, the energy demand of subsystem E_i is the sum of its energy demands during each process step $E_{i,ps}$, as shown in Eq. 2.

$$E_i = \sum_{ps=1}^{PS} E_{i,ps} \qquad (2)$$

Within each of these process steps, the pattern of the power curve is analyzed for each subsystem. Here, the pattern of the power curve can either be almost constant, fluctuate regularly or irregularly, or increase or decrease. Depending on the pattern, the further procedure is chosen. If a fluctuation occurs that is time-dependent or influenced by other process parameters, i.e., a regular fluctuation, or if the power curve increases or decreases once, for those subsystems several models need to be developed for each process step. For this, the process step is subdivided into several individual sections s, for which the power consumption $P_{i,ps,s}$ is individually modeled. The models are then summarized to represent the entire process step, as shown in Eq. 3 and Eq. 4.

$$E_{i,ps} = \sum_{s=1}^{S} E_{i,ps,s} \tag{3}$$

$$E_{i,ps,s} = P_{i,ps,s} \cdot t_{ps,s} \tag{4}$$

The individual sections are treated hereafter as the energy demand of the process steps. Therefore, for this case in the following, $P_{i,ps,s}$ is to be read and treated as $P_{i,ps}$. In the end, the energy demands of the sections $E_{i,ps,s}$ are summarized to the energy demand of the process steps E_i and then to the total energy demand E_{total}.

To determine $P_{i,ps,s}$, the procedure is the same as described below for power consumptions with a regular fluctuating or a constant pattern, instead of the process steps, however, the individual sections within the process steps are now examined.

The energy demand of a subsystem during a process step $E_{i,ps}$ can then be calculated based on Eq. 5. The power consumption of a subsystem within a process step or section $P_{i,ps}$ is approximated with the mean power consumption for each process step or section.

$$E_{i,ps} = P_{i,ps} \cdot t_{ps} \tag{5}$$

In the simplest case, the power consumption does not depend on any process parameter. Here Eq. 6 applies, as the power consumption of the subsystem can be approximated as constant.

$$P_{i,ps} = constant \tag{6}$$

If only one process parameter exists, the energy demand of the subsystem is determined by the power consumption depending on the setting of the process parameter $P_{i,ps,pp}$, as shown in Eq. 7.

$$E_{i,ps,pp} = E_{i,ps} = P_{i,ps,pp} \cdot t_{ps} \tag{7}$$

If the process parameter is cardinally scaled, the power consumption as a function of the set process parameter $P_{i,ps,pp}$ equals the power consumption of the entire subsystem $P_{i,ps}$. However, if the process parameter is only nominally or ordinally scalable, a case distinction has to be carried out. For this purpose, a sufficient number of different settings of the process parameter is defined for each process step or section of a process step. The averaged power consumption is then assigned for each process parameter setting. This case distinction, also shown in Eq. 8, makes it possible to approximate the power consumption for different process parameter settings.

$$P_{i,ps} = \begin{cases} P_{i,ps,pp,case\,1} = constant, \text{ if case } 1 \\ \qquad \cdots \\ P_{i,ps,pp,case\,C} = constant, \text{ if case } C \end{cases} \tag{8}$$

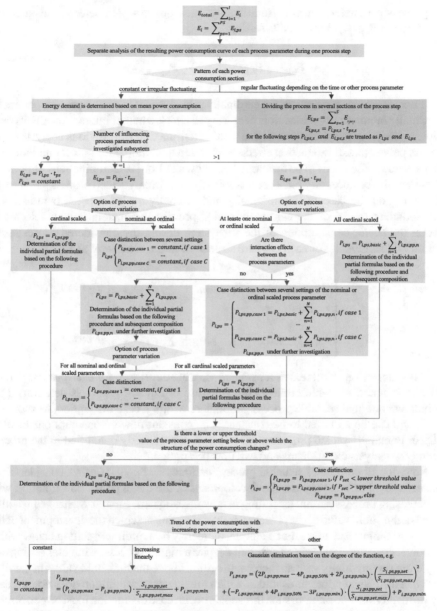

Fig. 6. Overview of the approach to create a model for energy prediction.

Similarly, if several process parameters influence the energy demand of a subsystem, it is necessary to examine whether they are scaled nominally, ordinally, or cardinally. If all process parameters are cardinally scaled, the power consumption of the subsystem $P_{i,ps}$ for a considered process step or section in a process step can be approximated by Eq. 9. Each subsystem has a basic power consumption P_{basic}, which is not changed by the process parameter settings. Additionally, each process parameter causes a power consumption $P_{i,ps,pp,n}$ based on its setting.

$$P_{i,ps} = P_{i,ps,basic} + \sum_{n=1}^{N} P_{i,ps,pp,n} \tag{9}$$

If not all process parameters are cardinally scaled, i.e. at least one process parameter is ordinally or nominally scaled, it is necessary to check whether interaction effects exist between the process parameters. These were determined in the previous analysis of the process parameters. If interaction effects exist, a separate model for power prediction for each assumed case needs to be developed, as shown in Eq. 10. Combinations of possible cases may also be necessary for several nominally or ordinally scaled process parameters. Depending on the case of the nominally and ordinally scaled process parameters, a corresponding model of the cardinally scaled process parameters is now selected to predict the power consumption of the subsystem during the considered process step. The individual models for each case are structured in the same way as Eq. 9.

$$P_{i,ps} = \begin{cases} P_{i,ps,pp,case\,1} = P_{i,ps,basic} + \sum_{n=1}^{N} P_{i,ps,pp,n}, \text{if } case\,1 \\ \qquad\qquad \ldots \\ P_{i,ps,pp,case\,C} = P_{i,ps,basic} + \sum_{n=1}^{N} P_{i,ps,pp,n}, \text{if } case\,C \end{cases} \tag{10}$$

If no interaction effects exist, separate models are developed for each process parameter and summed up, which follows the structure of Eq. 9. $P_{i,ps,pp,n}$ is determined for ordinal and nominal scalable process parameters within a case distinction, as shown in Eq. 8, and can be assumed to be constant. The constant power consumptions of ordinally or nominally scaled process parameters were previously quantified in the process parameter analysis experiments and can thus be applied.

Now, for all cardinally scalable process parameters $P_{i,ps,pp}$, further considerations are necessary. For this purpose, the development of the mean power consumption with an increasing process parameter setting is now investigated. First, it is checked whether there is a threshold value of the process parameter setting at which the development of the power consumption changes. For example, there may be a minimum setting below which the power consumption remains constant despite a higher process parameter setting and only increases gradually with an increasing setting above the threshold value. In parallel, there can also be an upper threshold at which the power consumption does not increase any further. When one or more such thresholds occur, case distinction must be performed. A separate model is created for each case, as shown in Eq. 11.

$$P_{i,ps} = \begin{cases} P_{i,ps,pp} = P_{i,ps,pp,case\,1}, \text{if } S_{p,ps,pp,set} < lower\ threshold\ value \\ P_{i,ps,pp} = P_{i,ps,pp,case\,2}, \text{if } P_{p,ps,pp,set} > upper\ threshold\ value \\ \qquad P_{i,ps,pp} = P_{i,ps,pp,n}, \text{else} \end{cases} \tag{11}$$

For the individual power consumptions of the subsystems during the process step or section depending on the setting of the process parameters, several types of cases can be distinguished by how the power consumption changes with increasing process parameter setting.The power consumption can be constant and thus independent of the process parameter in certain process steps, as shown in Eq. 12. This is often the case below lower threshold values or above upper threshold values.

$$P_{i,ps,pp} = constant \tag{12}$$

In most cases, however, the power consumption increases with an elevation of the process parameter setting. This can be linear or disproportionate. To obtain a regression for a linear slope, the slope of the function can be calculated by using the difference between the maximum power consumption and the minimum power consumption. The minimum power consumption also serves as the y-axis intercept. The x-axis intercept is then calculated by the proportion of the selected process parameter setting to the maximum process parameter setting, which is shown in Eq. 13.

$$P_{i,ps,pp} = \left(P_{i,ps,pp,max} - P_{i,ps,pp,min}\right) \cdot \frac{S_{i,ps,pp,set}}{S_{i,ps,pp,set,max}} + P_{i,ps,pp,min} \tag{13}$$

If there is no linear relationship, a function of any degree can be derived using the gauss elimination. For this, however, additional points must be integrated, whose values need to be determined within experiments. Equation 14 shows an example of the second-degree function resulting from the Gauss elimination.

$$
\begin{aligned}
P_{i,ps,pp} = & \left(2P_{i,ps,pp,max} - 4P_{i,ps,pp,50\%} + 2P_{i,ps,pp,min}\right) \cdot \left(\frac{S_{i,ps,pp,set}}{S_{i,ps,pp,set,max}}\right)^2 \\
& + \left(-P_{i,ps,pp,max} + 4P_{i,ps,pp,50\%} - 3P_{i,ps,pp,min}\right) \cdot \left(\frac{S_{i,ps,pp,set}}{S_{i,ps,pp,set,max}}\right) + P_{i,ps,pp,min}
\end{aligned}
\tag{14}
$$

The model is then validated using reference parts. This allows the identification of potential deficits in the model. If necessary, the model can then be improved by increasing the level of detail, e.g. by implementing additional caste distinctions or sections within a process step.

4 Example of an Application Using HS DED-LB

4.1 Capturing the Structure

For the investigated HS-DED system, a Ponticon pE3D[1], four independent subsystems can be distinguished, which have adjustable process parameters and tend to have a variable power consumption during the process. Other systems, such as the system control, can be grouped as peripheral subsystems.

The distinguished subsystems, which are also shown in Fig. 7, are:

- As laser generator, a Laserline LDF 8000–6 diode laser is used. Both adjustable process parameters, the set laser power and the laser spot diameter can be varied continuously between 504 W and 8400 W and 0.5 mm and 1.8 mm.

- The stirrer speed and the conveyor disc speed of powder feeder Twin-150-ARN216-OP by Oerlikon Metco can also be varied continuously up to 3300 rpm and between 0.2 and 10.0 rpm.
- For the suction system Dustomat 4–24 W3 eco + dry extractor by ESTA only the extracted volume can be adjusted. The extracted volume can be varied between 770 m^3/h and 2540 m^3/h in steps of 50 m^3/h.
- The trajectory system consists of three dynamic linear motors which move the build platform. Here, any scan path can be traveled at a continuously adjustable speed of up to 200 m/min.

In addition to power measurement sensors for the entire system, separate power measurement sensors were implemented for the laser generator, the powder feeder, and the suction system. The energy demand of the subsystems, which is assumed to be constant, and the trajectory system can then be calculated. As power measurement sensors, four current transformers and a corresponding EtherCAT Terminal from Beckhoff are used, which measure the consumed power at a frequency of 1 kHz, i.e. the consumed power per millisecond.

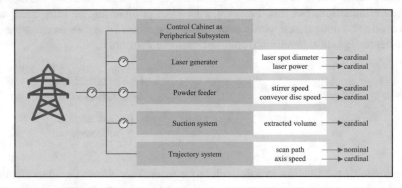

Fig. 7. Structure of the investigated HS DED-LB system

4.2 Process Analysis

Based on the initial experiments, a typical process workflow could be identified and the process was divided into three process steps [24]:

- **Pre-step:** At the beginning of the build cycle, a homogeneous powder gas flow is generated. For this, the two inert gas flows, the carrier gas flow, and the shielding gas flow, are first switched on. After a few seconds, the powder feeder is switched on and the powder is transported to the application head. After a few seconds, the powder cone has built up homogeneously. The trajectory system and the laser generator are on standby, and the suction operates constantly.
- **In-step**: The in-step is the actual additive manufacturing process. Thus, the part is additively manufactured. For this, the build platform moves along the scan path at a defined speed. Depending on the scan path, extra paths are necessary for decelerating

and reaccelerating the build platform. In such cases, the laser is not continuously melting the powder. Therefore, the laser only switches on at defined points and melts the powder to apply the material at the desired locations. The powder feeder and the suction system constantly run during this process step.

- **Post-step**: When the manufacturing process is finished, the laser and the powder feeder turn back into standby, and the inert gas flows stop. Depending on the setting, the build platform stops moving or can move at a slow speed to a previously defined position for better part removal. The post-step takes usually only a few seconds until the powder-gas cone has completely dissolved, and the build chamber can be entered.

For all subsystems, the power consumption varies only during their operation. In standby mode, experimental observations show that the power consumption remains constant and thus independent of any process parameters.

4.3 Analysis of the Process Parameters

For three of the investigated subsystems, there are two process parameters each. Accordingly, their influences on the power consumption of the subsystem and the interactions between parameters were investigated in full-factorial DOE. Additional experiments were then carried out for the relevant process parameters. The results presented here are based on research by Ehmsen et al. [24].

It was observed that the power consumption of the laser generator depends only on the set laser power and does not show any observable correlation to the selected laser spot diameter. Accordingly, only the laser power was investigated in further experiments. Here, it was determined that the power consumption increases approximately linearly with an increasing laser power setting. However, at which moment the laser is switched on and for how long depends on the part geometry and the resulting scan path.

In contrast, the power consumption of the powder feeder was dependent on both the stirrer speed and the conveyor disc speed. However, no interaction of the process parameters was observed. Both process parameters were then increased stepwise and independently of each other. For both, starting from a base power consumption, a linear increase with rising rotational speeds could be observed.

Since for the trajectory system the process parameter of the scan path is only nominally scaled, two cases were defined for the DOE: A circular scan path and a square scan path. Both the speed and the scan path affect the power consumption of the trajectory system, there are even interaction effects. For both cases of the scan path, the speed has now been increased successively. Furthermore, it was determined that for rather circular scan paths, the power consumption increases overproportionally. In contrast, for rather linear scan paths, the power consumption increases approximately linearly. However, for process safety reasons, for linear scan paths only a maximum speed of 100 m/min, i.e. 50%, could be set.

In the case of the suction system, only the power consumption with increasing extraction volume was investigated. Here, a lower threshold value was detected. Below an extraction rate of 36%, the power consumption is constant and corresponds to the standby level. Above an extraction rate of 36%, the power consumption increases overproportionally.

Thus, all process parameters except for the laser spot diameter must be considered in the model.

4.4 Creating the Model

To create the customized model to predict the energy demand of the HS DED-LB system, for each subsystem the procedure described in Sect. 3.4 was carried out.

The peripheral subsystem is constant throughout the entire process. Therefore, to calculate its energy demand, constant power is assumed, which was quantified previously in the experiments, and integrated over the entire process time.

The laser is on standby during the pre-step and the post-step and has constant power consumption here, which is independent of any process parameters. However, the power consumption during the in-step varies depending on the scan path. Thus, the process step is subdivided into further sections. There are sections in which the laser is in mode "ready to fire" and has a power consumption slightly above the standby level. The level is independent of the set laser power. In the sub-process steps, where the laser is on during scanning, the power consumption depends on the set laser power. A linear model was created here to predict the power consumption.

The powder feeder is only on standby during the post-step. Here, too, the power consumption was modeled by a constant. During the pre-step and the in-step, the powder feeder operates continuously. To capture the influence of both process parameters, independent linear models were developed for both, which were summarized with a constant base power input.

The suction system operates during the entire build cycle. Here, a case distinction between different extraction rates is necessary. Below an extraction rate of 36%, the power consumption is at the same level as during standby. Above an extraction rate of more than 36%, power consumption increases overproportionally with rising extraction rates. This increase can be approximated with a quadratic function obtained by the Gaussian elimination.

For the trajectory system, a case distinction was made for two different scan path patterns, a rather rotationally symmetric scan path, and a rather linear scan path. For both cases, a separate model was developed that predicts the power consumption as a function of the selected trajectory speed. Based on the results of the previous experiments, a linear model was developed for the case of the rather linear scan path. For the rather rotationally symmetric scan path, a third-degree function with the Gaussian elimination was received. During the pre-step and optionally also during the post-step, the trajectory system is not in motion and thus on stand-by. Therefore, it can be considered independent of any process parameters in these process steps.

Based on the modeled power consumption for each process step or sub-process step, the energy demand of each subsystem and thus the total energy demand of the HS DED-LB system can be determined. It may be necessary to model the build cycle time and the duration of the individual process steps as well. The exact models and their equations are presented in Ehmsen et al. [52].

4.5 Exemplary Application and Validation

To validate the accuracy of the model, a reference part was manufactured. The dimensions of the part and the corresponding process parameters are listed in Table 1. During manufacturing the part the power consumption was measured for each millisecond. The resulting power curve is illustrated in Fig. 8. Based on the power consumption, the energy demand which arises during the process was calculated and compared with the energy demand predicted by the model. As shown in Fig. 9, the predicted energy demand is only 2% lower than the measured energy demand.

Table 1. Process parameters of the reference part for validation

Part geometry and scan path		Process parameters	
Length	15.0 mm	Set laser power	3100 W
Width	10.2 mm	Trajectory speed	40 m/min
Height	2.96 mm	Extraction volume	1690 m^3/h
Pattern	linear	Conveyor disc speed	5.4 rpm
Number of tracks per layer	6	Stirrer speed	990 rpm
Number of layers	40		
Acceleration distance	70 mm		

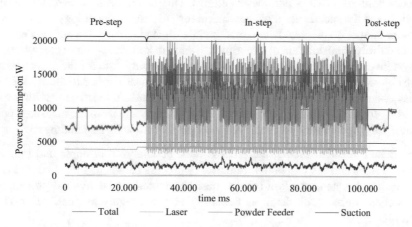

Fig. 8. Power curve of the manufacturing of the reference part

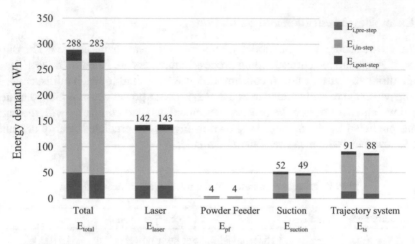

Fig. 9. Comparison of measured energy demand with modeled energy demand.

5 Conclusion

The research objective of the paper was to develop a methodology to create a customized model to predict the energy demand for an arbitrary AM system.

The model resulting from the methodology met the requirements to be detailed enough to identify potentials to reduce the energy demand and at the same time be able to model different process parameter settings. Due to the detailed description of the procedure and the specification of calculation formula, the methodology can be applied even with little or no prior knowledge of the investigated AM system and process. With a comparatively small amount of experiments, enough knowledge about the investigated AM system can be built up and at the same time, necessary data for the model creation can be collected. The functionality of the methodology was applied to the HS DED-LB process as an example, where a reference part was manufactured. The resulting model showed a very high prediction accuracy and deviate only 2% from the measured energy demand. This high quality was achieved through the following key findings:

- To obtain a detailed model, the process must be analyzed in depth and individual process steps and the individual influence of process parameters must be included.
- For each subsystem, the build cycle must be divided into individual process steps or sections in as much detail as necessary and as roughly as possible, to reduce complexity.
- For cardinally scaled process parameters, the development of the power consumption can be approximated by regression, whose function can be obtained based on a small number of process points.
- Nominal and ordinal scaled process parameters have to be approximated using a case distinction of averaged constant power consumption.

The results therefore indicate that the methodology is a powerful tool to develop a customized model and thus, predict the energy demand for different AM processes. In the future, the model will be applied to other AM processes such as powder bed fusion.

Furthermore, it will be important to investigate whether this methodology can also be used for formative or subtractive processes.

Acknowledgments. This research was funded by the Deutsche Forschungsgemeinschaft (DFG, German Research Foundation) – 252408385 – IRTG 2057.

References

1. Herrmann, C.: Ganzheitliches Life Cycle Management. Nachhaltigkeit und Lebenszyklusorientierung in Unternehmen. VDI-Buch. Springer, Berlin (2010)
2. Statista Research Department: Index zur Entwicklung des Industriestrompreises in Deutschland in den Jahren 1998 bis 2022 (2022). https://de.statista.com/statistik/daten/studie/12500/umfrage/entwicklung-der-industrie-strompreise-in-deutschland-seit-1998/. Accessed 28 Nov 2022
3. European Commission: Emissions cap and allowances (2022). https://climate.ec.europa.eu/eu-action/eu-emissions-trading-system-eu-ets/emissions-cap-and-allowances_en. Accessed 28 Nov 2022
4. Duflou, J.R., et al.: Towards energy and resource efficient manufacturing: a processes and systems approach. CIRP Ann. (2012). https://doi.org/10.1016/j.cirp.2012.05.002
5. Schmidt, C., Li, W., Thiede, S., Kara, S., Herrmann, C.: A methodology for customized prediction of energy consumption in manufacturing industries. Int. J. Precision Eng. Manuf.-Green Technol. **2**(2), 163–172 (2015). https://doi.org/10.1007/s40684-015-0021-z
6. Gebhardt, A.: Generative Fertigungsverfahren. Additive Manufacturing und 3D Drucken für Prototyping; Tooling; Produktion, 4th edn. Carl Hanser Fachbuchverlag, s.l. (2013)
7. Yi, L.: Eco-Design for Additive Manufacutring Using Energy Performance Quantification and Assessment. Dissertation. Produktionstechnische Berichte aus dem FBK. Lehrstuhl für Fertigungstechnik und Betriebsorganisation, Kaiserslautern (2021)
8. Huang, Y., Leu, M.C., Mazumder, J., Donmez, A.: Additive manufacturing: current state, future potential, gaps and needs, and recommendations. J. Manufact. Sci. Eng. (2015). https://doi.org/10.1115/1.4028725
9. Ribeiro, I., et al.: Framework for life cycle sustainability assessment of additive manufacturing. Sustainability **12**, 929 (2020)
10. DIN Normenausschuss Werkstofftechnologie (NWT): Additive Fertigung - Grundlagen - Terminologie (ISO/ASTM 52900:2017 (D)) (2017)
11. Wohlers Associates: Wohlers report 2019. 3D printing and additive manufacturing state of the industry. Wohlers Associates, Fort Collins, Colorado (2019)
12. DIN Normcnausschuss Werkstofftechnologie (NWT): Additive Fertigung – Grundlagen – Teil 2: Überblick über Prozesskategorien und Ausgangswerkstoffe, 2nd edn.(DIN EN ISO 17296-2) (2016)
13. Clayton, J.: Optimising metal powders for additive manufacturing. Met. Powder Rep. **69**, 14–17 (2014)
14. Gibson, I., Rosen, D., Stucker, B.: Additive manufacturing technologies. 3D printing, rapid prototyping and direct digital manufacturing. Springer, New York, Heidelberg, Dodrecht, London (2015)
15. Klocke, F.: Fertigungsverfahren 5. Gießen, Pulvermetallurgie, Additive Manufacturing, 4th edn. VDI-Buch. Springer, Heidelberg (2015)

16. Herzog, D., Seyda, V., Wycisk, E., Emmelmann, C.: Additive manufacturing of metals. Acta Mater. **117**, 371–392 (2016)

17. Yan, M., Yu, P.: An overview of densification, microstructure and mechanical property of additively manufactured Ti-6Al-4V — comparison among selective laser melting, electron beam melting, laser metal deposition and selective laser sintering, and with conventional powder. In: Lakshmanan, A. (ed.) Sintering Techniques of Materials. Intech (2015)

18. Li, T., et al.: Extreme high-speed laser material deposition (EHLA) of AISI 4340 steel. Coatings (2019). https://doi.org/10.3390/coatings9120778

19. Schaible, J., Sayk, L., Schopphoven, T., Schleifenbaum, J.H., Häfner, C.: Development of a high-speed laser material deposition process for additive manufacturing. J. Laser Appl. (2021). https://doi.org/10.2351/7.0000320

20. Schopphoven, T.: Experimentelle und modelltheoretische Untersuchungen zum Extremen Hochgeschwindigkeits-Laserauftragschweißen. Dissertation. Fraunhofer Verlag, Stuttgart (2020)

21. Küppers, W., Backes, G., Kittel, J.: Extremes Hochgeschwindigkeitslaserauftragsschweißver-fahren Patent DE 10 2011 100 456

22. Shen, B., Du, B., Wang, M., Xiao, N., Xu, Y., Hao, S.: Comparison on microstructure and properties of stainless steel layer formed by extreme high-speed and conventional laser melting deposition. Front. Mater. (2019). https://doi.org/10.3389/fmats.2019.00248

23. Schopphoven, T., Pirch, N., Mann, S., Poprawe, R., Häfner, C.L., Schleifenbaum, J.H.: Statis-tical/numerical model of the powder-gas jet for extreme high-speed laser material deposition. Coatings (2020). https://doi.org/10.3390/coatings10040416

24. Ehmsen, S., Glatt, M., Aurich, J.C.: Influence of process parameters on the power consumption of high-speed laser directed energy deposition. Procedia CIRP **116**, 89–94 (2023). https://doi.org/10.1016/j.procir.2023.02.016

25. Liu, Z., Jiang, Q., Cong, W., Li, T., Zhang, H.-C.: Comparative study for environmental perfor-mances of traditional manufacturing and directed energy deposition processes. Int. J. Environ. Sci. Technol. **15**(11), 2273–2282 (2017). https://doi.org/10.1007/s13762-017-1622-6

26. Bourhis, F.L., Kerbrat, O., Hascoet, J.-Y., Mognol, P.: Sustainable manufacturing: evalua-tion and modeling of environmental impacts in additive manufacturing. Int. J. Adv. Manuf. Technol. **69**(9–12), 1927–1939 (2013). https://doi.org/10.1007/s00170-013-5151-2

27. Slotwinski, J.A., Garboczi, E.J.: Metrology needs for metal additive manufacturing powders. JOM **67**(3), 538–543 (2015). https://doi.org/10.1007/s11837-014-1290-7

28. Morrow, W.R., Qi, H., Kim, I., Mazumder, J., Skerlos, S.J.: Environmental aspects of laser-based and conventional tool and die manufacturing. J. Clean. Prod. **15**, 932–943 (2007)

29. Ma, K., Smith, T., Lavernia, E.J., Schoenung, J.M.: Environmental sustainability of laser metal deposition: the role of feedstock powder and feedstock utilization factor. Procedia Manuf. (2017). https://doi.org/10.1016/j.promfg.2016.12.049

30. Muthu, S.S., Savalani, M.M.: Handbook of Sustainability in Additive Manufacturing. Volume 2. Environmental Footprints and Eco-design of Products and Processes. Springer, Singapore, s.l. (2016)

31. Singh, A., Kapil, S., Das, M.: A comprehensive review of the methods and mechanisms for powder feedstock handling in directed energy deposition. Addit. Manuf. (2020). https://doi.org/10.1016/j.addma.2020.101388

32. Chen, D., Heyer, S., Ibbotson, S., Salonitis, K., Steingrímsson, J.G., Thiede, S.: Direct digital manufacturing: definition, evolution, and sustainability implications. J. Clean. Prod. (2015). https://doi.org/10.1016/j.jclepro.2015.05.009

33. Kellens, K., Mertens, R., Paraskevas, D., Dewulf, W., Duflou, J.R.: Environmental impact of additive manufacturing processes: does AM contribute to a more sustainable way of part manufacturing? Procedia CIRP **61**, 582–587 (2017)

34. Yoon, H.-S., et al.: A comparison of energy consumption in bulk forming, subtractive, and additive processes: review and case study. Int. J. Precis. Eng. Manuf.-Green Technol. **1**(3), 261–279 (2014). https://doi.org/10.1007/s40684-014-0033-0
35. Colorado, H.A., Velásquez, E.I.G., Monteiro, S.N.: Sustainability of additive manufacturing: the circular economy of materials and environmental perspectives. J. Market. Res. (2020). https://doi.org/10.1016/j.jmrt.2020.04.062
36. Faludi, J., Baumers, M., Maskery, I., Hague, R.: Environmental impacts of selective laser melting: do printer, powder, or power dominate? J. Ind. Ecol. **21**, 144–156 (2017)
37. Kellens, K., Baumers, M., Gutowski, T.G., Flanagan, W., Lifset, R., Duflou, J.R.: Environmental dimensions of additive manufacturing: mapping application domains and their environmental implications. J. Ind. Ecol. **21**, 49–68 (2017)
38. Fredriksson, C.: Sustainability of metal powder additive manufacturing. Procedia Manuf. **33**, 139–144 (2019)
39. Baumers, M., Duflou, J.R., Flanagan, W., Gutowski, T.G., Kellens, K., Lifset, R.: Charting the environmental dimensions of additive manufacturing and 3D printing. J. Ind. Ecol. **21**, 9–14 (2017)
40. DIN Deutsches Institut für Normung e.V.: Umweltmanagement – Ökobilanz – Grundsätze und Rahmenbedingungen (ISO 14040:2006) (2009)
41. Jiang, Q., Liu, Z., Li, T., Cong, W., Zhang, H.-C.: Emergy-based life-cycle assessment (Em-LCA) for sustainability assessment: a case study of laser additive manufacturing versus CNC machining. Int. J. Adv. Manuf. Technol. **102**(9–12), 4109–4120 (2019). https://doi.org/10.1007/s00170-019-03486-8
42. Le Bourhis, F., Kerbrat, O., Dembinski, L., Hascoet, J.-Y., Mognol, P.: Predictive model for environmental assessment in additive manufacturing process. Procedia CIRP **15**, 26–31 (2014)
43. Kerbrat, O., Le Bourhis, F., Mognol, P., Hascoët, J.-Y.: Environmental impact assessment studies in additive manufacturing. In: Muthu, S.S., Savalani, M.M. (eds.) Handbook of Sustainability in Additive Manufacturing. EFEPP, pp. 31–63. Springer, Singapore (2016). https://doi.org/10.1007/978-981-10-0606-7_2
44. Serres, N., Tidu, D., Sankare, S., Hlawka, F.: Environmental comparison of MESO-CLAD® process and conventional machining implementing life cycle assessment. J. Clean. Prod. **19**, 1117–1124 (2011)
45. Xiong, Y., Lau, K., Zhou, X., Schoenung, J.M.: A streamlined life cycle assessment on the fabrication of WC–Co cermets. J. Clean. Prod. (2008). https://doi.org/10.1016/j.jclepro.2007.05.007
46. Wippermann, A., Gutowski, T.G., Denkena, B., Dittrich, M.-A., Wessarges, Y.: Electrical energy and material efficiency analysis of machining, additive and hybrid manufacturing. J. Clean. Prod. **251**, 119731 (2020)
47. Huang, R., et al.: Energy and emissions saving potential of additive manufacturing: the case of lightweight aircraft components. J. Clean. Prod. **135**, 1559–1570 (2016)
48. Jackson, M.A., van Asten, A., Morrow, J.D., Min, S., Pfefferkorn, F.E.: A comparison of energy consumption in wire-based and powder-based additive-subtractive manufacturing. Procedia Manuf. (2016). https://doi.org/10.1016/j.promfg.2016.08.087
49. Watson, J.K., Taminger, K.M.B.: A decision-support model for selecting additive manufacturing versus subtractive manufacturing based on energy consumption. J. Clean. Prod. (2015). https://doi.org/10.1016/j.jclepro.2015.12.009

50. Wegener, M.: Ressourceneffiziente Gestaltung von Prozessketten mit additiven Fertigungsverfahren. Dissertation, 1st edn. Prozesstechnologie, 2016, Band 42 (2016)
51. Dietmair, A., Verl, A.: A generic energy consumption model for decision making and energy efficiency optimisation in manufacturing. Int. J. Sustain. Eng. (2009). https://doi.org/10.1080/19397030902947041
52. Ehmsen, S., Yi, L., Linke, B.S., Aurich, J.C.: Reusable unit process life cycle inventory for manufacturing: high speed laser directed energy deposition. Prod. Eng. (2023). https://doi.org/10.1007/s11740-023-01197-4

Framework to Improve the Energy Performance During Design for Additive Manufacturing

L. Yi[1]([✉]), X. Wu[1], M. Glatt[1], B. Ravani[2], and J. C. Aurich[1]

[1] Institute for Manufacturing Technology and Production Systems, RPTU in Kaiserslautern, Kaiserslautern, Germany
li.yi@mv.uni-kl.de

[2] Department for Mechanical and Aerospace Engineering, UC Davis, Davis, USA

Abstract. Additive manufacturing (AM) is suitable for designing and producing complex components that are difficult or impossible to manufacture with conventional manufacturing processes. To ensure the design benefits of AM, novel design approaches such as structural topology optimization and cellular structure design are widely used and bring up the research domain of design for AM (DfAM). However, conventional DfAM approaches mainly focus on the geometry and manufacturability of AM components and rarely consider energy performance as an improvement objective. Given that the energy consumption in AM processes can be a great contributor to the overall environmental impact in the production stage with AM, the evaluation and improvement of the energy performance of AM should be considered in the DfAM approaches; otherwise, opportunities to improve the energy performance by changing product features are missed. To address this research question, we are proposing a framework that enables the evaluation and improvement of the energy performance of AM in the design stage resulting in a new method for DfAM. To validate the framework, two use cases are presented to illustrate the feasibility of developed methods and tools.

1 Introduction

Additive manufacturing (AM) has been used in a variety of industrial areas where high-value-added parts with complex geometries are needed, such as in aerospace technologies [1]. Since AM enables the creation of complex geometries that are difficult or impossible to manufacture with conventional milling or casting processes, novel design approaches such as topology optimization and cellular structures can be used in the design stage for AM, which further brings up the research field of design for additive manufacturing (DfAM) [2, 3]. On the part dimension, conventional DfAM usually aims at improving the functionality and manufacturability of AM parts [4]. Thus, to achieve these objectives, a number of previous works have focused on the development of innovative DfAM methods for structural topology optimization or lattice design for AM parts [4]. For example, in terms of material layout optimization, the multi-agent algorithm has been used in a method for designing, evaluating, and optimizing the manufacturability of AM parts [5]. Furthermore, the method based on constructive solid geometry is developed

© The Author(s) 2023
J. C. Aurich et al. (Eds.): IRTG 2023, *Proceedings of the 3rd Conference on Physical Modeling for Virtual Manufacturing Systems and Processes*, pp. 213–232, 2023.
https://doi.org/10.1007/978-3-031-35779-4_12

to generate multiple design variants and select the optimum design based on a genetic algorithm [6]. In terms of the cellular structure design, a method based on the moving asymptotes method is proposed [7]. In this area, another method for gradient lattice design using bidirectional evolutionary structural optimization has been developed [8], and the method based on the optimality criteria algorithm is proposed for the functional gradient lattice design [9]. In terms of support structure design, the method based on a genetic algorithm is developed to create and optimize tree-like support structures of AM parts [10].

Nevertheless, in assessing these works, it is seen that they do not consider other objectives in addition to the geometry and manufacturability, such as the energy performance that will be discussed in this paper. Therefore, conventional DfAM methods do not ensure the improvement of the environmental impact of AM in the product design stage. However, environmental issues have been another important topic in AM, in addition to the DfAM topics [11]. While the technology was emerging, AM was often considered to be generally more environment-friendly than conventional manufacturing processes. For example, Huang et al. have summarized the environmental benefits of AM, such as the absence of cutting tools or dies in AM leading to reduced resource usage, limited scraps during the build process leading to reduced wastes in AM, and the improved engineering performance of AM parts using lightweight leading to environment-related benefits during their usage phase [12]. However, the latest findings also point out that the environmental benefits of AM can only be ensured if they are considered during the design stage [11, 13]. Otherwise, the environmental benefit of AM (i.e. AM is more environment-friendly than conventional manufacturing) may be just an illusion [11]. Furthermore, recent studies have proven that energy consumption is the main contributor to the overall environmental impacts of the AM-based production phase [14], and therefore, the improvement of energy performance is currently an emerging research topic [15].

In existing methods, the energy performance is usually studied during the process planning or during the process chain planning for AM and is rarely considered in the design phase. Thus, chances that would help improve the energy performance of AM by varying product-related features are not fully exploited [16].

Aiming at the above background, this paper introduces the development of a new framework that considers energy performance as the optimization objective during the DfAM. The proposed framework consists of three key parts: structural topology optimization (1), tool-path length assessment (2), and multi-player competition algorithm (3). These three parts are combined in a holistic computational procedure, which enables the exploration of possible design variants from a given domain with the aim of finding out the design variant with the highest energy performance. To validate the framework, two use cases are presented to illustrate the feasibility of developed methods and tools.

2   Research Background

2.1   Energy Performance Issues in Additive Manufacturing

Energy performance is the key issue for ensuring the environmental sustainability of AM, considering that energy use may cause 67% to 75% of environmental impacts in the production phase [14]. In the past years, it was frequently argued that AM has more environmental benefits than conventional manufacturing for reasons that AM requires no cutting tools or dies, results in limited scrap, and enables design benefits leading to environmentally-related benefits during the lifetime of AM products [11–13]. Nevertheless, the latest findings have shown that these benefits can only be ensured if they have been carefully examined and validated during the design stage [11]. Thus, the energy performance of AM should be analyzed and assessed prior to the start of the build process of AM products. With respect to energy conversion [17], current studies on the energy issues of AM can be distinguished between primary energy, electricity (use energy), and thermal energy (final energy) issues.

The assessment and improvement of primary energy use of AM are usually discussed when determining the design solution related to a manufacturing process chain, supply chain, production network, or entire product life-cycle with AM. For example, primary energy demand can be used as an indicator to compare the use of AM versus conventional manufacturing in order to validate the benefits of AM (i.e. AM-based scenario has less primary energy demand than that of the conventional manufacturing-based scenario for providing the same function) [18–21]. For these works, the primary energy demand is usually quantified using the life-cycle assessment (LCA) or the cumulative energy demand (CED) method (e.g. [22]).

For electricity uses, energy performance quantification and improvement are performed when defining the parameters for a build process or a process chain. For example, the estimated electricity usage can be used as an indicator to compare the different design solutions with different process parameters or different manufacturing process chains based on different AM processes. For the electricity demand estimation, prediction software tools or analytical models are used [23–26]. For the improvement of energy performance in AM, optimization algorithms (e.g. genetic algorithm) can be used to optimize the process parameters [27]. Moreover, electricity issues are also widely analyzed using experiments, where process parameters (e.g. layer thickness and laser power) are varied, and the corresponding electricity consumption is measured using power meters. The relation between process parameters and electricity consumption can be analyzed using statistical methods (e.g. regression models), which can be further contributed to the design of AM build tasks [28–30].

For the thermal issues, the energy performance is improved when defining materials or energy input-related process parameters. For example, mixing copper powders with additive nanoparticles or defining a higher laser power will increase the energy absorption during the laser processing of laser powder bed fusion [31, 32].

In assessing the above works, only limited studies have addressed the DfAM and the energy issues in the same time, e.g. [20, 21, 23, 33–35]. The key issues of these approaches have been listed in Table 1. However, in these works, the authors consider the design activities (i.e. DfAM) separately from the energy performance evaluation activities, as depicted in Fig. 1. In those works, the product design (e.g. DfAM activities) is performed separately. Then, an AM build process or a manufacturing process chain based on AM is modeled to represent an AM-based production scenario. After that, the energy performance of the AM-based production scenarios is quantified and assessed, and further decision-making for product improvement is made. In such a way, the DfAM and the energy performance are rather two separate topics. This results in that the energy performance improvement is not integrated into the DfAM activities of current DfAM approaches. Subsequently, the time and effort for the design would be high due to the repeated DfAM and evaluation activities required by the improvement loop.

In our proposed work, the DfAM and energy performance assessment are integrated, as depicted in Fig. 1. The benefit of our approach is that it does not require repeated individual design and evaluation activities, which further saves time and cost in the design phase of AM. Moreover, if the time of a design cycle is reduced, AM designers are able to explore more different design possibilities, which also implies more chances to improve the functionality and utilization performance of AM products. To outline the difference between our approach from the existing approaches, the key issues in our approach have been described in italics and listed in Table 1.

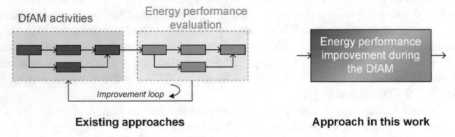

Fig. 1. Difference between the existing approaches and the approach in this work.

2.2   Research Target and Tasks for This Work

Based on the evaluation of the research background introduced in Sect. 2.1, it is clearly seen that the literature still lacks a methodology in which the energy performance evaluation is integrated with the DfAM. Therefore, this work aims at this research gap and contributes to a DfAM computational framework in which the energy performance evaluation and DfAM are iteratively executed in one algorithm. For this target, we have defined three research tasks. First, a method is proposed to describe the energy performance of AM during the DfAM. This is important since the description and quantification of the energy performance is the pre-request to the optimization of the energy performance. Second, the improvement of the energy performance of AM in the DfAM needs to be formulated as an optimization problem together with a computational technique to solve the problem. The third task is to implement and validate the feasibility of the proposed computational framework. In the remainder of the paper, the results of performing these three research tasks are discussed.

Table 1. List of the existing approaches and our approach.

References	DfAM issues	Energy performance evaluation issues	Collaboration between DfAM and evaluation
Ingarao et al. [20]	Propose different design geometries suitable for conventional and AM processes	Use primary energy to compare the different manufacturing process chains	Design the product first, map the process chains, and calculate primary energy in individual steps
Priarone et al. [34, 35]	Redesign a product using topology optimization to make them suitable for AM	Use primary energy to compare the full product life cycles with conventional and AM processes	Design the product first, map the process chain, and calculate the primary energy in individual steps
Yang et al. [21]	Part consolidation of an assembly into a single part to make it suitable for AM	Calculate the energy consumption of each phase with AM and conventional manufacturing routines	Design the product first, map the process chain, and model and calculate the energy consumption in individual steps
Tang et al. [33]	Redesign of product using topology optimization to make it suitable for AM	Calculate the energy consumption of each phase with AM and conventional milling process chains and compare their environmental impact	Design the product first, map the process chain, and calculate the energy consumption in individual steps
Yi et al. [23]	Use topology optimization to propose multiple designs for AM	Use electricity demand to compare different product and process designs	Design the product first, simulate the electricity demand in individual steps
Our approach	*Design product using topology optimization*	*Use tool-path length assessment to replace energy calculation*	*Topology optimization and evaluation of energy performance are integrated into one algorithm*

3 Framework of Energy Performance Improvement in DfAM

3.1 Overview of the Framework

The concept of the proposed computational framework developed in this work is illustrated in Fig. 2 and described in the following subsection. The computational framework includes three core parts.

- The first part is the structural topology optimization (TO) method, which enables the generation of a material layout with the best mechanical performance for a given design space and boundary conditions. In this framework, we have used the method of "smooth-edged material distribution for optimizing topology (SEMDOT)", which is a state-of-the-art TO algorithm suitable for AM [36]. In comparison to other TO algorithms, SEMDOT enables the creation of smooth geometric boundaries, which further ensures the manufacturability of the TO results for AM processes [36], and this is also the reason why this framework choses SEMDOT as the TO tool.
- The second core part is the "tool-path length assessment" method, in which the tool-path length to create the geometry using the AM process is estimated. The reason for choosing this method is that in the DfAM, the energy consumption calculation is difficult because the energy is a time-dependent process characteristic (i.e. energy is the time integral of power) instead of a product characteristic [16]. Therefore, it is necessary to define a parameter that is relevant to the product and energy at the same time. Thus, the evaluation of the energy can be replaced by the assessment of this parameter. In this work, the AM tool-path length has been regarded as such a parameter, which has a proportional relation to the energy consumption (i.e. longer tool-path leads to higher energy demand) and part geometry (i.e. more internal features lead to longer tool-path).
- The third part of the framework is the method of "multi-player competition algorithm", which is an iterative optimization technique to compare and select the optimum geometry variant with the minimum AM tool-path length (i.e. the highest energy performance in this work) [37]. The multi-player competition algorithm is used to iteratively execute the SEMDOT and the tool-path length assessment in one computational procedure. The reason for choosing this algorithm is that it compromises the computational efficiency and search quality for our problem.

The details of these three parts will be explained in the next subsections.

3.2 Structural Topology Optimization

In general, the design space for TO can be regarded as a discrete domain with a number of elements, as shown in Fig. 3. Thus, the material layout for this design space under certain boundary conditions can be considered as defining the relative material density X_e (a dimensionless value, not the density in g/cm^3) for each element in that discrete domain. Given a minimum relative material density ρ_{min}, the density of each element X_e should be a value between ρ_{min} and 1. For example, as depicted in Fig. 3, if X_e is 1, it means that this element is defined with a full material (i.e. black element), while if X_e is ρ_{min}, it means this element contains no material (i.e. while element). Thus, the

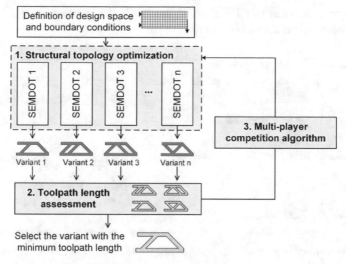

Fig. 2. Overview of the proposed framework.

objective of a TO problem based on varying the material density of elements can be described by Eq. 1 [36], where C represents the compliance; K, u, and f are the global stiffness matrix, the displacement vector, and the force vectors, respectively; V_e and V^* represent the volume of elements and pre-defined target volume, respectively; and M is the total number of elements.

$$
\begin{aligned}
minimize: \quad & C(X_e) = f^T u \\
subject\ to: \quad & K(X_e)u = f \\
& \frac{\sum_1^M X_e V_e}{\sum_1^M V_e} - V^* \le 0 \\
& 0 \le \rho_{min} \le X_e \le 1; e = 1, 2, \ldots M
\end{aligned}
\tag{1}
$$

However, element-based TO does not have a smooth edge, which does not meet the manufacturability requirement for AM. The SEMDOT method uses an approach involving inserting grid points within elements and then using a level-set function to represent the smooth edges of the geometry [36]. For each grid point, a grid density is defined ($\rho_{e,g}$), and therefore, the relation between X_e and $\rho_{e,g}$ can be expressed by Eq. 2 [36], where N represents the total number of grid points of an element.

$$
X_e = \frac{1}{N} \sum_{g=1}^{N} \rho_{e,g}
\tag{2}
$$

Finally, by using the Heaviside function, the smooth edges for the geometry can be generated [36]. During SEMDOT, the formulation of geometry is controlled by two pre-defined parameters: the radius of a circular filtering domain by elements (r_{filter}) and the heuristic radius of a circular filtering domain by grid nodes (Y_{filter}) [36]. Different combinations of these two parameters will result in different geometries, as shown by examples in Fig. 3. In general, the r_{filter} should be varied between 1 and 3.5, and Y_{filter}

should be varied between 1 and 3. Thus, in this work, the parameters (r_{filter}, Y_{filter}) are varied to propose different variants in the TO part.

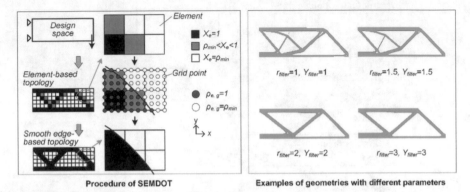

Procedure of SEMDOT   Examples of geometries with different parameters

Fig. 3. Computational procedure and results of SEMDOT algorithms (own illustration based on [36]).

3.3 Tool-Path Length Assessment

In general, the term 'energy performance' is regarded as a metric to express energy efficiency, energy use, or energy consumption of systems or processes [38]. Therefore, the prerequisite for the improvement of energy performance in AM is to quantify the energy required to perform a build task of AM. In general, the quantification of energy demand requires power and time information since energy is the time integral of power. However, the power and time information is related to the process parameters (e.g. laser power, laser speed, build orientation, and layer thickness), which directly affect the build time and power demands of AM machines. In DfAM, process parameters are not considered because they are mainly considered during the process planning stage in which product design has already been completed. Subsequently, missing process parameters in DfAM leads to difficulty in quantifying the energy demand for a build task in AM. To overcome this challenge, we use an equivalent evaluation indicator which should satisfy two requirements. First, the equivalent indicator should be related to the geometrical features of a product, and therefore, it can be described and investigated at the DfAM stage. Second, the indicator should be positively or inversely proportional to the energy consumption of AM, and therefore, the minimization of the energy consumption can be realized by the minimization or maximization of this equivalent indicator.

In this work, the AM tool-path length is regarded as the equivalent indicator since it is related to the geometrical features and energy consumption of AM at the same time. In general, the analogy of an AM tool can be considered as a means for processing the material in AM, and different AM processes have different AM tools. For example, in laser powder bed fusion, the AM tool is a laser beam, while in fused deposition modeling, the AM tool is a nozzle with a heating core. For each layer, the path of an AM tool can be generally distinguished between 'contour' and 'hatching', as depicted in Fig. 4 on the left.

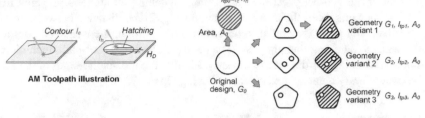

Fig. 4. Illustration of tool-path and its relation to the geometry variants.

The contour describes the path of the AM tool traveling on the edge of geometry, while the hatching refers to the path of the AM tool walking through the cross-sectional area of that geometry. Thus, the total tool-path length (l_{tp}) of the AM tool can be the sum of the path length for the contour (l_c) and the hatching (l_h), as expressed in Eq. 3, where L_C and L_H indicate the length of individual contour lines and hatching lines, respectively, and $N_{contour}$ and $M_{hatching}$ represent the total number of contour and hatching lines, respectively.

$$l_{tp} = \overbrace{\underbrace{\sum_{n=1}^{N_{contour}} L_{C(n)}}}^{l_c} + \overbrace{\underbrace{\sum_{m=1}^{M_{hatching}} L_{H(m)}}}^{l_h} \tag{3}$$

Given a geometric design (G_0) with a specific area (A_0), the tool-path to create this geometry can be denoted as l_{tp0}. If this geometry can be modified to other shapes (G_k) with the same area size A_0, the tool-path length for the new geometry would be different since there may be many or fewer internal holes and more or less complex curves in the new geometry. Given the constant power of AM tool during the material processing, the energy demand of the AM tool to create a geometry is related to the tool-path for reasons that the length of AM tool determines the processing time. If the AM tool-path lengths of different geometries with the same area A_0 are different, the energy required to produce them would be different, as shown in Fig. 4 on the right. Thus, the problem of finding out a design variant with the least required energy consumption can be understood as being equivalent to the problem of finding out a design variant with the shortest l_{tp}, as expressed in Eq. 4, where G represents a geometry variant generated by the SEMDOT method, S_G represents the set of all geometries for a given population K, and G^* represents the optimum geometry to be found. This method is denoted as the tool-path length assessment (TLA) in this work.

$$l_{tp}(G^*) = \underset{G^* \in S_G}{\text{argmin}} \left\{ G^* \in S_G : l_{tp}(G_k) \geq l_{tp}(G^*) \text{ for } G_k \in S_G, k \in [1, K] \right\} \tag{4}$$

Finally, to implement the TLA, we have used an image processing technique, as shown in Fig. 5, and explained in the following. First, the result of the SEMDOT method is an optimized geometry, which can be exported in standard image file format (e.g. JPG files). After that, we have used the python library OpenCV to detect and extract the contour of the geometry. At the same time, a hatching template is prepared in the form of another image. In the example shown in Fig. 5, the distance between any two neighbor hatching lines is defined to be 1 mm. Furthermore, the extracted contour is infilled with the hatching line template. The result of that is the tool-path, including the contour and hatching for the given geometry. Finally, the tool-path length of the image can be conveniently detected and calculated using OpenCV.

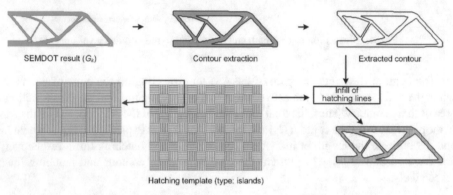

Fig. 5. Image processing technique to implement TLA.

3.4 Multi-player Competition Algorithm

To solve the problem formulated in Eq. 4, we propose the multi-player competition algorithm (MPCA), which is inspired by sport games in which multiple players compete in multiple rounds, and finally, only one player wins the game (e.g. table tennis and running). Considering our problem as a game, the scenario would be that multiple players are selecting geometry variants from a design domain, and the one who picks up the geometry with the shortest l_{tp} wins the game.

For a better understanding of the MPCA, the computational procedure is illustrated in a pseudo code (as depicted in Fig. 6) and explained in the following. First, it is assumed that there are a certain number of players (K), which is expressed in a set $S_{player} = \{1, 2, 3, \ldots, K\}$, and the game will be repeated in n_{max} rounds. In the first round, each of them should pick up a parameter combination (r_{filter}, Y_{filter}) stochastically based on respective value ranges (i.e. $[r_{min}, r_{max}]$ and $[Y_{min}, Y_{max}]$). Thus, all parameter combinations can be summarized in a new set $S_{rY} = \left\{ \left(r_{filter}^{(1)}, Y_{filter}^{(1)} \right), \ldots \left(r_{filter}^{(K)}, Y_{filter}^{(K)} \right) \right\}$. After that, the method SEMDOT is applied based on each parameter combination in the set S_{rY}, and the result is a set of geometry variants, denoted as $S_G = \{ G^{(1)}, G^{(2)}, \ldots, G^{(K)} \}$. By applying the TLA for each geometry variant in the set S_G, a new set, including tool-path lengths for all geometry variants, is obtained $S_{ltp} = \left\{ l_{tp}^{(1)}, l_{tp}^{(2)}, \ldots, l_{tp}^{(K)} \right\}$. Finally, all l_{tp} values from the set S_{ltp} are ranked in ascending order, and the player with the shortest l_{tp} is regarded as the winner in this round. For the next round, the total number of players K is updated by $K = K(1 - \eta_{eli})$, where η_{eli} is a parameter between 0 and 1 and indicates the percentage of players that should be eliminated from the game. This is intended to enable the convergence of the computational procedure. Moreover, it is to mention that only $K \cdot \eta_{eli}$ players with the lowest ranking positions (i.e., players who have longer l_{tp}) are eliminated from the game, whereas the surviving players will continue to the next round. As an example, Fig. 6 shows a case in which 10 players are in the game. Assuming the η_{eli} of 0.3, three players will be eliminated from the game (i.e. the players marked with red colors).

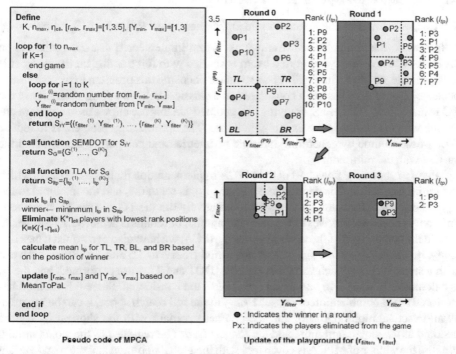

Fig. 6. Pseudo code and update logic for (r_{filter}, Y_{filter}) to explain MPCA.

Moreover, to enable convergence, the pick-up domain of (r_{filter}, Y_{filter}) is scaled down for each round following the strategy of "area selection by minimum mean l_{tp} (MeanToPaL)". The MeanToPaL strategy is illustrated in Fig. 6 and explained in the following. First, the value ranges for (r_{filter}, Y_{filter}) are initially defined to $r_{filter} \in [1, 3.5]$ and $Y_{filter} \in [1, 3]$. . Therefore, it can be regarded as a square playground, where each play should pick up a point indicating a combination of (r_{filter}, Y_{filter}). Assuming 10 initial players in the game and player P9 wins Round 0, the playground will be split in four regions: the regions at top-left (TL), top-right (TR), bottom-left (BL), and bottom-right (DR). One of these four regions will be selected for a new competition round. In the MeanToPaL strategy, the mean l_{tp} of the players present in each region is calculated, and the region with the lowest mean l_{tp} is selected as the region for a new competition round. In the example shown in Fig. 6, player P9 wins Round 0, and the region TR has the minimum man l_{tp}, and therefore, the other three regions are excluded in the next round (i.e. Round 1). In the next two rounds, the region keeps narrowing down, and finally, player P9 wins the game.

By iteratively using TO and TLA in multiple competition rounds, the winner of the last round is regarded as the final winner of the game.

4 Use Cases

4.1 Use Case 1: 2D Optimization Problem

4.1.1 Description of the Scenario and Implementation Procedure

In this work, the computational framework is first implemented in a 2D optimization problem, where a simply supported beam is studied with force acting on the lower side in the middle, as shown in Fig. 7. The reason for choosing this problem in the use case is because this problem has been considered as a classical benchmarking geometry in the TO research field. Therefore, studying this geometry in this use allows us to compare our results with the results in the existing literature. The optimization problem is to reduce 70% of the volume by generating multiple geometries and comparing them to find the variant with the minimum l_{tp}.

Figure 7 shows the flowchart of the MPCA implementation for this use case, in which the initial population K is defined to be 30, the η_{eli} is set to 0.2, and the maximal round n_{max} is set to 5. Afterward, a loop is set to iteratively run the TO and the TLA methods until only one player survives or the maximal number of rounds is achieved. Moreover, to enable the comparison of the method without MPCA, we have also performed a baseline study, in which r_{filter} and Y_{filter} values are varied from 1 to 3.5 and 1 to 3, respectively, with a step of 0.1. For each variation, the SEMDOT and TLA are performed, and the l_{tp} is calculated. In total, 546 variations (i.e. 21·26) are considered. Since this method has explored every combination of (r_{filter}, Y_{filter}) in the full search space, it can be regarded as an exact solution approach, and the geometry variant with the shortest l_{tp} can be denoted as the global optimum, and the longest l_{tp} can be regarded as the worst variant. Finally, the winner of MPCA is compared with the global optimum and the worst variant for the 2D problem.

4.1.2 Results of the Computational Framework

Figure 7 shows the results of the winner of the MPCA, the global optimum, and the worst variant. The corresponding l_{tp} values of them are listed in Table 2. In comparing the geometries of the global optimum with that of the winner of the MPCA, it is observed that they look very similar. Moreover, this is also supported by their l_{tp} values, as shown in Table 2. The l_{tp} of the global optimum and the winner of the MPCA are 3338.8 mm and 3340.2 mm, respectively, with a difference of 1.4 mm. Thus, although MPCA fails to capture the global optimum, the difference is negligible. In comparing the geometries of the winner of MPCA with the worst variant, it is observed that the worst variant contains five large internal contours, which enlarges the l_{tp} length. To verify the effectiveness of the energy consumption reduction by the MPCA, the geometries of the global optimum, winner of MPCA, and worst variant are printed with a material extrusion printer

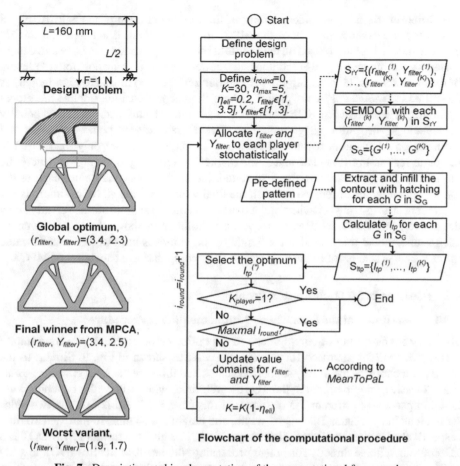

Fig. 7. Description and implementation of the computational framework.

(the printer Ultimaker 3). The energy consumption is measured using a power meter, and the results are also listed in Table 2. In comparing the results, a reduction rate of approximately 6% is observed for this case, which verifies the feasibility of MPCA for improving the energy performance during the DfAM in a 2D problem.

Table 2. List of the l_{tp} and energy required to produce the geometries.

Geometries	Global optimum	Winner of MPCA	Worst variant (ref.)
l_{tp} in mm	3338.8	3340.2	3405.4
E in kJ	1028	1033	1097

Moreover, for a better understanding of the functional logic of the MPCA, Fig. 8 shows the convergence process of the MPCA approach. First, it is noted that the 3D contour surface is plotted based on the results of the full exploration (i.e. l_{tp} of 546 design variants). The markers on the 3D contour surface describe the pick-up positions of players in different rounds. In assessing the 3D contour surface, two areas are highlighted, as described in "mountains" and "plains", where the deepest position implies the global optimum, as seen in in Fig. 8. In Round 0, it is seen that the markers are almost equally distributed in the entire playground. The markers can be seen both in montain and plain regions. For the next round (i.e. Round 1), the markers are only distributed in the plain region. For Rounds 2 and 3, the markers are moving to the region, where the global optimum is located. In the final round, although the players do not capture the global optimum, the difference between the final winner and the global optimum is not significant. The moving behaviors of the markers on the 3D contour surface clearly show the convergence of the MPCA. Moreover, the violin chart shows the l_{tp} of players in each round, where it is seen that the height of the violins is narrowed to smaller value ranges. This is also proof of the convergence and computational capability of MPCA.

4.2   Use Case 2: 3D Optimization Problem

4.2.1   Description of the Scenario and Implementation Procedure

To verify the capability of our approach in 3D optimization problems, we have implemented our MPCA approach for a beam use case, as shown in Fig. 9. Similar to the reasons for choosing the deeply supported beams that this case has been chosen because it is a classic design problem in the TO research field, which allows for benchmarking and comparison of different TO methods. In this case, the beam is fixed on two sides with a load in the center. The length, width, and height are 60 mm, 10 mm, and 10 mm, respectively. Unlike the 2D optimization problem, in which the result of SEMDOT is a 2D contour suitable directly for image processing, the result of 3D SEMDOT is a 3D mesh, which cannot be directly used for image processing. Thus, in this 3D optimization problem, the mesh is first sliced into 50 layers, and each layer is further inserted with the hatching patterns. To evaluate the l_{tp} of a 3D mesh, the sum of the l_{tp} for all 50 layers is calculated and compared. In addition to the slicing steps, the remaining steps are kept the same as the computational procedure in the 2D use case, as shown in Fig. 7. Moreover, the full exploration has also been carried out as the baseline study.

4.2.2   Results of the Computational Framework

Figure 9 shows examples of the hatching image of a layer for the three meshes. Table 3 summarizes the l_{tp} values of the components for the global optimum, final winner of MPCA, and worst variant. The mesh of the winner from MPCA is finally converted into an STL file, which can be directly used for AM.

In assessing the point and contour chart in Fig. 9, it is observed that the final winner of the MPCA is not the global optimum. Nevertheless, the difference between the global optimum and the winner of MPCA can be neglected. The (r_{filter}, Y_{filter}) values of the global optimum, winner of MPCA, and the worst variant are, (1.5, 1.1), (1.2, 1.7), and (3, 1.1), respectively. Moreover, it is also observed in the contour chart that the pick-up

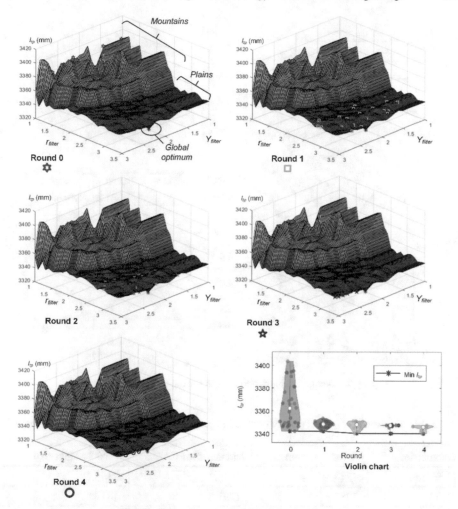

Fig. 8. Convergence process of the approach.

points of players are converged to the plain region, where the global optimum is allocated. Moreover, in assessing the violin chart depicting the l_{tp} of players, it is seen that the distribution of the l_{tp} picked up by players are successively narrowed into a smaller range. Based on these observations, it is concluded that our MPCA is converged.

Furthermore, to verify the energy saving of MPCA, the STL files are used to print the respective geometry using the same material extrusion printer as in the case of 2D problem. The energy consumption measured is summarized in Table 3. In comparing the energy consumptions, it is observed that the reduction rates for the global optimum and the winner of MPCA are 1.8% and 1.5%, respectively. This implies that the MPCA is suitable for the 3D problem.

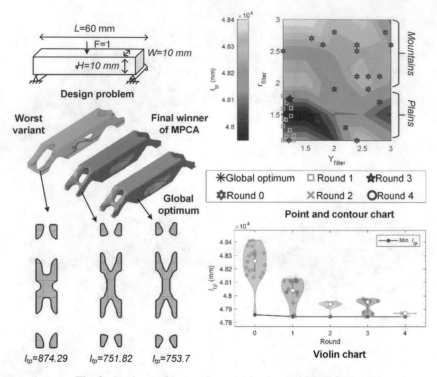

Fig. 9. Design problem and results for the case MBB-beam.

Table 3. List of the l_{tp} and energy required to produce the geometries in 3D problems.

Geometries	Global optimum	Winner of MPCA	Worst variant (ref.)
l_{tp} in mm	47844.1	47949.3	48446.9
E in kJ	1785.6	1791.8	1819.8

5　Discussion

Based on the use cases, two issues should be further discussed. First, in comparing the 2D design problem with the 3D design problem, it is observed that the reduction rate of 2D problem (6%) is higher than that of the 3D problem (1.5%). This implies the fact that the energy reduction performance of our MPCA is case-specific. Nevertheless, in terms of the effectiveness of the energy consumption reduction for the MPCA, it is still concluded that the MPCA enables the energy-saving in 3D design problems for AM.

Second, in terms of the result quality of MPCA, it is seen that in both use cases, the global optimum is not achieved. This can be reflected by the contour charts in Fig. 8 and Fig. 9 that none of any pick-up positions has been overlapped with the position of the global optimum. The reason is that the MPCA approach is not an exact algorithm that explores every possible pick-up position in the playground. The MPCA approach is an approximation method that searches the playground according to several pre-defined rules (e.g. MeanToPal for updating the search domain), which also means that missing the optimal solution is always possible. On the contrary, the full exploration approach, which is considered the baseline for both use cases, is a typical exact solution that ensures the capture of global optimum. However, ensuring the result quality comes at the cost of the computational time. For example, in the 2D problem, the computational time for the full exploration is approximately 37 h, while the time for the MPCA approach is only approximately 11 h using a laptop with an i7 CPU and 32 G RAM. This means that the MPCA approach has saved almost 70% computational time. Based on this observation, it is still concluded that the MPCA is suitable for the energy-saving of AM processes in the DfAM stage.

Third, the MPCA approach is proposed in this work by us and has never been compared with other existing optimization methods yet, such as the genetic algorithm and the particle swarm optimization. Therefore, although the effectiveness of the MPCA approach has been confirmed in this work, it still requires to be compared with conventional optimization methods in future work.

6   Conclusion and Outlook

In summary, this paper introduces the development and validation of a computational framework in which the energy performance of AM is evaluated and improved during the DfAM. Based on the use cases, three conclusions are drawn. First, it is confirmed that AM tool-path length can be used as an equivalent indicator during the DfAM to approximate the energy performance assessment since the DfAM lacks process parameters for precise energy consumption quantification. Second, in a 2D problem, it is confirmed that the proposed MPCA enables a reduction rate of approximately 6% for energy savings, while in a 3D problem, the reduction rate of energy consumption has been observed to be approximately 2%. Third, the MPCA approach is a suitable approximation approach to balance the result quality with the computational time in the proposed optimization problem.

In terms of future works, the following three topics are suggested. The first one is to keep improving the computational efficiency considering that the current computational time of the MPCA can be up to hours. The second future work is to compare the MPCA with other optimization techniques. Finally, future work should consider multi-objective optimization scenarios to include more objectives (e.g. mechanical performance and manufacturing cost), whereas this work only considers the energy performance improvement as a single-objective problem.

Acknowledgment. This work is funded by Deutsche Forschungsgemeinschaft (DFG, German Research Foundation) – 252408385 – IRTG 2057.

References

1. Gibson, I., Rosen, D., Stucker, B.: Additive Manufacturing Technologies. Springer, New York (2015). https://doi.org/10.1007/978-1-4939-2113-3
2. Thompson, M.K., et al.: Design for additive manufacturing: trends, opportunities, considerations, and constraints. CIRP Ann. **65**, 737–760 (2016). https://doi.org/10.1016/j.cirp.2016.05.004
3. Vaneker, T., Bernard, A., Moroni, G., Gibson, I., Zhang, Y.: Design for additive manufacturing: framework and methodology. CIRP Ann. **69**, 578–599 (2020). https://doi.org/10.1016/j.cirp.2020.05.006
4. Tang, Y., Zhao, Y.F.: A survey of the design methods for additive manufacturing to improve functional performance. Rapid Prototyping J. **22**, 569–590 (2016). https://doi.org/10.1108/RPJ-01-2015-0011
5. Dhokia, V., Essink, W.P., Flynn, J.M.: A generative multi-agent design methodology for additively manufactured parts inspired by termite nest building. CIRP Ann. **66**, 153–156 (2017). https://doi.org/10.1016/j.cirp.2017.04.039
6. Wang, Z., Zhang, Y., Bernard, A.: A constructive solid geometry-based generative design method for additive manufacturing. Addit. Manuf. **41**, 101952 (2021). https://doi.org/10.1016/j.addma.2021.101952
7. Wu, J., Aage, N., Westermann, R., Sigmund, O.: Infill optimization for additive manufacturing-approaching bone-like porous structures. IEEE Trans. Vis. Comput. Graph. **24**, 1127–1140 (2018). https://doi.org/10.1109/TVCG.2017.2655523
8. Tang, Y., Kurtz, A., Zhao, Y.F.: Bidirectional Evolutionary Structural Optimization (BESO) based design method for lattice structure to be fabricated by additive manufacturing. Comput. Aided Des. **69**, 91–101 (2015). https://doi.org/10.1016/j.cad.2015.06.001
9. Li, D., Liao, W., Dai, N., Xie, Y.M.: Anisotropic design and optimization of conformal gradient lattice structures. Comput. Aided Des. **119**, 102787 (2020). https://doi.org/10.1016/j.cad.2019.102787
10. Zhang, Y., Wang, Z., Zhang, Y., Gomes, S., Bernard, A.: Bio-inspired generative design for support structure generation and optimization in Additive Manufacturing (AM). CIRP Ann. **69**, 117–120 (2020). https://doi.org/10.1016/j.cirp.2020.04.091
11. Baumers, M., Duflou, J.R., Flanagan, W., Gutowski, T.G., Kellens, K., Lifset, R.: Charting the environmental dimensions of additive manufacturing and 3D printing. J. Ind. Ecol. **21** (2017). https://doi.org/10.1111/jiec.12668
12. Huang, R., et al.: Energy and emissions saving potential of additive manufacturing: the case of lightweight aircraft components. J. Clean. Prod. **135**, 1559–1570 (2016). https://doi.org/10.1016/j.jclepro.2015.04.109
13. Kellens, K., Mertens, R., Paraskevas, D., Dewulf, W., Duflou, J.R.: Environmental impact of additive manufacturing processes: does AM contribute to a more sustainable way of part manufacturing? Proc. CIRP **61**, 582–587 (2017). https://doi.org/10.1016/j.procir.2016.11.153
14. Faludi, J., Baumers, M., Maskery, I., Hague, R.: Environmental impacts of selective laser melting: do printer, powder, or power dominate? J. Ind. Ecol. **21**, S144–S156 (2017). https://doi.org/10.1111/jiec.12528
15. Peng, T., Kellens, K., Tang, R., Chen, C., Chen, G.: Sustainability of additive manufacturing: an overview on its energy demand and environmental impact. Addit. Manuf. **21**, 694–704 (2018). https://doi.org/10.1016/j.addma.2018.04.022
16. Yi, L., Ravani, B., Aurich, J.C.: Energy performance-oriented design candidate selection approach for additive manufacturing using tool-path length comparison method. Manuf. Lett. **33**, 5–10 (2022). https://doi.org/10.1016/j.mfglet.2022.06.001

17. Kaltschmitt, M., Streicher, W., Wiese, A.: Renewable Energy. Springer, Heidelberg (2007). https://doi.org/10.1007/3-540-70949-5
18. Faludi, J., Bayley, C., Bhogal, S., Iribarne, M.: Comparing environmental impacts of additive manufacturing vs traditional machining via life-cycle assessment. Rapid Prototyping J. **21**, 14–33 (2015). https://doi.org/10.1108/RPJ-07-2013-0067
19. Morrow, W.R., Qi, H., Kim, I., Mazumder, J., Skerlos, S.J.: Environmental aspects of laser-based and conventional tool and die manufacturing. J. Clean. Prod. **15**, 932–943 (2007). https://doi.org/10.1016/j.jclepro.2005.11.030
20. Ingarao, G., Priarone, P.C., Deng, Y., Paraskevas, D.: Environmental modelling of aluminium based components manufacturing routes: additive manufacturing versus machining versus forming. J. Clean. Prod. **176**, 261–275 (2018). https://doi.org/10.1016/j.jclepro.2017.12.115
21. Yang, S., Min, W., Ghibaudo, J., Zhao, Y.F.: Understanding the sustainability potential of part consolidation design supported by additive manufacturing. J. Clean. Prod. **232**, 722–738 (2019). https://doi.org/10.1016/j.jclepro.2019.05.380
22. Ehmsen, S., Yi, L., Aurich, J.C.: Process chain analysis of directed energy deposition: energy flows and their influencing factors. Proc. CIRP **98**, 607–612 (2021). https://doi.org/10.1016/j.procir.2021.01.162
23. Yi, L., et al.: An eco-design for additive manufacturing framework based on energy performance assessment. Addit. Manuf. **33**, 101120 (2020). https://doi.org/10.1016/j.addma.2020.101120
24. Giudice, F., Barbagallo, R., Fargione, G.: A Design for Additive Manufacturing approach based on process energy efficiency: electron beam melted components. J. Clean. Prod. **290**, 125185 (2021). https://doi.org/10.1016/j.jclepro.2020.125185
25. Yi, L., Glatt, M., Thomas Kuo, T.-Y., Ji, A., Ravani, B., Aurich, J.C.: A method for energy modeling and simulation implementation of machine tools of selective laser melting. J. Clean. Prod. **263**, 121282 (2020). https://doi.org/10.1016/j.jclepro.2020.121282
26. Yi, L., Ravani, B., Aurich, J.C.: Development and validation of an energy simulation for a desktop additive manufacturing system. Addit. Manuf. **32**, 101021 (2020). https://doi.org/10.1016/j.addma.2019.101021
27. Ma, F., Zhang, H., Hon, K., Gong, Q.: An optimization approach of selective laser sintering considering energy consumption and material cost. J. Clean. Prod. **199**, 529–537 (2018). https://doi.org/10.1016/j.jclepro.2018.07.185
28. Baumers, M., Tuck, C., Wildman, R., Ashcroft, I., Hague, R.: Energy inputs to additive manufacturing: does capacity utilization matter? University of Texas at Austin (2011)
29. Baumers, M., Tuck, C., Wildman, R., Ashcroft, I., Rosamond, E., Hague, R.: Transparency built-in. J. Ind. Ecol. **17**, 418–431 (2013). https://doi.org/10.1111/j.1530-9290.2012.00512.x
30. Lunetto, V., Galati, M., Settineri, L., Iuliano, L.: Unit process energy consumption analysis and models for Electron Beam Melting (EBM): effects of process and part designs. Addit. Manuf. **33**, 101115 (2020). https://doi.org/10.1016/j.addma.2020.101115
31. Jadhav, S.D., et al.: Influence of carbon nanoparticle addition (and impurities) on selective laser melting of pure copper. Materials **12**, 2469 (2019). https://doi.org/10.3390/ma12152469
32. Ye, J., et al.: Energy coupling mechanisms and scaling behavior associated with laser powder bed fusion additive manufacturing. Adv. Eng. Mater. **21**, 1900185 (2019). https://doi.org/10.1002/adem.201900185
33. Tang, Y., Mak, K., Zhao, Y.F.: A framework to reduce product environmental impact through design optimization for additive manufacturing. J. Clean. Prod **137**, 1560–1572 (2016). https://doi.org/10.1016/j.jclepro.2016.06.037
34. Priarone, P.C., Ingarao, G., Lunetto, V., Di Lorenzo, R., Settineri, L.: The role of re-design for additive manufacturing on the process environmental performance. Proc. CIRP **69**, 124–129 (2018). https://doi.org/10.1016/j.procir.2017.11.047

35. Priarone, P.C., Lunetto, V., Atzeni, E., Salmi, A.: Laser powder bed fusion (L-PBF) additive manufacturing: on the correlation between design choices and process sustainability. Proc. CIRP **78**, 85–90 (2018). https://doi.org/10.1016/j.procir.2018.09.058

36. Fu, Y.-F., Rolfe, B., Chiu, L.N.S., Wang, Y., Huang, X., Ghabraie, K.: SEMDOT: smooth-edged material distribution for optimizing topology algorithm. Adv. Eng. Softw. **150**, 102921 (2020). https://doi.org/10.1016/j.advengsoft.2020.102921

37. Yi, L., Wu, X., Nawaz, A., Glatt, M., Aurich, J.C.: Improving energy performance in the product design for additive manufacturing using a multi-player competition algorithm. J. Clean. Prod. **391**, 136173 (2023). https://doi.org/10.1016/j.jclepro.2023.136173

38. ISO: ISO 50006:2014(en) Energy management systems—Measuring energy performance using energy baselines (EnB) and energy performance indicators (EnPI)—General principles and guidance (2014). https://www.iso.org/obp/ui/#iso:std:iso:50006:ed-1:v1:en

Investigation of Micro Grinding via Kinematic Simulations

N. Altherr[✉], B. Kirsch, and J. C. Aurich

Institute for Manufacturing Technology and Production Systems, RPTU in Kaiserslautern,
Kaiserslautern, Germany
nicolas.altherr@rptu.de

Abstract. With increasing demand for micro structured surfaces in hard and brittle materials, the importance of micro grinding increases. The application of micro pencil grinding tools (MPGTs) in combination with ultra precision multi axes machine tools allow an increased freedom of shaping. However, small dimensions of the grinding tools below 500 μm substantiate high rotational speeds and low feed rates to enable the machining process. Besides, the abrasive grits of the tool can be large in comparison to the tool dimensions. All factors will influence the resulting surface topography of the workpiece. But some of the topography properties are no longer accessible for optical measurements, making process evaluations and improvements difficult.

In the present contribution the measurement results are supplemented by the results of a kinematic simulation model. The built up of such a kinematic simulation is described, which considers real process and tool properties. The results received by the simulation are compared to measurements to validate the model and point out the advantages of the simulations. In a further step, a principle is shown how the simulation can be used to make the undeformed chip thickness accessible, a process result which cannot be measured within the real machining process.

1 Introduction

Micro grinding is an abrasive process which is suitable to machine brittle materials such as hardened steel, silicon [1], or glass [2]. Micro structuring such materials can be used to create special surface properties and structures which are necessary for the application in optical or electronic industry. Tools for this process are MPGTs which usually have a diameter below 500 μm [3]. However, such tools have many statistical characteristics concerning the grit geometry and the location of the grits on the tool surface. For this reason, the distribution of the abrasive grits of each tool is unique. In peripheral grinding this property is also transferred to the resulting workpiece topography leading to unique scratch-structures on the surface [4]. Especially when using larger sizes for the abrasive grits, which are beneficial concerning tool life and cutting characteristics [5], the scratches significantly influence the resulting surface properties. For the investigation of the micro grinding process, this major influence of the scratches leads to a disadvantage: The topography is no longer mainly influenced by the process parameters. Instead, the

J. C. Aurich et al. (Eds.): IRTG 2023, *Proceedings of the 3rd Conference on Physical Modeling for Virtual Manufacturing Systems and Processes*, pp. 233–259, 2023.
https://doi.org/10.1007/978-3-031-35779-4_13

tools and their grit distribution become a prominent factor influencing the final surface. However, the real tools cannot be used for multiple experiments since they are affected by wear. This hinders the comparison between two surfaces, received by different process parameters and different tools: The influence of the process parameter cannot be separated from the influence of the tool statistics.

At this point, using kinematic simulations can simplify the investigation of the micro grinding process. In its basic idea, a kinematic simulation model does not consider material behavior, such as material deformations or machine stiffness [6]. Instead, the focus of the simulation is on the influence of kinematic process parameters such as feed rate or the tilt angle of the spindle on the resulting workpiece shape and surface topography.

In general, a kinematic simulation model consists of a tool model and a workpiece model [7]. Such simulations are time discrete, which means that the state of the model is calculated after discrete time steps [8]. Hence, the relative motion between both models is determined for discrete time steps evaluating kinematic equations [9]. For each time step, the intersection between the virtual workpiece and the tool model is calculated [9]. The intersecting volumes are removed from the workpiece model to update the workpiece shape and to represent the virtual machining process.

For kinematic simulation, the workpiece model is spatially discretized. One suitable method is the dexel model, established by van Hook [10]. In the model, usually two dimensions of the workpiece are divided into equidistant calculation points. In the perpendicular third dimension, the surface height is stored on each calculation point as a numerical value. Several authors already used comparable approaches to visualize surfaces received by grinding [6, 11, 12].

The tool model is assumed to be a rigid body which means that it does not change during the simulation. Hence, tool wear is usually not considered [13]. In grinding, different approaches exist to model the tool with focus on the abrasive grits. The challenge is that each grinding tool is unique regarding the arrangement and shape of the abrasive grits. One approach which is based on digitization of real abrasive grits was developed by Klocke et al. [14]. They used a micro computer tomograph to receive three-dimensional volume models of the abrasive grits [14]. However, such digitization techniques go along with high measurement and data processing effort [14].

As an alternative, statistical properties of the grits on the grinding tool are used as basis for modeling [15]. In this case, simplified grit geometries, based on standard geometric bodies are used to approximate the real grit shape. In literature, spheres [16], cuboids, octahedrons [17], tetrahedrons [12], cones [18], ellipsoids, or general polyhedrons [19] were used as basic geometries for the virtual grit representations. To improve the compliance of the basic geometries with the real grits, further adaptations were applied: The intersection of hexahedrons with tetrahedrons in different size ratios leads to grit shapes fitting to the ideal crystal morphology of cubic boron nitride (cBN) grits [14, 17]. Further adaptions of the resulting virtual grit representations were done by Warnecke et al. by fitting planes with different orientations, randomly cutting areas of the grit volume [17]. Compared to this, Chen et al. used tetrahedrons to model the grits, of which the corner points were randomly cut off, resulting in further triangular facet on virtual grit representation [12]. In general, the choice of the virtual grit representation depends not only on the material of the abrasive grits, but also on the manufacturing process, since the grit shapes of, for instance, cBN depends on the temperature and pressure during the synthetic production [20]. Due to the different shapes, the size of the abrasive grits is not the same for all grits within a sample and they differ from the nominal value. Koshy et al. took the sieving process of the grits as basis: The lower and upper mesh size of the sieves were used to calculate the mean value and standard deviation [21], assuming a normal distribution of the grit sizes [22]. Another approach is based on measurements of the abrasive grits. Warnecke et al. used scanning electron microscopy (SEM) images of the grits to measure the length of the short and long axis of the grits [17]. This approach was also adopted by Chen et al. who measured the edge length of the grits via SEM images [12]. For analyzing powders in general, laser diffraction is a suitable method using the light scattering properties of the powders and solving the Maxwell equations [23]. This method was already used for the characterization of abrasive grits [24]. Two theories to evaluate the light diffraction and calculate the particle size exist: The Fraunhofer approximation and the Mie theory [23, 25]. For the Fraunhofer approximation, no values for the optical properties of the particles are necessary, however it is only suitable for larger diameters and does not consider light transmission through the particles [23]. For the Mie theory, the theoretical fundamentals were set by Gustav Mie [25]. This theory does not only use light diffraction, but also transmission through the particles [26] leading to more reliable results for particles below 10 μm [27]. However, the Mie theory requires the value of the complex refraction index of the analyzed powder [23]. Concerning the results, the particle size distribution of real powders is often elongated to larger dimensions [28]. This effect is not well described by the normal distribution, but by the lognormal distribution [28].

For placing the virtual grit representations on the virtual tool surface, the single grits are randomly rotated around their axis first [29]. The protrusion height of the grits on the tool surface is modeled using a normal distribution with a mean value equal to the mean value of the grit size [30]. The position of each virtual grit representation on the virtual tool surface can be set with a uniform distribution, until a given grit density on the tool is reached [6]. Another method was introduced by Chen et al. [31]. In their model, all grit representations were initially located on an equidistant mesh on the virtual tool surface to predefine the number of grits within the model [31]. Then, a slight deviation of the position of each grit was added to randomize the grit positions [31]. Liu et al. denoted the method as "shaking" of the virtual grit representations [18].

However, with this principle, overlapping of the grits with each other is possible. To avoid this, the placing on the tool surface was restricted: When spheres are used as virtual grit representations, Koshy et al. formulated a mathematical condition to assure that the center points of two grits have at least a distance equal to the sum of both grit radii [21]. If more complex grit shapes are used, the condition can also be applied by calculating the minimum enclosing sphere of each virtual grit representation [12].

In previous works we already performed kinematic simulations for grinding processes. Adjustments to the principle were made to determine forces and chip thickness in the conventional grinding process [32]. With the help of kinematic simulations, we investigated the undeformed chip thickness of single grit scratch tests in combination with the pile-up effect [33]. We also performed investigations concerning micro grinding with MPGTs. The generation of the bottom surface of ground micro channels were investigated via kinematic simulations [13]. We found that the radial position of the abrasive grit on the tool face causes a step shaped workpiece topography [13].

The investigation of peripheral grinding using kinematic simulations for electroless plated MPGTs was not yet performed. To fill this research gap, in the present contribution the kinematic simulation is applied to micro grinding and the built up of the simulation model is described. One advantage of the kinematic simulation is, that the same tool model can be used for an arbitrary number of virtual experiments. Furthermore, errors appearing due to the digitization of the real ground surface topographies can be avoided within simulations. The simulations were based on real grinding experiments with MPGTs and implemented in Matlab[1]. The work originated within the IRTG 2057 on Physical Modeling for Virtual Manufacturing Systems and Processes.

2 Properties of the MPGTs

The MPGTs considered within this contribution were self-manufactured. The tool blanks were made of high speed steel and a cylindrical shape with a nominal height of 400 μm and a diameter of 360 μm was manufactured on the tool tip. This cylindrical shape was the base for the abrasive body of the MPGT. The abrasive body was produced via an electroless plating process, a solely chemical process we developed previously [34]. The abrasive grits were made of cBN with a nominal grit size between 20 μm and 30 μm. For the plating process, the abrasive grits were added to the plating solution in the form of loose powder. During the plating process a nickel-phosphorous compound was deposited, which embedded the abrasive grits in one layer on the cylindrical shape of the tool blank [35]. The final nominal diameter of the tool was 400 μm, which is composed

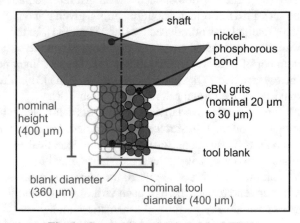

Fig. 1. Geometric properties of the MPGTs

---

[1] Naming of specific manufactures is done solely for the sake of completeness and does not necessarily imply an endorsement of the named companies nor that the products are necessarily the best for the purpose.

of the blank diameter (360 μm) and twice the smallest nominal grit size (20 μm). The properties of the tools researched within this contribution are depicted in Fig. 1.

3   Model of the MPGT for Kinematic Simulations

For modeling the tool, geometric properties of the real MPGTs are necessary. A digitization of the complete micro grinding tool was not applicable due to the size and curvature of the tools. As an alternative, statistical methods were applied to describe the properties of the tools and to get the geometric information. For this purpose, values for the statistics of grit size and grit shape distribution were used for modeling. Both statistics are received by analysis of the real cBN grits and the real MPGTs. In the following sections it will be shown how the results of the analyses can be used to generate a virtual representation of the MPGT.

3.1   Analysis of the Grit Size Distribution

The cBN particles used as abrasive grits for the tools were purchased in the form of powder with a nominal size between 20 μm and 30 μm. However, solely the range is not suitable to be used in the simulations, since within this range a uniform distribution of the grit sizes would be assumed. Instead, the distribution of the grit sizes was needed to receive realistic sizes of the virtual grit representation.

A method was necessary which delivered repeatable results, without subjective evaluation. This can efficiently be done via particle analysis. The 3P instruments Bettersizer S3 plus (See Footnote 1) was applied which uses the principle of laser diffraction. A large amount of grits can be analyzed within a short period of time and in a repeatable way. The original cBN powder was used to measure the grit size distribution. For a measurement of the grits, which were actually embedded on the MPGTs, hundreds of tools would have had to be chemically dissolved to receive a suitable quantity of grits.

Within the particle analysis, water was used as dispersant medium and the dispersant stirrer had a rotational speed of 2500 rpm. According to the standard measurement procedure, cBN grits were added to the dispersant medium until an obscuration of 12% within the measurement system was displayed. Before the grit size analysis was started, ultrasound was activated for one minute to remove bubbles within the dispersant medium. During the measurement a total amount of 10,000 grits were analyzed. The evaluation of the measurement was done using the Mie theory since grit sizes partly below 10 μm appeared in the cBN powder sample and the transmission of the light through the particles is possible. For using this evaluation method, the complex refraction index was necessary, a parameter which was not exactly known for cBN. Hence, the option of the measurement system to calculate its value was used, leading to a value of 2.12-0.001i. Using the calculated value of the complex refraction index, a measurement residual of 6.945% for the particle size analysis was reached. The histogram of the grit sizes is given in Fig. 2. It can be seen that the distribution of the grit sizes has no symmetrical shape, but the left flank is steeper than the right flank. Besides, ten percent of the grits have a size below the d_{10} value of 12.57 μm and 90 percent of the grits are below the d_{90} value of 27.77 μm. However, grit sizes below 12.57 μm and above 27.77 μm also occurred in the abrasive grit sample: The smallest grit size identified by the measuring system was about 3.55 μm and the largest grit size about 51.82 μm.

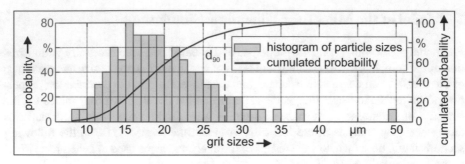

Fig. 2. Histogram of the grit size distribution received by the particle size analysis.

The finite number of analyzed grits cause that no continuous distribution function of the grit sizes can be determined by the particle analysis. As a consequence, values between the discrete measured values are also possible values for the grit size with a certain probability. For this reason, a distribution function was fitted into the discrete data using the Matlab (See Footnote 1) function "fitdist". Due to the asymmetric shape of the grit size distribution, the normal distribution did not fit well. A better correlation was received using the log-normal distribution function. Fitting a log-normal distribution function into the measured data of the particle size analysis led to a distribution which is described by a mean value of 2.9375 and a standard deviation of 0.3133. Both values were used as input parameters for the simulation.

3.2 Analysis of the Grit Shape

In case of the grit shape, it cannot be assumed that all grit shapes included in the original powder of cBN are also found on the tool. Grits with an unfavorable geometry which does not allow a form locked join within the bond are not embedded. For instance, this can be grits which only have a small volume fraction inside the bond leading to a weak connection between grit and bond. Hence, the shape of the grit had to be evaluated by analyzing the abrasive body of the manufactured tools as it is shown in Fig. 3. Since the protrusion height of the grits, which is the distance between the outermost point of the grit and the bond material, is not evaluated within this contribution, three dimensional topography measurements to receive the height information were not necessary. Hence, SEM images of the peripheral surface of real MPGTs were analyzed which have a higher lateral resolution in comparison to optical topography measurements. The procedure for the evaluation of the SEM image is depicted in Fig. 3 and was applied to ten different MPGTs for statistical validation. In detail, an area of 200 μm times 200 μm in the middle of the depicted tool surface was considered and a uniform number of 20 grits for each tool were analyzed. The number of 20 grits was available within the evaluation area for all analyzed tools. Furthermore, using a constant value of measured grits instead of all available grits avoided overrating of individual MPGTs with larger grit density.

Each grit shape was manually allocated to the best fitting basic grit shape. For the allocation, two classes were defined for grits with the following properties:

- Elongated grits with large aspect ratios or predominantly sharp corner points and acute angles.
- Bulky grits with aspect ratios similar to cuboids or predominantly obtuse angles.

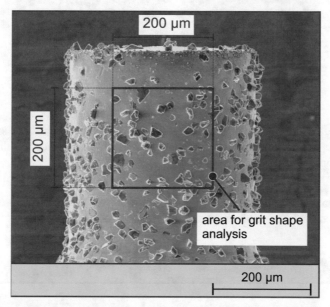

Fig. 3. Area for evaluation of the grit shapes on a SEM image of the MPGT.

Subsequently to the allocation, the ratios of elongated and bulky shape were calculated in relation to the total number of evaluated grits in order to estimate the probability of the grit classes. The result of the evaluation is depicted in Fig. 4. It can be seen that both grit classes appear approximately with the same probability. The probability values are used as input parameters for the simulation, what will be described in later sections.

3.3 Requirements and Assumptions for the Tool Model

For the application within a kinematic simulation of the peripheral micro grinding process, the following requirements and assumptions for the tool model were defined:

1) Due to the comparatively large abrasive grits for micro grinding in combination with the associated low grit density, each single grit has a high impact on the final surface topography. Hence, the grit shape is represented in the resulting surface topography. Therefore, detailed virtual grit representations are necessary for the simulation.
2) The exact number of grits on the real tool is not known. Hence, a predefined number of grits on the tool model is not suitable. Instead, the goal is to implement a virtual saturation process of virtual grit representations on the virtual tool surface.

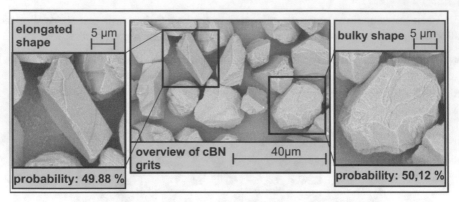

Fig. 4. Definition of the two grit shape classes

3) Input data for tool modeling are supposed to be determined with less effort in a repeatable way and the procedure should be transferable to other tool diameters and abrasive grit sizes.
4) As for the real tool, overlapping grits should be avoided within the tool model.

The superordinated procedure which is used to model the tool is depicted in Fig. 5. In the next sections, the single steps are described in detail.

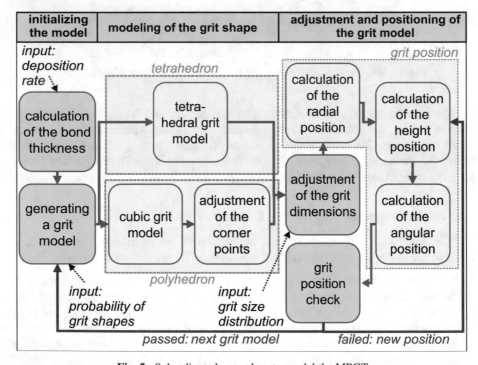

Fig. 5. Subordinated procedure to model the MPGT.

3.4 Modeling of the Virtual Bond of the Tool Model

The bond is the material in which the grits are embedded with a certain volume fraction. The embedded volume fraction depends on the thickness of the bond material.

For the tool model, the bond is considered to be an ideal cylinder. Within this cylinder, the bond is limited by the blank material, which is also considered to be an ideal cylinder with a smaller radius. The radius of the blank is set by the manufacturing process to 180 μm. The difference between the blank radius and the radius of the bond surface is the thickness of the bond as it is depicted in Fig. 6. In the real manufacturing process of the MPGTs, this value depends on the plating time. In previous work, we found a deposition rate of about 21 μm/h [35]. For the tools, a total plating time of 35 min was applied leading to a bond thickness of 12.25 μm. Adding the value to the radius of the tool blank provides the radius of the bond. Both, the radius of the blank and the radius of the bond are used for the tool model.

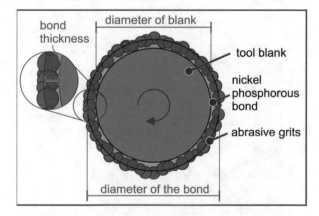

Fig. 6. Definition of the bond thickness for MPGTs.

3.5 Validation of the Bond Thickness

Measurements on real tools were used to check whether the analytical value for the bond thickness is suitable for modeling. For the analysis, the faces of the tools were considered, and the diameter of the bond was measured using SEM images of the tools. Before the SEM images were performed, the tools were cut to make the bond visible and improve the identification of the bond diameter. For cutting, the MPGT was rotated, and a dicing blade was used. The feed rate of the dicing blade was set to a value of 1 mm/min for a gentle cutting process. Dough mixed with cBN powder was used to remove the emerging burr.

The electroless plating process does not produce ideally smooth surfaces and minor variations in bond thickness may occur. Therefore, the determined value for the diameter also varies depending on the measuring position. For this reason, the measurement of the bond diameter was statistically validated. Within the measurement procedure, rectangles were set around the tool with each side touching the bond of the tool image. The rectangle led to two perpendicular measurements for the diameter of the bond, using both dimensions of the fitted rectangle. For the statistical validation, the procedure was performed three times using rectangles with an angle of $0°$, $+120°$ and $-120°$ as it is depicted in Fig. 7. Thus, for each tool six values for the bond diameter were available. The measurement principle was applied to the SEM images of six different MPGTs. To get the value for the bond thickness, the measured values are halved to receive the values for the bond radius. Subtracting the radius of the tool blank leads to the value of the bond thickness. The mean value of the measurements was 13.03 μm with a standard deviation of 2.91 μm.

Even though large standard deviations appear, which can be explained by material deformations due to the cutting process of the tool faces, the mean value fits well to the value received by using the deposition rate to calculate the bond thickness.

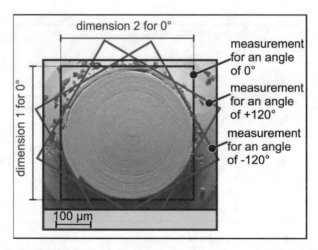

Fig. 7. Evaluation of the bond diameter using rectangles with different tilt angles.

3.6 Modeling of the Abrasive Grits

For the virtual grit representations, the previously defined two classes of grits are approximated with basic geometries. In order to meet the requirement of realistic virtual grit representations, those basic geometries were considered which are constructed of plane facets and then fit to the real crystal shape of the grits. Thus, spheres and ellipsoids are excluded.

The implementation of the virtual grit representations into the tool model was performed considering the limited computational power: To keep the memory requirement low, only the corner coordinates of each virtual grit representation is stored. Besides the points, a list is created, in which the corner points are allocated to the grit facets.

At the beginning of the generation of a virtual grit representation, the corner points are set so that the size of the virtual grit in each dimension is equal to one. Thus, multiplying the size of the grit by the coordinates of the corner points, the virtual grit representation is stretched to its final size.

For the elongated grit class, tetrahedrons were selected as virtual grit representation as it is shown in Fig. 8. Tetrahedrons consist of spatially arranged triangular faces, which create sharp corner points as it was defined for the elongated grit class. Besides, tetrahedrons appear in the ideal crystal morphology of cBN [36].

The bulky shape is represented using a polyhedron (see Fig. 8). Starting point for modeling the polyhedron is a cuboid. Since the real grits allocated to the bulky class did not have sharp corner points, the corner points of the cuboid are cut off at a randomized position. It is done by spanning another triangular facet which substitutes the corner point of the cuboid.

To span the new facet, the corner point of the original cuboid is used. The corner point of the original cuboid is stored three times to receive three individual points at the same position. Each of the three corner points is displaced in the direction of one edge of the cuboid. The displacement is done with randomized values, leading to arbitrary triangular facets. The procedure is depicted in Fig. 8.

The decision which virtual grit representation geometry is chosen is done by the probability evaluated for the grit shape distribution. Subsequently, the virtual grit representation size was adjusted. For this purpose, random sizes are calculated according to a log-normal distribution. The distribution function is defined using the mean value and standard deviation received by the particle size analysis. In a last step, the grit is randomly rotated around its three rotational axes.

3.7 Positioning of the Virtual Grit Representations on the Virtual Tool

In the next step of the tool model generation algorithm, the virtual grit representations are placed on the virtual tool surface. Since peripheral grinding is considered, the virtual grit representations are only located on the peripheral surface of the virtual tool.

Using cylinder coordinates, the radial position of the grit determines the protrusion height. The position is determined by a normal distribution. However, the grits are supposed to be inside the virtual bond. Thus, the values for the radial position are checked: It was avoided that the grit is located inside the blank material and outside of the bond material.

For the first virtual grit representation, all height and angular positions on the tool peripheral surface are equally probable. However, for all subsequent grits the positions which are already occupied by virtual grits are no longer available. In order to represent this effect in the simulation, a virtual saturation process of the tool model surface with grits was replicated: The randomly calculated position of the virtual grit representation is compared with the position of the grits already placed on the tool model. However, due to the randomized shape of the virtual grit representations, computationally extensive calculation would have been necessary to determine the intersection between two grits. To circumvent such calculations, the comparison between the grit positions is abstracted: Since only the height and angular positions of the grits are relevant for a possible interaction with other grits, the checking of the grit positions was reduced to two dimensions (angular and height direction). Furthermore, the grit shape is reduced to a quadrangle enclosing the virtual grit representation. Using this procedure, it is possible to calculate

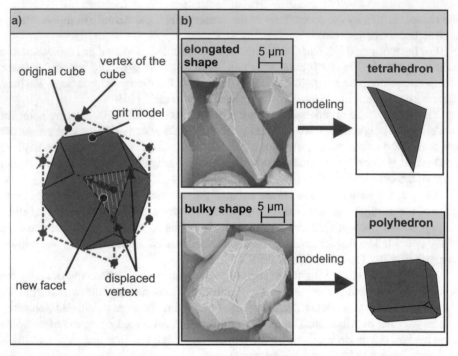

Fig. 8. Modeling of the polyhedron (a) and allocation of the grit models to the grit classes (b)

the intersections between the grits using two-dimensional geometry: A valid position is only present if the smallest or largest angle respectively height position of the new grit is not inside the squares of an already existing grit.

Two special cases are taken into account which could lead to overlapping grits event though the previously mentioned condition was met:

- The new grit model completely encloses the quadrangle of an already existing virtual grit representation.
- The angular position of the new grit is close or equal to 360°. Here, the maximum angular position of the grit could reach values above 360°, leading to overlapping with existing grits with minimum angular positions close to 0°.

If an overlapping of the virtual grit representations is detected, the calculated position of the grit is discarded. Consequently, the algorithm skips back to the calculation of the grit position and a new position is generated. Especially with increasing amount of virtual grit representations which are already placed on the virtual tool, the procedure can end in an infinite program loop. To avoid this and to limit the total computing time, the maximum number of loops was limited to 100 repetitions. When no valid position for the virtual grit representation was found within this period, the virtual grit representation is discarded and a new one is generated. With increasing computing time, the grit density on the virtual tool is enlarged. At the same time, the probability of a new grit to be located on the tool model decreases which is comparable to the saturation process of the real tool.

For the tool models used within the present contribution, a number of 1,000 virtual grit representations were predefined. After generating the tool model, between 580 and 680 virtual grit representations were actually integrated into the final tool model depending on the grit sizes. However, the amount of virtual grit representations on the virtual tool is not a static value but it can change due to the statistical characteristics of the grinding tool. The process fits to the real electroless plating process: Only if enough space is left on the tool surface, a further grit can be embedded. Hence, also for the real process a saturation point for the amount of grits on the tool is reached. A comparison between the tool model and the real MPGT will be given in the next section.

3.8 Evaluation of the Grit Size on the Real Tool

During the plating process, the chemical solution is set into rotation to swirl up the cBN grits. However, the rotation of the solution, including the cBN causes centrifugal forces acting on the abrasive grits. Due to the centrifugal forces, especially larger and therefore heavier cBN grits tend to move away from the middle of the plating solution. Such grits cannot reach the tool surface and are therefore not embedded. Thus, the coating process operates as grit size filter, what means, that not all grit sizes of the original cBN powder are actually present on the final tool.

For the evaluation of this effect, the values of the particle size analysis were compared to values of grit measurements of the real tool. Optical measurements using a confocal microscope were not suitable due to the same reasons mentioned in the section on the grit shape analysis. Thus, the analyses were done using SEM images of tool peripheral surface. To avoid the influence of the curvature of the tool, only the grits which were located inside a 200 μm × 200 μm square in the middle of the tool image were considered as it was already done for the grit shape analysis. The measurement of each grit was done two times: One measurement in the longest and one in the shortest expansion of the grit. The dimension was measured using two parallel lines enclosing the grit. It was applied to the SEM images of ten different tools and for each tool, the size of 20 grits were analyzed. A log-normal distribution was fitted into the resulting grit sizes, as it was also done for the results of the particle size analysis. The corresponding mean value and the standard deviation are given in Table 1 in comparison to the values received by the particle size analysis.

Table 1. Mean value and the standard deviation of the fitted distribution functions for the grit sizes received via particle size analysis in comparison to the statistical values received by grit size measurements in SEM images.

Value	Particle size analysis	Measurement via SEM
Mean value	2.9375	2.8211
Standard deviation	0.3133	0.2928

The values show that the measurements via the SEM images have a lower mean value and a lower standard deviation. This supports the previously mentioned filtering effect of the coating process of the MPGTs. To make the difference between both statistics visible, both are used to generate a tool model. The comparison is depicted in Fig. 9. On the left, the tool model using the complete data of the particle size analysis is shown (tool model 1 (TM-1)). When comparing it to the image of the real tool in the middle, it is again recognizable that the grit sizes are too large. On the right side, a tool model calculated with the grit sizes (comparison tool model (TM-C)) measured via the SEM images is shown, which fits much better to the real tool. For further analysis, both tool models are compared. With the virtual tools, parameters are available which are not directly accessible by measurements of the real tool. The first one is the total amount of virtual grit representations. The tool modeled via the data received by the SEM image (TM-C) has 680 grits on its peripheral surface, which is 100 grits more than for the other tool model. This fact also corresponds to the smaller grit sizes of the tool model on the right in Fig. 9: Since the virtual grit representations are smaller, more models can be placed on the surface until the virtual saturation is reached. Furthermore, the protrusion heights of the virtual grit representations were calculated virtual grit representation. For the tool modelled via the particle size analysis (TM-1), the mean value of the protrusion heights is 10.86 μm with a standard deviation of 6.89 μm. For the tool models created via the SEM measured values (TM-C), the mean value of 7.80 μm is significantly smaller which is also true for the standard deviation of 5.72 μm.

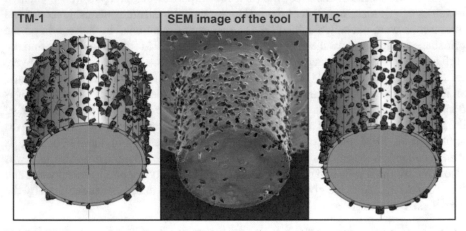

Fig. 9. Comparison of the tool model 1 (TM-1) calculated via the data of the particle size analysis, the tool model received by the data of the grit size measurement (TM-C) and the SEM image of the real tool.

3.9 Adaption of the Grit Sizes for the Tool Model

In the previous section it was shown that the grit sizes of the real tool differ from the grit sizes measured with the particle analysis. To feasibly use the data from the particle size analysis within the modelling procedure for the MPGT, the statistical distribution received via the particle size analysis was adjusted in the following way: As larger grit sizes do not appear on the final MPGT, values of larger grit sizes were excluded from the distribution function. For this purpose, the threshold value was set to the d_{90} value of the particle size distribution. It is a standard parameter [23], which is used to characterize the particle size distributions. Its value is directly exported by the particle measurement system. The implementation in the algorithm for the tool model is done by a testing loop of the grit size: According to the determined distribution function of the particle size analysis, a value for the size of the virtual grit representation is calculated. If the value is above d_{90} it is discarded, and a new value is generated. The resulting tool model, which will be called TM-2, is depicted in Fig. 10 in comparison to the tool model using the original particle size distribution (TM-1), and to the model generated with the statistical values received via the measurement of the SEM images (TM-C). Comparing TM-2 to TM-1, the amount of grits are increased as expected due to the decreasing average grit sizes. The amount of modelled grits is also closer to the number of grits in TM-C. Furthermore, the mean value and standard deviation of the grit protrusion heights of TM-2 are much closer to the corresponding values of TM-C. An exact match does not have to be achieved, since the parameters mentioned are subject to statistical fluctuations even in real tools. In a last step, the grits were given an identification number to allocate the corner points to the corresponding virtual grit representation.

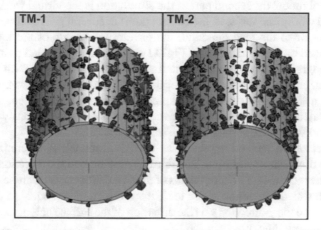

Fig. 10. Comparison between tool model 1 (TM-1) and tool model 2 (TM-2)

3.10 Conclusion on Tool Modeling

A principle was presented to model MPGTs using data of particle size analysis of the abrasive grits and the deposition rate of the tool bond during manufacturing.

For the tool model, polyhedrons with adapted corner points and tetrahedrons were used as basic virtual grit representations. The virtual grit representations were placed randomly on the tool surface and overlapping of the grits was avoided. The amount of virtual grit representations on the virtual tool surface was not predefined, but a virtual saturation process was used which better emulates the real manufacturing process of the MPGTs. To adjust the size of the virtual grit representations, particle size analyses were used, rapidly delivering statistically validated and repeatable results without subjective assessments. However, when comparing the values of the particle size analysis with the values received by measurements of the embedded grits in basis of SEM images of the MPGTs, it was found that the real grits on the tools are smaller. In the modelling procedure of the tool, this was considered by excluding virtual grit representation size values larger than the d_{90} value of the particle size analysis. This resulted in virtual grit representation sizes which fitted to the real grits on the tool.

All in all, the modeling procedure of the MPGT led to a realistic virtual representation of the tool which is suitable for kinematic simulations.

4   Setup of the Simulation

For the application of the tool model within a simulation, a simulation model is necessary. The steps for the setup of this model are described in the following sections.

4.1   Workpiece Representation Within the Simulation

Besides the tool model the second part of the simulation is the virtual workpiece. In the case of ground workpiece surfaces, the contact paths of several grits overlap. This causes that perpendicular to the feed direction, the enveloping profile of the tool peripheral surface is imaged on the workpiece. In Fig. 11 this profile is depicted as a blue line. In addition, the engagement of a grit is not continuous, but periodic with each tool rotation. For this reason, an analytical description of the surface would lead to complex equation systems. As an alternative, discrete workpiece models can be applied. Such models facilitate the calculation of the tool-workpiece interaction by continuously updating the resulting virtual workpiece surface.

The discretization method applied was the dexel model. A discretization of the surface in two dimensions was used. The third dimension was the direction of the dexel which has higher accuracy. The coordinates of the simulation model were set in a way that the x-direction was the direction of the feed, the z-direction was equal to the rotational axis of the tool and the y-direction was the direction of the surface heights, corresponding to the peripheral grinding processes [37].

Since optical surface measurements were used to digitize the real ground surfaces, the dexel model was adapted to the measurement to ensure comparability: In the lateral directions of the measurement system, the resolution was 0.7 μm. This value was also used for the resolution of the dexel model in discretized x- and z-directions. The vertical resolution is much higher and was used to identify the surface heights. The principle for discretization is depicted in Fig. 11 [37]. In Matlab (See Footnote 1), the dexel grid was implemented as a matrix. Each matrix entry depicts the height of one dexel. Hence, the dexels can also be interpreted as discrete calculation points for the surface height of the virtual workpiece.

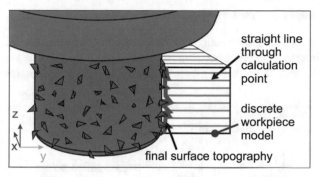

Fig. 11. Discretization of the workpiece model [37]

4.2 Kinematics and Time Discretization

For the calculation, the relative movements between the tool model and the workpiece model are determined. The feed motion and the tool rotation are considered. Effects due to process-machine-interactions, such as material deformations, tool deflections, or limited machine stiffness were not taken into account.

The simulation is time discrete. Hence, the tool movement and its intersection with the workpiece is calculated after equal time steps. In connection with the rotational speed of the tool, the value of the time steps determines which angle of the tool rotation is included within each calculation step. Due to the small diameters, high rotational speeds are required for micro machining. Thus, a low value for the time steps was necessary. For the simulations with a constant rotational speed of 30,000 1/min, a value of 10 μs was used for the time steps to ensure, that the rotational position of the smallest virtual grit representation in two consecutive time steps are at least touching each other and a continuous engagement of the grit into the virtual workpiece is approximated. Within each time step the grit passes a distance of 6.12 μm which is equal to the dimension of the smallest grit within the tool model. On the other hand, the value of 10 μs of the time step led to an acceptable calculation time of the simulation [37].

To calculate the kinematics, each corner points of the virtual grit representation is virtually moved. For each time step, the current position is calculated in relation to the initial position of the models to reduce a chain of errors due to the numerical calculation of the trigonometry. Within the simulation, the feed motion is considered as a translatory displacement of the virtual tool model. The distance for the displacement is calculated using the feed rate multiplied by the value of the time step. The rotational motion of the virtual grit representations is realized using a rotation matrix around the z-axis. Analogous to the feed rate, the angle is calculated via the rotational speed of the tool multiplied by the value of the time step.

The complete motion data is saved for the following calculations of the tool workpiece interactions. However, a large part of the data could be excluded from further calculations because they cannot geometrically reach the virtual workpiece surface. For instance, this condition is true for grits which are turned away from the grinding zone within the current time step. Excluding such grit provides the advantage that the calculation time for the tool workpiece intersections is reduced [37].

4.3 Calculation of the Tool-Workpiece Intersection

For the calculation of the intersection between tool and workpiece model, the main task is to find the intersection point between dexels and the virtual grit representations and to determine the new height value for each dexel. The principle is depicted in Fig. 12.

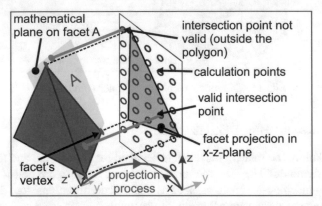

Fig. 12. Principle for the calculation of the surface heights [37]

For the calculation, a straight line is set for each dexel in the direction of the surface heights. Each of the straight lines is intersected with the facets of the virtual grit representations. To determine the intersection, a vectorial plane, spanned by the corner point vectors, is placed on the facet and the geometric intersection point is calculated. Since the points could also be outside of the facet, the facet and the intersection point are projected into the x-z-plane leading to a polygon in two dimensions. If the projected point is inside the two dimensional polygon, a valid intersection point is present. In this case, the y-coordinate is used to define the new length of the dexel on condition that the new length of the dexel is shorter than the original length [37].

5   Application of the Simulation Model to the Investigation of Micro Grinding

In the following sections the simulation model is applied to investigate the kinematic influence of two different feed rates on the resulting surface topography and to investigate the undeformed chip thickness, which is a parameter, that cannot be measured during experiments.

5.1   Influence of the Feed Rate on the Resulting Surface Topography

The simulation model was used to investigate the influence of the feed rate on the resulting surface topography when performing peripheral micro grinding. The advantage in comparison to experiments is that solely the influence of the feed rate can be considered, independent from other effects on the surface topography that appear in real experiments.

5.1.1 Experimental Setup

Micro grinding experiments were performed to compare the real surface topographies with the topographies received by the simulations. The workpiece for the experiments was made of 16MnCr5, hardened to 650 HV 30. The workpiece was tilted to 45° leading to V-shaped grooves due to the horizontal feed direction of the tool. This setup, which is depicted in Fig. 13 enabled to make the workpiece surface, generated by the peripheral tool surface, accessible for optical measurements. The experiments were performed on a LT Ultra MMC600H ultra precision machine tool with five axes. Two different feed rates $v_f = 5$ mm/min and $v_f = 1$ mm/min were applied at a constant rotational speed of 30,000 rpm and a constant axial depth of cut of 150 μm which was achieved within one single pass. For both feed rates the feed travel was 10 mm.

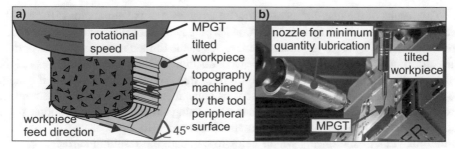

Fig. 13. Experimental setup a) kinematic properties, b) setup within the machine tool [37]

For the evaluation of the experimentally generated surfaces, a three-dimensional topography measurement system Nanofocus μsurf Explorer (See Footnote 1) was used to digitize the surfaces. Processing of the data was done according to DIN EN ISO 25178-2 [38] using the MountainsMap (See Footnote 1) software. Within the software, the complete surfaces were aligned, a filter with a cutoff wavelength of 1.5 μm was applied to remove surface fractions with short wavelength and another one was applied with a cutoff wavelength of 24 μm to remove larger wavelength fractions of the surface. For each experiment, three measurements were performed: One at the start, one in the middle and one at the end of the groove. From the results, the arithmetic mean value Sa and the root mean square Sq of the surface heights were calculated. Further details on the measurement procedure can be found in [37]. The procedure for processing of the data was used for both, the measured data and the data received by the simulations to ensure comparability [37].

5.1.2 Results of the Experiments on the Feed Rate

The results of the experiments showed the expected structure of the micro ground surface: The abrasive grits of the tool led to continuous scratches in feed direction. Perpendicular to the feed direction, the enveloping profile of the grits on the tool were imaged on the surface.

The evaluated surface parameters varied within one experiment for the different measurement positions. For the experiments with a feed rate of 1 mm/min the values for both, Sa and Sq, showed a decreasing trend from the beginning of the machined groove towards the end. This can be explained by the wear of the tool. With increasing feed travel, the abrasive grits are subjected to wear. Hence, the envelope of the tool changes leading to reduced surface parameters.

For the experiments using a feed rate of 5 mm/min the surface parameters are generally larger than those for a feed rate of 1 mm/min. In the experiments using a feed rate of 5 mm/min, the surface parameters for the middle measurement decreased which was explained due to wear. However, they increased again for the last measurement position on the left. The effect can appear due to severe wear of the tools: Due to wear the abrasive grits become smaller or even break out. This also partly changes the enveloping profile of the tool especially at the edge between tool face and peripheral surface. This leads locally to less material removal and hence increasing values of the surface parameters [37].

The results of the measurements show an advantage of the simulations: Especially the large grits cause rough surface topographies perpendicular to the feed direction which are characterized by steep gradients. Hence, due to the limited resolution of the optical measurement system, parts of the surface were not digitized correctly and missing height values within the measured data appeared. Since this is a physical constraint, it cannot be avoided by repeating the measurement or adjustments to the optical measurement procedure. These challenges do not occur when using simulations. Furthermore, since tool wear is not considered within the simulations, the tool model can be used for numerous virtual experiments which helps to compare the resulting surfaces. Hence, the investigation of the kinematic influence of the feed rate on the resulting surface topography is accessible when considering the results of the kinematic simulations [37].

Figure 14 exemplarily depicts the experimental results at the end of the groove and for both feed rates in comparison to the results received by the simulations. It is evident that the surface heights of the simulations fit to the real surface, showing characteristic scratches in feed direction due to the abrasive grits.

Regarding the simulations the discretization of the model led to minimized but not completely eliminable deviations. However, it was found that the values for both surface parameters only had a divergence below 10 nm. This difference can be neglected since the surface topography created by the grits is much more prominent. Hence, regarding the described tools with a nominal diameter of 400 µm and nominal grit sizes between 20 µm and 30 µm it could be stated that both simulated surfaces for the two different feed rates are almost identical when using the spatial discretization equal to the lateral resolution of the optical measurement system. On the other hand, the experiments showed larger surface roughnesses for the larger feed rate. Comparing the experimental results with the results of the simulation it can be concluded that the differences do not occur due to the tool kinematics. Further effects such as wear and material behavior influenced the experimental results [37].

Further details on the evaluation can be found in [37].

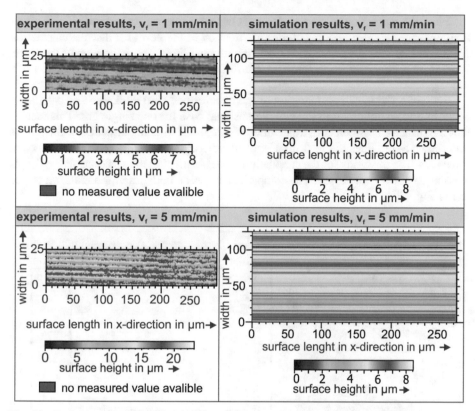

Fig. 14. Experimental results of the surface heights exemplarily for the end of the groove. For two different feed rates the comparison to the simulation results is given according to [37]

5.2 Calculation of the Undeformed Chip Thickness

One process parameter which cannot be measured during machining is the undeformed chip thickness. It depicts the height of the removed material perpendicular to the trajectory of the cutting edge respectively to the grit. The undeformed chip thickness is an important measure to describe the grinding process, e.g. for determining the material separation mode when grinding brittle materials according to the theory of Bifano et al. [39].

Within the simulation, the undeformed chip thickness can be calculated using the difference in height values of the dexels for two consecutive time steps. However, the dexels are always aligned parallel to each other instead of perpendicular to the trajectory of the associated grit. Hence, an estimation of the error was performed. For the error estimation, the maximum feed rate was set to 5 mm/min and the minimum rotational speed to 30,000 rpm, depicting a parameter combination which is limiting for micro grinding with the described MPGTs. Lower feed rates or larger rotational speed would decrease the undeformed chip thickness and thus promote material separation in ductile mode. Using these parameters, estimated and approximated values for the undeformed chip thickness in the two dimensional case were calculated as follows:

First, the estimation of the undeformed chip thickness was calculated using the vertical distance between two trajectories in direction of the dexel for discrete calculation points. In Fig. 15a) the principle is marked with blue lines. The estimation of the undeformed chip thickness is depicted by the difference of the dexel heights of the two trajectories.

Secondly, the ideal consideration of the undeformed chip thickness was calculated according to its definition in the two dimensional case for one single grit. This means, that the distance between both trajectories is calculated in radial direction of the tool as it is shown as red lines in Fig. 15a). Hence, the direction is perpendicular to the second trajectory. The positions for the calculation were identical to the positions of the dexels enabling a direct comparison. The comparison between both principles is shown on the left in Fig. 15.

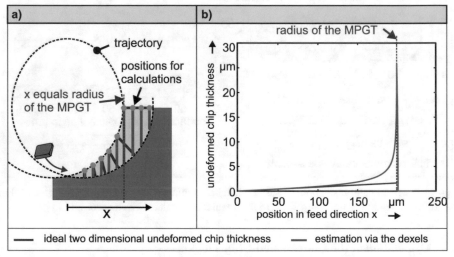

Fig. 15. a) general principle for the calculation and estimation of the undeformed chip thickness (not to scale) and b) the corresponding results using the limiting process parameters

It shows the values for the undeformed chip thickness as a function of the position in cutting direction x. For x-values below approximately 60 μm, the difference between the calculation of undeformed chip thickness and the estimation via the dexel data is not visible in the diagram. For the x-values below 60 μm, undeformed chip thicknesses below 54 nm are estimated. The relative error was determined by calculating the difference between the ideal calculation of the undeformed chip thickness for the simplified two dimensional case, and the estimation via the dexels, divided by the value of the ideal two dimensional undeformed chip thickness. For the x-values smaller than 60 μm, the relative error was below 5%. Furthermore, the relevant x-positions can be limited: Since a large part of the resulting surface is machined by further grits, only the chip thicknesses that occur at small x-values are relevant for the final workpiece surface. In the extreme case, assuming that only one grit is located on the tool, the grit would penetrate the workpiece with each tool revolution. Hence, the maximum distance between two workpiece-grit-contacts would equal the feed per revolution, which has a value of 0.167 µm and can be set as maximum x-value. In this case, the relative error would be below 0.00004%.

As a conclusion, the vertical dexel data is suitable to approximate the undeformed chip thickness in micro grinding. The limiting values for the feed rate and the rotational speed represent parameters suitable for micro machining. However, for machining in ductile mode, the feed rate will be much lower and the rotational speed higher than the limiting values. Hence, the error between the approximation and the ideal two dimensional calculation of the undeformed chip thickness will be further reduced.

The determination of the undeformed chip thickness via the vertical dexel data was done by calculating the difference between the dexel heights in two consecutive time steps. Hence, for each dexel not only a single value for the undeformed chip thickness is received, but one for each time step. As a consequence, the value of the last time step which is not equal to zero was chosen. The reason is that only the final surface is considered which is generated with the last adjustment of the dexel height.

The undeformed chip thickness was analyzed using the same parameters as for the investigation of the feed rate with 5 mm/min. Figure 16 shows that the majority of the surface has a chip thickness in the expected range of nanometers. The large value for the undeformed chip thickness at low x-values appear due to run-in effects which will be avoided in further simulations.

The patterns in feed direction (x-direction) result from the necessary discretization within the simulation model. According to the simulation, the undeformed chip thickness is not equal for the complete surface: Some areas of the surface have very low undeformed chip thicknesses below 1 µm whereas they tend to be in the area of one-digit micrometers for other areas, especially for larger values of the z-direction. The reason is the grit distribution on the tool. Depending on the grit sequence and the grit protrusion height on the tool, the undeformed chip thickness increases or decreases. The effect especially appears when comparably large grit sizes are used. The conclusion of the simulation is that for a complete material separation in ductile mode, the process parameters have to be optimized.

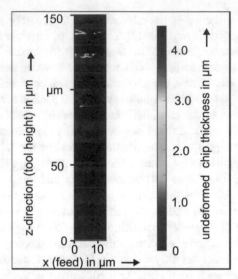

Fig. 16. Result for the undeformed chip thickness of the final surface estimated via the dexel data of the simulation.

6 Conclusion and Outlook

Within the IRTG 2057 a kinematic simulation model was built up to investigate micro grinding. A method was presented to virtually represent the MPGT using input data generated by particle size analyses and knowledge about the deposition rate of the electroless plating process. The tool model considered realistic grit geometries which made a detailed analysis of the resulting surface possible. Furthermore, the detailed grit geometries enabled the calculation of the undeformed chip thickness which's values depend on the grit shape. The modeling of the tool was supported by analysis of the real MPGTs.

The kinematic model was validated by comparing the surfaces received by the simulations with topography measurements of real micro ground surfaces. Both surfaces depicted the characteristic scratches in feed direction originated by the abrasive grits. When using the lateral resolution of the measurement system as the distances of the dexels, the simulations showed that differences between the surface topography generated by the different feed rates were not based on kinematic effects.

The model also made the undeformed chip thickness accessible, a parameter that cannot be measured experimentally. It was found that for micro grinding applications with MPGTs the undeformed chip thickness can vary strongly due to a comparable wide range of abrasive grit sizes.

In future investigations, the kinematic simulation model will be used for further analysis of micro grinding regarding process results which are not available via measurements. This includes the detailed study of the influence of the feed rate on the resulting surface, as well as the study of the chip thickness regarding the machining of glass in ductile mode.

Acknowledgements. Funded by the Deutsche Forschungsgemeinschaft (DFG, German Research Foundation) – 252408385 – IRTG 2057.

References

1. Carrella, M., Aurich, J.C.: Micromachining of silicon - study on the material removal mechanism. In: AMR, vol. 1018, pp. 167–174 (2014)
2. Aurich, J.C., Engmann, J., Schueler, G.M., Haberland, R.: Micro grinding tool for manufacture of complex structures in brittle materials. CIRP Ann. **58**(1), 311–314 (2009)
3. Masuzawa, T.: State of the art of micromachining. CIRP Ann. **49**(2), 473–488 (2000)
4. Setti, D., Arrabiyeh, P.A., Kirsch, B., Heintz, M., Aurich, J.C.: Analytical and experimental investigations on the mechanisms of surface generation in micro grinding. Int. J. Mach. Tools Manuf. **149**, 103489 (2020)
5. Kirsch, B., Bohley, M., Arrabiyeh, P.A., Aurich, J.C.: Application of ultra-small micro grinding and micro milling tools: possibilities and limitations. Micromachines **8**(9), 261–279 (2017)
6. Chakrabarti, S., Paul, S.: Numerical modelling of surface topography in superabrasive grinding. Int. J. Adv. Manuf. Technol. **39**(1–2), 29–38 (2008). https://doi.org/10.1007/s00170-007-1201-y
7. Böß, V., Nespor, D., Samp, A., Denkena, B.: Numerical simulation of process forces during re-contouring of welded parts considering different material properties. CIRP J. Manuf. Sci. Technol. **6**(3), 167–174 (2013)
8. Page, B.: Diskrete Simulation. Eine Einführung mit Modula-2. Springer, Heidelberg (1991). https://doi.org/10.1007/978-3-642-76862-0
9. Denkena, B., Tönshoff, H.K.: Spanen. Grundlagen, 3rd edn. Springer, Heidelberg (2011). https://doi.org/10.1007/978-3-642-19772-7
10. van Hook, T.: Real-time shaded NC milling display. SIGGRAPH Comput. Graph. **20**(4), 15–20 (1986)
11. Salisbury, E.J., Domala, K.V., Moon, K.S., Miller, M.H., Sutherland, J.W.: A three-dimensional model for the surface texture in surface grinding, part 1: surface generation model. J. Manuf. Sci. Eng. **123**(4), 576–581 (2001)
12. Chen, H., Yu, T., Dong, J., Zhao, Y., Zhao, J.: Kinematic simulation of surface grinding process with random cBN grain model. Int. J. Adv. Manuf. Technol. **100**(9–12), 2725–2739 (2019). https://doi.org/10.1007/s00170-018-2840-x
13. Setti, D., Kirsch, B., Arrabiyeh, P.A., Aurich, J.C.: Visualization of geometrical deviations in micro grinding by kinematic simulations. In: Volume 4: Processes. ASME 2018 13th International Manufacturing Science and Engineering Conference. College Station, Texas, USA, 18–22 June 2018. American Society of Mechanical Engineers (2018)
14. Klocke, F., et al.: Modelling of the grinding wheel structure depending on the volumetric composition. Proc. CIRP **46**, 276–280 (2016)
15. Brinksmeier, E., Aurich, J.C., Govekar, E., Heinzel, C., Hoffmeister, H.-W., Klocke, F., et al.: Advances in modeling and simulation of grinding processes. CIRP Ann. **55**(2), 667–696 (2006)
16. Zahedi, A., Azarhoushang, B.: FEM based modeling of cylindrical grinding process incorporating wheel topography measurement. Proc. CIRP **46**, 201–204 (2016)
17. Warnecke, G., Zitt, U.: Kinematic simulation for analyzing and predicting high-performance grinding processes. CIRP Ann. **47**(1), 265–270 (1998)

18. Liu, Y., Warkentin, A., Bauer, R., Gong, Y.: Investigation of different grain shapes and dressing to predict surface roughness in grinding using kinematic simulations. Precis. Eng. **37**(3), 758–764 (2013)
19. Chen, X., Li, L., Wu, Q.: Effects of abrasive grit shape on grinding performance. In: 2017 23rd International Conference on Automation and Computing (ICAC), Huddersfield, United Kingdom, 07–08 September 2017, pp. 1–5. IEEE (2017)
20. Linke, B.S.: A review on properties of abrasive grits and grit selection. Int. J. Abras. Technol. **7**(1), 46–58 (2015)
21. Koshy, P., Jain, V.K., Lal, G.K.: Stochastic simulation approach to modelling diamond wheel topography. Int. J. Mach. Tools Manuf. **37**(6), 751–761 (1997)
22. Hou, Z.B., Komanduri, R.: On the mechanics of the grinding process – part I. Stochastic nature of the grinding process. Int. J. Mach. Tools Manuf. **43**(15), 1579–1593 (2003)
23. ISO 13320: Particle size analysis. Laser diffraction methods (2020)
24. Nosenko, V.A., Aleksandrov, A.A.: The relation between the geometric parameters of grinding powders grains measured by laser diffraction and light-microscopical methods. In: MATEC Web of Conferences, vol. 224, p. 1129 (2018)
25. Mie, G.: Beiträge zur Optik trüber Medien, speziell kolloidaler Metallösungen. Ann. Phys. **330**(3), 377–445 (1908)
26. Eshel, G., Levy, G.J., Mingelgrin, U., Singer, M.J.: Critical evaluation of the use of laser diffraction for particle-size distribution analysis. Soil Sci. Soc. Am. J. **68**(3), 736–743 (2004)
27. Etzler, F.M., Sanderson, M.S.: Particle size analysis: a comparative study of various methods. Part. Part. Syst. Charact. **12**(5), 217–224 (1995)
28. Brittain, H.G.: Particle-size distribution, part I: representations of particle shape, size, and distribution. Pharm. Technol. **25**(12), 38–45 (2001)
29. Wang, W.S., Su, C., Yu, T.B., Zhu, L.D.: Modeling of virtual grinding wheel and its grinding simulation. In: KEM, vol. 416, pp. 216–222 (2009)
30. Zhou, X., Xi, F.: Modeling and predicting surface roughness of the grinding process. Int. J. Mach. Tools Manuf. **42**(8), 969–977 (2002)
31. Chen, X., Rowe, W.: Analysis and simulation of the grinding process. Part I: generation of the grinding wheel surface. Int. J. Mach. Tools Manuf. **36**(8), 871–882 (1996)
32. Aurich, J.C., Kirsch, B.: Kinematic simulation of high-performance grinding for analysis of chip parameters of single grains. CIRP J. Manuf. Sci. Technol. **5**(3), 164–174 (2012)
33. Setti, D., Kirsch, B., Aurich, J.C.: Experimental investigations and kinematic simulation of single grit scratched surfaces considering pile-up behaviour: grinding perspective. Int. J. Adv. Manuf. Technol. **103**(1–4), 471–485 (2019). https://doi.org/10.1007/s00170-019-03522-7
34. Arrabiyeh, P., Raval, V., Kirsch, B., Bohley, M., Aurich, J.C.: Electroless plating of micro pencil grinding tools with 5–10 μm sized cBN grits. In: AMR, vol. 1140, pp. 133–140 (2016)
35. Arrabiyeh, P.A., Kirsch, B., Aurich, J.C.: Development of micro pencil grinding tools via an electroless plating process. In: Proceedings of the 2016 Manufacturing Science and Engineering Conference (2016)
36. Bailey, M.W., Hedges, L.: Die Kristallmorphologie von Diamant und ABN. Industrie-Diamanten-Rundschau **1995**(3/95), pp. 126–129 (1995)
37. Altherr, N., Lange, A., Zimmermann, M., Kirsch, B., Aurich, J.C.: Kinematic simulation model for micro grinding processes using detailed tool models. In: Procedia CIRP Conference on Modeling of Machining Operations (Accepted)
38. DIN EN ISO 25178-2: Geometrical product specifications (GPS) - Surface texture: Areal - Part 2: Terms, definitions and surface texture parameters (2012)
39. Bifano, T.G., Dow, T.A., Scattergood, R.O.: Ductile-regime grinding: a new technology for machining brittle materials. J. Eng. Ind. **113**(2), 184–189 (1991)

Molecular Dynamics Simulation of Cutting Processes: The Influence of Cutting Fluids at the Atomistic Scale

S. Schmitt[1], S. Stephan[1(✉)], B. Kirsch[2], J. C. Aurich[2], H. M. Urbassek[3], and H. Hasse[1]

[1] Laboratory of Engineering Thermodynamics (LTD), RPTU, Kaiserslautern, Germany
{sebastian.schmitt,simon.stephan}@rptu.de
[2] Institute for Manufacturing Technology and Production Systems (FBK), RPTU, Kaiserslautern, Germany
[3] Physics Department and Research Center OPTIMAS, RPTU, Kaiserslautern, Germany

Abstract. Molecular dynamics simulations are an attractive tool for studying the fundamental mechanisms of lubricated machining processes on the atomistic scale as it is not possible to access the small contact zone experimentally. Molecular dynamics simulations provide direct access to atomistic process properties of the contact zone of machining processes. In this work, lubricated machining processes were investigated, consisting of a workpiece, a tool, and a cutting fluid. The tool was fully immersed in the cutting fluid. Both, a simple model system and real substance systems were investigated. Using the simplified and generic model system, the influence of different process parameters and molecular interaction parameters were systematically studied. The real substance systems were used to represent specific real-world scenarios. The simulation results reveal that the fluid influences mainly the starting phase of an atomistic level cutting process by reducing the coefficient of friction in this phase compared to a dry case. After this starting phase of the lateral movement, the actual contact zone is mostly dry. For high pressure contacts, a tribofilm is formed between the workpiece and the cutting fluid, i.e. a significant amount of fluid particles is imprinted into the workpiece crystal structure. The presence of a cutting fluid significantly reduces the heat impact on the workpiece. Moreover, the cutting velocity is found to practically not influence the coefficient of friction, but significantly influences the dissipation and, therefore, the temperature in the contact zone. Finally, the reproducibility of the simulation method was assessed by studying replica sets of simulations of the model system.

1 Introduction

Understanding the fundamental mechanisms of lubricated tribological processes on the atomistic scale is crucial for many technical applications such as cutting and grinding in manufacturing. Cutting fluids are used for two reasons: a) reduce the friction and b) reduce the heat impact and temperature of the workpiece. The fundamental mechanisms

© The Author(s) 2023
J. C. Aurich et al. (Eds.): IRTG 2023, *Proceedings of the 3rd Conference on Physical Modeling for Virtual Manufacturing Systems and Processes*, pp. 260–280, 2023.
https://doi.org/10.1007/978-3-031-35779-4_14

in the small contact zone between a tool, a workpiece, and a lubricant behind these are today not fully understood. However, understanding these processes on the atomistic scale can be helpful for improving the macroscopic process, e.g. for modern micro- and precision machining technology. In-situ experimental investigations of the fundamental mechanisms on the atomistic scale are not possible today. As an alternative, classical molecular dynamics (MD) simulations can be used to gain insights into lubricated tribological processes on the atomistic scale. Due to the strong physical basis, molecular simulations can be applied in two general ways: a) the prediction of thermophysical properties of matter and b) modelling nanoscopic processes. In this work, the focus is on the modeling of nanoscopic cutting processes using molecular dynamics simulation.

Molecular simulations are based on solving Newtons equation of motion for an atomistic many particle system – considering boundary condition imposed by the simulation scenario, e.g. the movement of a tool. The interactions between the particles on the atomistic level are defined by force fields. Force fields aim at representing the molecular interactions and structure of a given real substance and, hence, mostly provide a reliable representation of the behavior matter. Besides using force fields for real substances, model systems can be favorably used in molecular simulation to study the link between molecular interaction parameters and macroscopic properties for obtaining generic information on processes [36, 35, 17, 4, 14, 33, 28]. Thereby, model systems can be used to study the fundamental mechanisms of complex processes in detail [8, 18, 13, 19, 9]. The Lennard-Jones model system is the most popular and widely used model system [34, 37]. In this work, both real substance systems and Lennard-Jones model systems were used for studying the atomistic processes of lubricated cutting.

In the literature, there are several MD studies available investigating the deformation of a workpiece and material removal caused by a tool, e.g. Refs. [36, 11, 1, 25, 27, 32, 39, 38]. However, in most studies, only dry contact processes (no cutting fluid) are considered, e.g. on the influence of the shape of the tool [2, 10] and the solid material [17, 11, 3, 42, 16]. The influence of cutting fluids is considered in only few studies, e.g. Refs. [39, 45, 26, 30, 31, 6].

In this work, the influence of cutting fluids on cutting processes were investigated on the atomistic scale. Thereby, different aspects of a cutting process were investigated such as the behavior of the fluid in the lubrication gap [36, 32], the formation of a tribofilm on the workpiece surface [32, 39], the influence of the cutting speed on the process [27], the thermal balance of the system, and the temperature field in the contact zone [36, 27]. Also, the reproducibility of cutting simulations was investigated [38].

This chapter is organized as follows: First, the simulation scenarios, the molecular models, as well as the observables used for characterizing the system are introduced. Then, the results are presented and discussed, which includes mechanical properties, the workpiece deformation, lubrication and the formation of a tribofilm, thermal properties, and the evaluation of the reproducibility.

2 Methods

2.1 Simulation Scenario

The simulation scenario used in this work is depicted in Fig. 1. It consists of a workpiece, a (cutting) tool and, in the lubricated cases, of a fluid. This models the contact zone of a machining process, e.g. the tip of an abrasive particle in micro grinding or an asperity contact. In the lubricated simulations, the tool was fully submersed in the fluid. In the dry cases, the tool was in a vacuum. The workpiece surface is in the x-y plane.

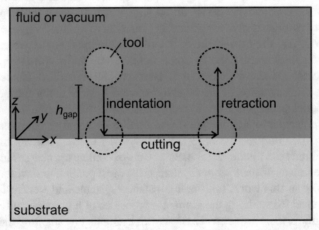

Fig. 1. Sketch of the simulation scenario. The scenario consists of a workpiece (grey), a tool (green), and a fluid or vacuum (blue).

The simulation box had periodic boundary conditions in x- and y-direction. The fluid was confined in the box by a soft repulsive wall at the top. The workpiece was locked in position by prescribing the at least three atom layers at the bottom of the box. Dissipated heat is removed from the system by imposing the initial system temperature (specified below for the different systems) to at least four atom layers above the fixed layer. In the course of the simulation, the tool carries out three consecutive movements: the indentation (negative z-direction), the cutting (positive x-direction), and the retraction (positive z-direction). Hence, the movement of the tool is predefined. Three different tool shapes were used: a sphere [27, 32], a spherical cap [39], and a cylinder [36, 38]. In the latter case, a quasi-2D scenario was considered. Moreover, different cutting gaps h_{gap} were considered for the cutting, cf. Figure 1. The simulations contained up to 9.2 $\times 10^6$ particles per simulation. For the model system simulations, 5.43×10^6 particles were used in the simulations: 3.65×10^6 for the workpiece, 7.9×10^5 for the tool, and 1.78×10^6 for the fluid. For the real substance simulations, at least 8×10^5 particles were used for the workpiece, 7×10^5 for the tool, and 8.68×10^5 for the fluid. The molecular simulation software LAMMPS [24] was used for the simulations. Details on the simulation scenario can be found in Refs. [36, 27, 32, 39, 38].

2.2 Molecular Model

Simulations were carried out using a simplified model system as well as a real substance system, which are briefly introduced in the following. Details on the model system simulations are given in Refs. [36, 27, 38]. Details on the real substance systems in Refs. [32, 39].

2.2.1 Model Systems

The Lennard-Jones truncated and shifted (LJTS) model system is a generic and simplified system. Yet, this provides a reasonable representation of a real substances as the basis molecular interactions, i.e. repulsive and dispersive, are captured [35, 12, 44]. Hence, the simulations with the LJTS model system do not aim to model a specific real cutting process but to investigate general mechanisms of cutting processes and the influence of molecular interaction parameters. Despite its simplicity, the LJTS system provides a physically robustness model backbone and is, at the same time, computationally relatively cheap. In the presented model system, all occurring interactions were modelled by the LJTS potential (cf. Eq. 2.1).

$$u_{LJ}(r) = 4\varepsilon \left[\left(\frac{\sigma}{r}\right)^{12} - \left(\frac{\sigma}{r}\right)^{6} \right] \text{ and}$$

$$u_{LJTS}(r) = \begin{cases} u_{LJ}(r) - u_{LJ} & r \le r_c \\ 0 & r > r_c \end{cases} \text{ with } r_c = 2.5\sigma \tag{2.1}$$

In Eq. (2.1), u_{LJ} indicates the full Lennard-Jones potential. For the LJTS potential, interactions are truncated at r_c and the potential energy u is shifted such that no discontinuity appears at the truncation radius. Each component (workpiece, fluid, and tool) has two molecular interaction parameters: the energy parameter ε and the size parameter σ. The size parameter σ as well as the particle mass m of all components were the same in all cases. The solid-solid interactions between the workpiece and the tool were described by the LJTS potential with a cut-off radius of $r_{cut} = 2^{1/6}\sigma$ such that they only interact repulsively. The workpiece energy parameter was chosen to represent iron [12] and the fluid energy parameter was chosen to represent methane [44]. All quantities for the LJTS model systems are given in reduced units (cf. Table 1) with respect to the fluid particle interaction parameters ε_F and σ_F. The solid-fluid interaction energy was systematically varied in the range $0 \le \varepsilon_{SF}^* \le 1.7$ to study its influence on the cutting process. The initial temperature was $T^* = 0.8$ and the initial pressure was $p^* = 0.014$.

2.2.2 Real Substance Systems

In the real substance simulations, the workpiece was modeled as a single crystal iron block, the tool was modeled as a diamond single crystal with a spherical tip shape, and the fluid was either methane or decane [32, 39]. The iron workpiece was described by an embedded atom model (EAM) [21]. The fluid was either modelled as methane by a BZS force field [44, 43] or as n decane by the TraPPE force field [22], which provide an excellent description of the fluid bulk phase properties [29]. The diamond tool was modelled by a Tersoff potential [41]. For methane, the initial temperature was either 100 K (for simulations with a spherical indenter) or 130 K (for simulations with a spherical cap indenter). The initial pressure was 0.1 MPa and 50 MPa, respectively.

Table 1. Definition of physical quantities in reduced units. Reduced quantities are marked by (*). The Boltzmann constant is indicated as k_B.

Length	$L^* = L/\sigma_F$
Temperature	$T^* = T/(\varepsilon_F/k_B)$
Force	$F^* = F/(\varepsilon_F/\sigma_F)$
Time	$t^* = t/(\sigma_F\sqrt{m_F/\varepsilon_F})$
Velocity	$v^* = v/\sqrt{\varepsilon_F/m_F}$
Energy	$E^* = E/\varepsilon_F$

2.3 Definition of Observables

Different quantities were computed from the simulation trajectories, which are briefly introduced in the following. The forces on the tool in tangential (x) and normal (y) direction, F_t and F_n, respectively, were calculated as the sum of all forces acting on the particles of the tool. The forces were calculated every 1000 timesteps. Mean values have been calculated during the cutting phase in the stationary part of the process. The coefficient of friction is defined as $\mu = F_t/F_n$. The coefficient of friction was calculated by the mean values of the forces from the stationary cutting phase. The internal energy of the fluid U_F and the workpiece U_S were calculated as the sum of the interaction energies between all respective particles. They were also sampled every 1000 timesteps. The energy dissipated by the thermostat in the substrate ΔU_{thermo} was computed as the energy difference after and before the thermostatization in each time step. The following quantities were calculated based on the configurational data of the particles that were written out at least every 10^4 timesteps. The number of fluid particles in the gap between the tool and the substrate were computed as N_{gap}. The gap was geometrically defined as depicted in Fig. 2 as the volume between the tool and the workpiece surface in front of the tool center and below the undeformed surface height. The surfaces were analyzed using the *alpha shape* algorithm [7]. Based on the calculated workpiece surface, the number of fluid particles below the surface that form a tribolayer were computed as N_{tribo}. The temperature profile in the x-z plane was sampled by averaging the per-atom temperature binwise. The bins are defined as depicted in Fig. 2. An estimation of the statistical uncertainty and significance of the observations is possible by the separate reproducibility study [38]. Details on the definition and sampling of the observables are given in Refs. [36, 27, 32, 39].

3 Results

3.1 Mechanical Properties

Figure 3. Shows the time evolution of the tangential and the normal force in cutting simulations for a dry case and a lubricated case. During the indentation, the normal force increases drastically when the tool penetrates into the workpiece and causes elastic and plastic deformation. In that phase, the tangential force fluctuates around $F_t = 0$. In the

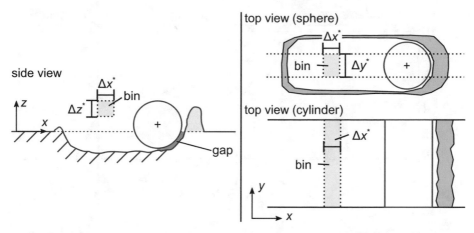

Fig. 2. Sketch illustrating the definition of observables. Left: Side view with the gap between the tool and the substrate and the bin to sample the temperature in x-y plane. Right: Top view with the bin to sample the temperature profile in x-y plane for the case of a spherical (top) and a cylindrical tool (bottom).

cutting phase, the normal force decreases and the tangential force increases until a steady state is reached. During the retraction, the both forces decay to zero.

Overall, the results from the dry and lubricated case are similar. Yet, the normal force on the tool is slightly affected by the cutting fluid during the indentation and the starting phase of the cutting. The peaks observed during the indentation in the dry case (caused by dislocation movement) are damped by the presence of the cutting fluid molecules in the gap. During the indentation, the vast majority of fluid particles is squeezed out of the gap. During the starting phase of the cutting, until a steady state is reached, the tangential force is slightly increased in the lubricated case compared to the dry case, which is due to an effective enlargement of the tool due to adsorbed fluid particles [32], i.e. the tool causes more elastic and plastic deformation of the workpiece in the lubricated cases. The coefficient of friction increases strongly in the starting phase of the cutting. In the starting phase, the lubricated simulations yield slightly lower coefficients of friction compared to the dry simulations due to lubricant molecules remaining in the gap between the tool and the substrate. In the stationary phase, the coefficient of friction is very similar in the dry and the lubricated case with values between 0.4 and 0.6. Since the differences between the dry and lubricated case are small and might in general be within the scattering of the results, the reproducibility of the findings discussed here was studied (and confirmed) using the model system, cf. Sect. 3.4.

The influence of the solid-fluid interaction energy ε_{SF}^* on the cutting process was systematically investigated using the LJTS model system [36]. The results are shown in Fig. 4. In the starting phase of the cutting process, the coefficient of friction in the lubricated cases is reduced by about 25% compared to the dry case. This is due to an increased normal force and a slightly decreased tangential force in the starting phase [36]. In the stationary phase, the coefficient of friction is increased in the lubricated case compared to the dry case by approximately 15%, which is due to individually fluid

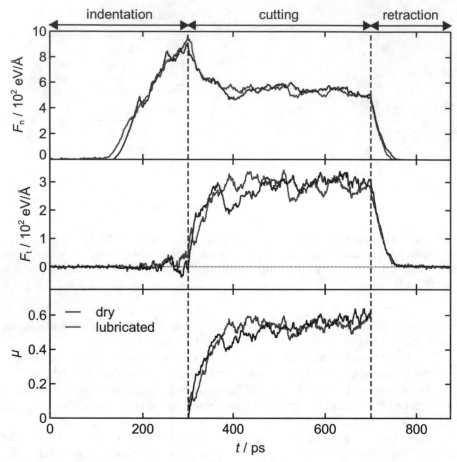

Fig. 3. Normal (top) and tangential (bottom) forces on the tool as a function of time for a dry and a lubricated simulation [32]. The tool had a spherical shape. The temperature was $T = 100$ K. The workpiece was an iron single crystal, the tool was a diamond, and the fluid was methane.

particles being imprinted into the workpiece surface. Interestingly, these findings do not dependent on the solid-fluid interaction energy. The coefficient of friction is reduced in the starting phase due to fluid molecules remaining in the gap and effectively increasing the tool size. These fluid molecules are squeezed out mostly with ongoing cutting process. The squeeze out process is discussed in detail in Sect. 3.3.

Figure 5. Shows the results for the influence of the cutting speed on the coefficient of friction. For both cases (dry and lubricated), no significant influence of the cutting speed on the coefficient of friction is observed in the considered velocity range. For the smallest considered cutting speed $v^* = 0.66$, the coefficient of friction is nearly the same for the dry and the lubricated simulations with $\mu^* \approx 0.85$. For the velocities $0.1 < v^* < 0.3$, the lubricated simulations yield smaller values for the coefficient of friction compared to the dry simulations. These differences are probably within the uncertainty of the data. This is supported by the result for the highest considered velocity $v^* = 0.332$, which

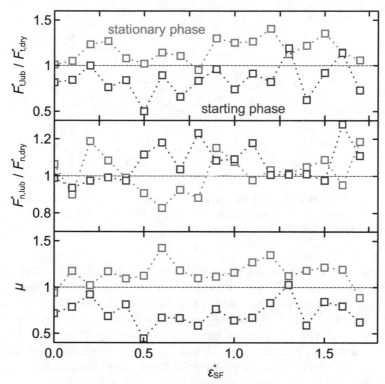

Fig. 4. Tangential force (top), normal force (middle), and coefficient of friction (bottom) of the lubricated simulations related to the dry simulation as function of the solid-fluid interaction ε_{SF}^* [36]. The tool had a cylindrical shape. The temperature was $T^* = 0.8$. The workpiece, the tool, and the fluid were modeled by the LJTS potential. Mean values for the starting phase (red, $142 < t^* < 335$) as well as the stationary phase (blue, $335 < t^* < 625$).

show the opposite trend. Using the potential parameters for methane and iron [12, 44] for the fluid and workpiece, respectively, the cutting speed range corresponds to 20 - 100 m/s in SI units, which is typical for cutting and grinding processes [20, 5]. In the literature, relative velocities between two solid bodies up to 400 m/s were considered using molecular simulation (which is not representative for common manufacturing cutting processes). Nevertheless, in this high-speed regime, the coefficient of friction was reported to decrease with increasing velocity [23, 46].

Using the real substance system simulation setup, the influence of the cutting depth, i.e. the z-position of the tool with respect to the workpiece surface, was systematically investigated [39]. Here, also configurations with no direct contact between tool and substrate were considered, i.e. hydrodynamic lubrication (HL). In Fig. 6, the forces on the tool as well as the coefficients of friction are shown as function of the cutting depth h. Results are given for decane as fluid. Three different lubrication regimes can be identified for different cutting depth: hydrodynamic lubrication (HL), mixed lubrication (ML), and boundary lubrication (BL). The normal and tangential force on the tool increase with increasing cutting depth as expected. The normal forces are larger than the tangential

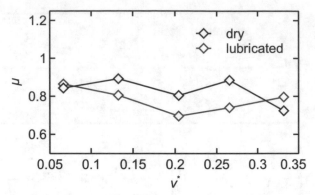

Fig. 5. Coefficient of friction μ as a function of the tool velocity v^* for a dry and lubricated cases [27]. The tool had a spherical shape. The temperature was $T^* = 0.8$. The workpiece, the tool, and the fluid were modeled by the LJTS potential.

forces in general. The coefficient of friction behaves differently in the three regimes. In the HL regime, the coefficient of friction is lowest – as expected. Here, a stable fluid film separates the two solids and leads to a good lubrication with small values of the coefficient of friction. In the ML regime, the coefficient of friction increases with increasing cutting depth as the normal force strongly increases due to the direct contact between the tool and the workpiece, which is transferred via the thin lubrication film. The coefficient of friction reaches a maximum of $\mu \approx 0.21$ at the border between the ML and BL regime. For larger cutting depth (BL regime), the coefficient of friction decreases with increasing cutting depth. The behavior in the BL regime is mainly determined by the formation of a tribofilm, the squeeze-out and the resistance of the lubrication film. Therefore, the tangential force increases more strongly compared to the normal force, which leads to a decrease of the coefficient of friction.

3.2 Workpiece Deformation

In the following, phenomena related to the surface of the workpiece and to the formation of dislocations in the workpiece are discussed. Figure 7 shows the results for the total dislocation length as a function of the simulation time for a dry and a lubricated case. Overall, the results for the dry and the lubricated case are similar. In the indentation phase, the presence of the cutting fluid leads to a slightly earlier formation of dislocations compared to the dry simulation, which was also confirmed by replica simulations of a model systems in an earlier work of our group [38]. This is probably due to fluid adsorption layers on both the workpiece and the tool surface that effectively increases the size of the tool and lead to an earlier starting of elastic and plastic deformation on the workpiece. During the cutting phase, the dislocation length slightly increases in the dry case and the lubricated case. The presence of the fluid has only minor effects on the dislocation behavior – also considering the statistical uncertainties of the data [38]. After the retraction, an annihilation of some of the dislocations can be observed as the total dislocation length decreases.

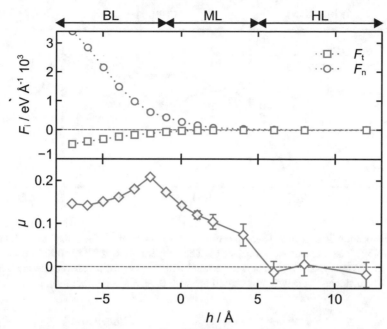

Fig. 6. Normal and tangential forces on the tool (top) and coefficient of friction (bottom) as a function of the cutting depth for simulations with decane [39]. The tool was a spherical cap. The temperature was $T = 350$ K. The workpiece was an iron single crystal, the tool was a diamond, and the fluid was decane. Three different regimes were distinguished: boundary lubrication (BL), mixed lubrication (ML), and hydrodynamic lubrication (HL). The statistical uncertainties were estimated from the fluctuation of block average values in the quasi-stationary cutting phase.

Figure 8 shows the deformed substrate surface at the end of the cutting process. Results are shown for two different cutting depths for both methane and decane as fluid. For a cutting depth of $h = -2$Å, the cutting process roughens the surface over the entire cutting length, which is mostly due to the formation of a tribofilm.

No distinct chip formation is observed for $h = -2$Å, cf. Fig. 8. A distinct chip formation is obtained for $h = -6$Å, cf. Fig. 8 (bottom). The chip forms primarily at the sides of the tool which is built-up by the workpiece atoms that are thrust aside of the tool. Comparing the methane results with the decane results, the chip is more pronounced and the atoms at the workpiece surface are less disordered in the methane case. This is due to the different characteristics of the adsorption layers of methane and decane [39] including the particles trapped in the gap between the tool and the substrate.

3.3 Lubrication and Formation of Tribofilm

The number of fluid particles trapped in the gap between the tool and the workpiece (cf. Fig. 2) were computed at the end of the indentation phase. These results are shown as a function of the solid-fluid interaction in Fig. 9. For the case of no attractive interactions between the solid and the fluid particles, the number of particles in the gap N_{gap} at the

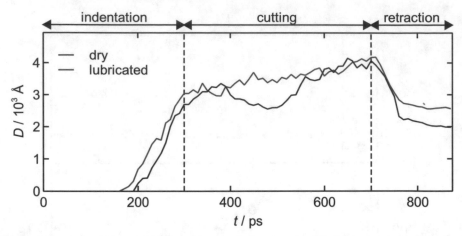

Fig. 7. Dislocation length in the workpiece as a function of time for a dry and a lubricated simulation [32]. The tool had a spherical shape. The temperature was $T = 100$ K. The workpiece was an iron single crystal, the tool was a diamond with a spherical tip shape, and the fluid was methane. The dislocation analysis was carried out with the DXA algorithm [40].

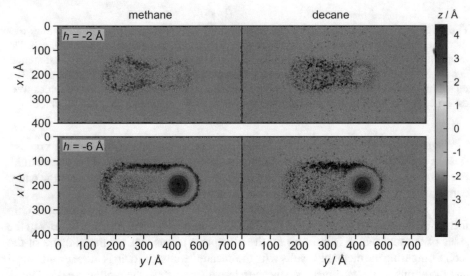

Fig. 8. Top view on the workpiece surface at the end of the cutting phase for the cutting depths $h = -2, -6$Å (top to bottom) [39]. Results are shown for simulations with methane (left) and decane (right) as fluids. The tool was a spherical cap. The workpiece was an iron single crystal and the tool was a diamond. The visualizations were created using Paraview [15].

end of the indentation is lowest and close to zero. With increasing solid-fluid interaction energy, more particles remain trapped in the gap until a steady plateau is reached for approximately $\varepsilon_{SF}^* \approx 0.75$. For larger values, N_{gap} stays approximately constant. Interestingly, this behavior is in line with the contact angle behavior of the studied LJTS system [4], i.e. total wetting is reached at approximately $\varepsilon_{SF}^* \approx 0.75$. This means the number of particles increases with decreasing contact angle until total wetting is reached.

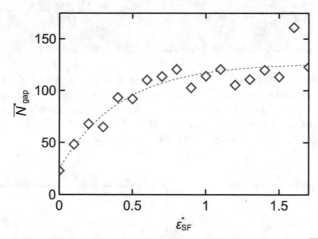

Fig. 9. Number of fluid particles in the gap between the tool and the workpiece $\overline{N}^*_{\text{gap}}$ (cf. Fig. 2) as function of the solid-fluid interaction energy ε_{SF}^* [36]. The tool had a cylindrical shape. The temperature was $T^* = 0.8$. The workpiece, the tool, and the fluid were modeled by the LJTS potential. The dashed line is an empirical correlation of the form $\overline{N}^*_{\text{gap}} = 126.5 - 101.7e^{-\varepsilon_{SF}^*/0.422}$.

The properties of the surface are strongly influenced by single fluid molecules that are imprinted into the workpiece surface. This is observed for both the methane as well as the decane case. Yet, both systems show different characteristics due to the different molecular shape of the fluid molecules. A screenshot of each system is shown in Fig. 10. The imprinted fluid molecules can be interpreted as a tribofilm that forms in the upper part of the workpiece near the surface due to the extreme load. In the case of methane, the tribofilm is mainly build-up by methane atoms occupying regular lattice sites of the iron crystal. The overall lattice structure of the workpiece remains undamaged in this case. Due to decane being a long linear chain molecule, the workpiece lattice structure is significantly broken up in the decane case. Decane molecules are also imprinted into the substrate surface, but the imprinted molecules destroy the regular lattice and cause an unstructured formation of the upper atom layers of the workpiece.

The formation of a tribofilm is further analyzed in Fig. 11, which shows the number of fluid particles imprinted in the substrate surface N_{tribo} as a function of the z-coordinate for three different cutting depths.

The formation of the tribofilm is mainly observed at large cutting depth as the number of imprinted fluid particles is significantly lower. For both studied fluids, the number of fluid particles is largest for $z \approx -2\text{Å}$ and the number decreases with decreasing z. For methane, the imprinted sites reach depths of up to $z = 12\text{Å}$. The decane sites only reach depths of up to $z = -9\text{Å}$. The different characteristics observed in the screenshots (cf. Fig. 10) can be confirmed by the histograms shown in Fig. 11. The methane particles are accumulated at specific depths, i.e. distinct peaks forming in the histogram. This is due to the methane particles occupying regular lattice sites of the workpiece lattice.

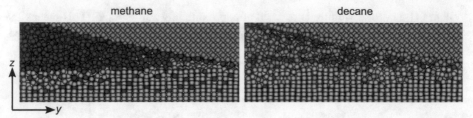

Fig. 10. Screenshots of the rear side of the contact zone (tool moves to the right) at cutting depth $h = -6$Å [39]. The tool was a spherical cap. The workpiece was an iron single crystal, the tool was a diamond, and the fluid was methane (left) or decane (right). Green particles indicate tool particles, grey particles the workpiece particles, dark blue CH_4 (left) or CH_3 (right) sites (end group of decane), and light blue CH_2 sites (middle group of decane).

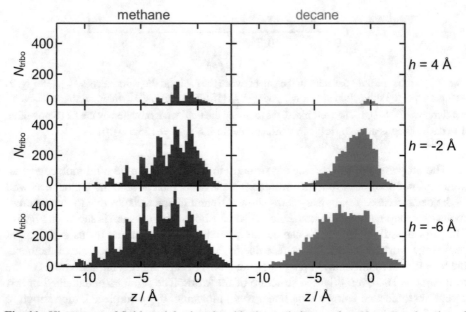

Fig. 11. Histograms of fluid particles imprinted in the workpiece surface N_{tribo} as a function of the z-coordinate [39]. Results for $h = 4, -2, -6$Å. The tool was a spherical cap. The workpiece was an iron single crystal, the tool was a diamond, and the fluid was methane (left) or decane (right).

For the decane simulations, these peaks cannot be observed which is in line with the unstructured tribofilm observed in Fig. 10.

3.4   Thermal Properties

Thermal properties of a cutting process such as the temperature in the contact zone and the heat flux absorbed by the workpiece, are crucial for the product quality and the manufacturing process. The MD simulations carried out in this work were evaluated in detail regarding the thermal properties. Figure 12 shows the spatial temperature distribution in the $x^* - y^*$ plane by snapshots at a cutting length of $L^* = 77$.

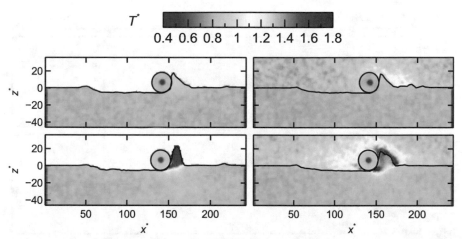

Fig. 12. Temperature profile in the $x^* - y^*$ plane at the cutting length $L^* = 77$ for dry (left) and lubricated (right) simulations [27]. Results shown for two different velocities, $v^* = 0.066$ (top) and $v^* = 0.332$ (bottom). The tool had a spherical shape. The bulk temperature was $T^* = 0.8$. The workpiece, the tool, and the fluid were modeled by the LJTS potential.

The evaluation procedure is depicted in Fig. 2. Simulation results for two different cutting speeds, $v^* = 0.066$ and $v^* = 0.332$, are shown. Results from a dry case are compared to the results from a lubricated case. The temperature increases for both cases with increasing tool velocity. In the dry simulation, the temperature increases mainly in the chip, i.e. the energy dissipates in the direct vicinity of the contact zone – as expected. From the chip, the heat is transported to the bulk of the substrate in the dry case. In the simulation scenario, the tool is thermostatted which is why the temperature in the tool keeps constant. If a fluid is present, the increase of the temperature in the contact zone is reduced as the heat is also transported into the fluid. Therefore, the temperature of the cutting fluid near the contact zone is increased compared to the bulk fluid temperature. Hence, the fluid has important cooling capabilities.

In Fig. 13, the results for the energy balance of the system are shown. The energy balance includes all sources and sinks of the system. Energy is added to the system by the work done by the tool. Energy is removed from the system by a thermostat in the workpiece (cf. Fig. 13). In the cutting simulations, the thermostat acts as a heat sink and removes dissipated energy from the system such that a quasi-stationary state is established. The energy which is not removed by the thermostat heats up the substrate and (if present) the fluid, i.e. their internal energy increases. The change of the internal energy of the workpiece U_W and the fluid U_F, energy removed by the thermostat ΔU_{thermo}, and the work done by the tool W_T are shown in Fig. 13 as a function of time. Therein, U_W and U_F indicate the changes of the total energy (kinetic and potential) of the workpiece and fluid particles, respectively. The energy removed from the system ΔU_{thermo} was computed from the rescaled kinetic energy imposed by the thermostat. The work done

274   S. Schmitt et al.

by the tool W_T was computed from the integral of the total force (in moving direction) during the process. The fluid strongly influences the energy balance of the process.

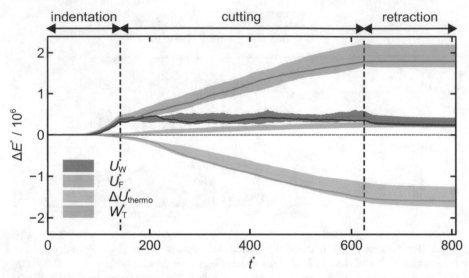

Fig. 13. Energy balance of the system including internal energy changes of the workpiece U_W and the fluid U_F, energy removed by the thermostat ΔU_{thermo}, and the work done by the tool W_T [36]. The shaded areas include all lubricated simulation cases with different solid-fluid interactions energies. The solid lines represent the dry simulation case. The tool had a cylindrical shape. The initial temperature was $T^* = 0.8$. The workpiece, the tool, and the fluid were modeled by the LJTS potential.

The heat removed by the thermostat as well as the change of the internal energy of the substrate is reduced up to 20% by the presence of a fluid. This cooling effect depends on the solid-fluid interaction energy [36]. The fluid reduces the friction during the cutting phase, and part of the dissipated energy in the contact zone heats up the fluid, which directly cools the contact zone. Nevertheless, the main part of the energy added to the system by the cutting process is dissipated. The energy dissipated is significantly larger than energy required for the defect generation and plastic deformation.

3.5 Reproducibility

The statistical uncertainties and the reproducibility of the simulation method was assessed using a set of eight replicas. The single simulations of the set only differ in their initial velocities that are assigned before the equilibration of the simulation box. Based on the eight replica simulations, the standard deviation σ was calculated for several observables. In Fig. 14, the results for the normal force are shown. The time evolution of the normal force agrees in general with the results given in Fig. 3. The higher normal force in the case with a fluid compared to the dry case is confirmed by these results. The standard deviation of the normal force among the replica simulations is relatively small in the indentation phase.

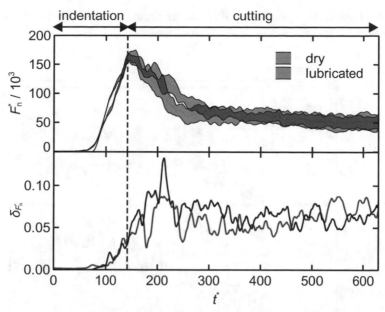

Fig. 14. Normal force on the tool as a function of time for the dry and the lubricated case (top) and the corresponding instantaneous standard deviation (bottom) [38]. The shaded area includes the entire range of the results of all eight replica simulations. The tool had a cylindrical shape. The temperature was $T^* = 0.8$. The workpiece, the tool, and the fluid were modeled by the LJTS potential.

The standard deviation increases with progressing indentation. In the starting phase of the cutting (cf. Sect. 3.1), the lubricated simulations show significantly higher statistical uncertainties. At the beginning of the stationary phase ($t^* \approx 300$), the normal forces in the dry and the lubricated simulations agree within the scattering of the replica sets. The standard deviation is slightly higher in the dry simulations in the starting phase of the cutting compared to the lubricated simulations.

In Fig. 15, the coefficient of friction calculated from the eight replicas is shown with its corresponding standard deviation. In the starting phase ($t^* < 300$), the coefficient of friction of the dry simulations is significantly larger compared to the simulations with a fluid. The difference exceeds the scattering of the replica sets, which indicates that the differences between a lubricated and a dry case in the starting phase are significant. The coefficient of friction is reduced in the lubricated case in the starting phase of the cutting due to fluid particles trapped in the gap between tool and the workpiece (cf. Sect. 3.3). Until the fluids are squeezed out of the gap, the tool experiences a larger normal force, which decreases the coefficient of friction in a lubricated case compared to a dry case. In the stationary phase of the cutting process, the coefficient of friction is slightly increased in the lubricated case, which is due to individual fluid particles being imprinted into the workpiece surface, which requires additional work done by the tool in the lubricated case. In general, no systematic differences in the reproducibility of dry and lubricated simulations were found. Moreover, the time dependency of the standard deviation for the normal force and the coefficient of friction indicate that no differences between the simulations of a replica build up with ongoing process.

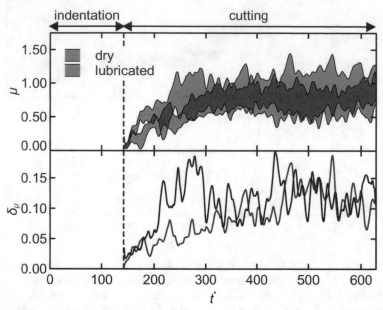

Fig. 15. Coefficient of friction as a function of time for the dry and the lubricated case (top) and the corresponding instantaneous standard deviation (bottom) [38]. The shaded area includes the entire range of the results of all eight replica simulations. The tool had a cylindrical shape. The temperature was $T^* = 0.8$. The workpiece, the tool, and the fluid were modeled by the LJTS potential.

4 Conclusions

In this work, the influence of lubrication on the contact zone of cutting processes was studied on the atomistic scale using classical MD simulation. Thereby, new insights were obtained on the fundamental mechanisms of lubrication and cooling provided by the presence of cutting fluids. Different simulations scenarios were used including model systems as well as real substance systems. The mechanical properties of the contact were studied in means of the normal and tangential force as well as the coefficient of friction. It was found that the presence of a fluid has an important influence in the starting phase of the atomistic cutting process, i.e. decreases the coefficient of friction, which is due to fluid molecules trapped in the gap between the workpiece and the tool. The influence of the cutting depth on the cutting process was investigated using two different fluids: methane and decane. Based on the results, three different lubrication regimes were identified for different cutting depths: the hydrodynamic lubrication regime for very small cutting depths, the mixed lubrication regime for cutting depths that correspond approximately to the size of fluid molecules, and the boundary lubrication regime cutting for cutting depths that yield significant elastic and plastic deformation of the workpiece. Within these different regimes, the coefficient of friction shows different characteristics.

The presence of a fluid has important thermal effects on the atomistic cutting process, i.e. it is found to reduce the maximum temperature in the contact zone and reduce the heat impact of the workpiece. The heat absorbed by the workpiece is reduced by up to 20% by the presence of a fluid lubricant. For future for, the simulation scenario should be refined, e.g. considering a rough surface topography, such that the reality of tribological contact processes is captured in more detail.

Acknowledgement. The authors gratefully acknowledge access to the ELWE supercomputer at Regional University Computing Center Kaiserslautern (RHRK) under the grant TUKL-MTD.

References

1. An, R., Huang, L., Long, Y., Kalanyan, B., Lu, X., Gubbins, K.E.: Liquid-solid nanofriction and interfacial wetting. Langmuir **32**, 743–750 (2016). https://doi.org/10.1021/acs.langmuir.5b04115
2. AlabdAlhafez, I., Brodyanski, A., Kopnarski, M., Urbassek, H.M.: Influence of tip geometry on nanoscratching. Tribol. Lett. **65**(1), 1–13 (2017). https://doi.org/10.1007/s11249-016-0804-6
3. AlabdAlhafez, I., Urbassek, H.M.: Scratching of hcp metals: a molecular-dynamics study. Comput. Mater. Sci. **113**, 187–197 (2016). https://doi.org/10.1016/j.commatsci.2015.11.038
4. Becker, S., Urbassek, H.M., Horsch, M., Hasse, H.: Contact angle of sessile drops in Lennard-Jones systems. Langmuir **30**, 13606–13614 (2014). https://doi.org/10.1021/la503974z
5. Brinksmeier, E., Meyer, D., HuesmannCordes, A.G., Herrmann, C.: Metalworking fluids – mechanisms and performance. CIRP Ann. **64**, 605–628 (2015). https://doi.org/10.1016/j.cirp.2015.05.003
6. Dai, L., Sorkin, V., Zhang, Y.-W.: Effect of surface chemistry on the mechanisms and governing laws of friction and wear. ACS Appl. Mater. Interfaces. **8**, 8765–8772 (2016). https://doi.org/10.1021/acsami.5b10232
7. Edelsbrunner, H., Kirkpatrick, D., Seidel, R.: On the shape of a set of points in the plane. IEEE Trans. Inf. Theory **29**, 551–559 (1983). https://doi.org/10.1109/TIT.1983.1056714
8. Fertig, D., Hasse, H., Stephan, S.: Transport properties of binary Lennard-Jones mixtures: insights from entropy scaling and conformal solution theory. J. Mol. Liq. **367**, 120401 (2022). https://doi.org/10.1016/j.molliq.2022.120401
9. Fertig, D., Stephan, S.: Influence of dispersive long-range interactions on transport and excess properties of simple mixtures. Mol. Phys. e2162993 (2023). https://doi.org/%2010.1080/00268976.2022.2162993
10. Gao, Y., Lu, C., Huynh, N.N., Michal, G., Zhu, H.T., Tieu, A.K.: Molecular dynamics simulation of effect of indenter shape on nanoscratch of Ni. Wear **267**, 1998–2002 (2009). https://doi.org/10.1016/j.wear.2009.06.024
11. Gao, Y., Ruestes, C.J., Urbassek, H.M.: Nanoindentation and nanoscratching of iron: atomistic simulation of dislocation generation and reactions. Comput. Mater. Sci. **90**, 232–240 (2014). https://doi.org/10.1016/j.commatsci.2014.04.027
12. Halicioğlu, T., Pound, G.M.: Calculation of potential energy parameters from crystalline state properties. Phys. Stat. Sol. (a) **30**, 619–623 (1975). https://doi.org/10.1002/pssa.2210300223

13. Heier, M., Stephan, S., Liu, J., Chapman, W.G., Hasse, H., Langenbach, K.: Equation of state for the Lennard-Jones truncated and shifted fluid with a cut-off radius of 2.5 σ based on perturbation theory and its applications to interfacial thermodynamics. Mol. Phys. **116**, 2083–2094 (2018). https://doi.org/10.1080/00268976.2018.1447153

14. Heier, M., Stephan, S., Diewald, F., Müller, R., Langenbach, K., Hasse, H.: Molecular dynamics study of wetting and adsorption of binary mixtures of the Lennard-Jones truncated and shifted fluid on a planar wall. Langmuir **37**, 7405–7419 (2021). https://doi.org/10.1021/acs.langmuir.1c00780

15. Henderson, A.: Paraview guide, a parallel visualization application. Kitware Inc. (2007). http://www.paraview.org

16. Lautenschlaeger, M.P., et al.: Effects of lubrication on the friction in nanometric machining processes: a molecular dynamics approach. AMM. **869**, 85–93 (2017). https://doi.org/10.4028/www.scientific.net/AMM.869.85

17. Lautenschlaeger, M.P., Stephan, S., Horsch, M.T., Kirsch, B., Aurich, J.C., Hasse, H.: Effects of lubrication on friction and heat transfer in machining processes on the nanoscale: a molecular dynamics approach. Proc. CIRP **67**, 296–301 (2018). https://doi.org/10.1016/j.procir.2017.12.216

18. Lautenschlaeger, M.P., Hasse, H.: Transport properties of the Lennard-Jones truncated and shifted fluid from non-equilibrium molecular dynamics simulations. Fluid Phase Equilib. **482**, 38–47 (2019). https://doi.org/10.1016/j.fluid.2018.10.019

19. Lautenschlaeger, M.P., Hasse, H.: Shear-rate dependence of thermodynamic properties of the Lennard-Jones truncated and shifted fluid by molecular dynamics simulations. Phys. Fluids **31**, 063103 (2019). https://doi.org/10.1063/1.5090489

20. Klocke, F.: Fertigungsverfahren 2: Zerspanung mit geometrisch unbestimmter Schneide. Springer, Berlin Heidelberg, Berlin, Heidelberg (2017)

21. Mendelev, M.I., Han, S., Srolovitz, D.J., Ackland, G.J., Sun, D.Y., Asta, M.: Development of new interatomic potentials appropriate for crystalline and liquid iron. Phil. Mag. **83**, 3977–3994 (2003). https://doi.org/10.1080/14786430310001613264

22. Martin, M.G., Siepmann, J.I.: Transferable potentials for phase equilibria. 1. united-atom description of n-Alkanes. J. Phys. Chem. B. **102**, 2569–2577 (1998). https://doi.org/10.1021/jp972543+

23. Noreyan, A., Amar, J.G.: Molecular dynamics simulations of nanoscratching of 3C SiC. Wear **265**, 956–962 (2008). https://doi.org/10.1016/j.wear.2008.02.020

24. Plimpton, S.: Fast parallel algorithms for short-range molecular dynamics. J. Comput. Phys. **117**, 1–19 (1995). https://doi.org/10.1006/jcph.1995.1039

25. Rentsch, R., Inasaki, I.: Effects of fluids on the surface generation in material removal processes-molecular dynamics simulation. CIRP Ann. **55**, 601–604 (2006). https://doi.org/10.1016/S0007-8506(07)60492-2

26. Ren, J., Zhao, J., Dong, Z., Liu, P.: Molecular dynamics study on the mechanism of AFM-based nanoscratching process with water-layer lubrication. Appl. Surf. Sci. **346**, 84–98 (2015). https://doi.org/10.1016/j.apsusc.2015.03.177

27. Schmitt, S., et al.: Molecular simulation study on the influence of the scratching velocity on nanoscopic contact processes. In: 2nd International Conference of the DFG International Research Training Group 2057 – Physical Modeling for Virtual Manufacturing (iPMVM 2020), vol. 89, pp. 17:1–17:16 (2021). https://doi.org/10.4230/OASIcs.iPMVM.2020.17

28. Schmitt, S., Vo, T., Lautenschlaeger, M.P., Stephan, S., Hasse, H.: Molecular dynamics simulation study of heat transfer across solid–fluid interfaces in a simple model system. Mol. Phys. **120**, e2057364 (2022). https://doi.org/10.1080/00268976.2022.2057364

29. Schmitt, S., Fleckenstein, F., Hasse, H., Stephan, S.: Comparison of force fields for the prediction of thermophysical properties of long linear and branched alkanes. J. Phys. Chem. B **127**(8), 1789–1802 (2023). https://doi.org/10.1021/acs.jpcb.2c07997

30. Shi, J., Zhang, Y., Sun, K., Fang, L.: Effect of water film on the plastic deformation of monocrystalline copper. RSC Adv. **6**, 96824–96831 (2016). https://doi.org/10.1039/C6RA17 126E

31. Sivebaek, I.M., Persson, B.N.J.: The effect of surface nano-corrugation on the squeeze-out of molecular thin hydrocarbon films between curved surfaces with long range elasticity. Nanotechnology **27**, 445401 (2016). https://doi.org/10.1088/0957-4484/27/44/445401

32. Stephan, S., Lautenschlaeger, M.P., Alhafez, I.A., Horsch, M.T., Urbassek, H.M., Hasse, H.: Molecular dynamics simulation study of mechanical effects of lubrication on a nanoscale contact process. Tribol. Lett. **66**(4), 1–13 (2018). https://doi.org/10.1007/s11249-018-1076-0

33. Stephan, S., Liu, J., Langenbach, K., Chapman, W.G., Hasse, H.: Vapor−liquid interface of the lennard-jones truncated and shifted fluid: comparison of molecular simulation, density gradient theory, and density functional theory. J. Phys. Chem. C **122**, 24705–24715 (2018). https://doi.org/10.1021/acs.jpcc.8b06332

34. Stephan, S., Thol, M., Vrabec, J., Hasse, H.: Thermophysical properties of the Lennard-Jones fluid: database and data assessment. J. Chem. Inf. Model **59**, 4248–4265 (2019). https://doi.org/10.1021/acs.jcim.9b00620

35. Stephan, S., Horsch, M., Vrabec, J., Hasse, H.: MolMod–an open access database of force fields for molecular simulations of fluids. Mol. Simul. **45**, 806–814 (2019). https://doi.org/10.1080/08927022.2019.1601191

36. Stephan, S., Dyga, M., Urbassek, H.M., Hasse, H.: The Influence of lubrication and the solid-fluid interaction on thermodynamic properties in a nanoscopic scratching process. Langmuir **35**, 16948–16960 (2019). https://doi.org/10.1021/acs.langmuir.9b01033

37. Stephan, S., Staubach, J., Hasse, H.: Review and comparison of equations of state for the Lennard-Jones fluid. Fluid Phase Equilib. **523**, 112772 (2020). https://doi.org/10.1016/j.fluid.2020.112772

38. Stephan, S., Dyga, M., AlabdAlhafez, I., Lenhard, J., Urbassek, H.M., Hasse, H.: Reproducibility of atomistic friction computer experiments: a molecular dynamics simulation study. Mol. Simul. **47**, 1509–1521 (2021). https://doi.org/10.1080/08927022.2021.1987430

39. Stephan, S., Schmitt, S., Hasse, H., Urbassek, H.M.: Molecular dynamics simulation of the Stribeck curve: boundary lubrication, mixed lubrication, and hydrodynamic lubrication on the atomistic level. Friction (2023). https://doi.org/10.1007/s40544-023-0745-yinpress

40. Stukowski, A., Albe, K.: Extracting dislocations and non-dislocation crystal defects from atomistic simulation data. Modell. Simul. Mater. Sci. Eng. **18**(8), 085001 (2010). https://doi.org/10.1088/0965-0393/18/1/01501202. http://www.ovito.org/

41. Tersoff, J.: Modeling solid-state chemistry: interatomic potentials for multicomponent systems. Phys. Rev. B. **39**, 5566–5568 (1989). https://doi.org/10.1103/PhysRevB.39.5566

42. Wu, C.-D., Fang, T.-H., Lin, J.-F.: Atomic-scale simulations of material behaviors and tribology properties for FCC and BCC metal films. Mater. Lett. **80**, 59–62 (2012). https://doi.org/10.1016/j.matlet.2012.04.079

43. Vrabec, J., Stoll, J., Hasse, H.: A set of molecular models for symmetric quadrupolar fluids. J. Phys. Chem. B. **105**, 12126–12133 (2001). https://doi.org/10.1021/jp012542o
44. Vrabec, J., Kedia, G.K., Fuchs, G., Hasse, H.: Comprehensive study of the vapour–liquid coexistence of the truncated and shifted Lennard-Jones fluid including planar and spherical interface properties. Mol. Phys. **104**, 1509–1527 (2006). https://doi.org/10.1080/002689706 00556774
45. Zheng, X., Zhu, H., Kosasih, B., KietTieu, A.: A molecular dynamics simulation of boundary lubrication: the effect of n-alkanes chain length and normal load. Wear **301**, 62–69 (2013). https://doi.org/10.1016/j.wear.2013.01.052
46. Zhu, P.-Z., Qiu, C., Fang, F.-Z., Yuan, D.-D., Shen, X.-C.: Molecular dynamics simulations of nanometric cutting mechanisms of amorphous alloy. Appl. Surf. Sci. **317**, 432–442 (2014). https://doi.org/10.1016/j.apsusc.2014.08.031

Visual Analysis and Anomaly Detection of Material Flow in Manufacturing

E. Kinner[1]([⊠]), M. Glatt[2], J. C. Aurich[3], and C. Garth[4]

[1] Scientific Visualization Lab, RPTU - University of Kaiserslautern-Landau, Kaiserslautern, Germany
`ekinner@rhrk.uni-kl.de`
[2] Institute for Manufacturing Technology and Production Systems, RPTU - University of Kaiserslautern-Landau, Kaiserslautern, Germany
[3] Chair of Institute for Manufacturing Technology and Production Systems, RPTU - University of Kaiserslautern-Landau, Kaiserslautern, Germany
[4] Chair of Scientific Visualization Lab, RPTU - University of Kaiserslautern-Landau, Kaiserslautern, Germany

Abstract. The automated tracking of objects in factories via real-time locating systems (RTLS) is gaining increased attention due to its improved availability, technical sophistication, and most of all, its plethora of applications. The tracking of workpieces through their production process, for example, unlocks a detailed understanding of timings, patterns, and bottlenecks. While research mostly focuses on technological advancements, the analysis of the generated data is often left unclear. We propose a visual analysis framework based on ultra-wide-band (UWB) RTLS tracking data of material flow for this purpose. With this, we present an analysis and define a practical approach for how factory-level data can be analyzed. Advanced algorithms adapted from non-adjacent research domains are used to process and detect anomalies in the data, which would otherwise be hidden behind oversimplified analysis methods. Our approach considers different levels of granularity for the analysis in its visualization and, therefore, scales with increasing data sizes effortlessly. We also generated a ground truth dataset of RTLS UWB data with labeled anomaly cases. Combined, we provide a full, end-to-end, efficient processing and multi-visualization analysis pipeline for self-contained yet generalizable UWB RTLS data.

1 Introduction

The manufacturing of products usually requires multiple steps to complete, often split among several working stations. In general, it is true that the more complex a product is, the more steps are needed to manufacture it. Additionally, it is not uncommon that a complex product requires other sub-products and might have some variations that address the specific needs of a particular customer. This introduces a lot of complexity into the manufacturing process. The path a product takes through the factory is linked to that complexity. In order to make any statements about the performance of a factory layout or production process pipeline, a general overview and understanding of the material flow are necessary. When and where a workpiece is produced and in what time

J. C. Aurich et al. (Eds.): IRTG 2023, *Proceedings of the 3rd Conference on Physical Modeling for Virtual Manufacturing Systems and Processes*, pp. 281–293, 2023.
https://doi.org/10.1007/978-3-031-35779-4 15

frame is critical information for the assessment of productivity and efficiency. For that, real-time-locating-system (RTLS)-based approaches reduce the work required to gather this information. Instead of labor-intensive manual tracking of workpieces or expensive full-scale robotic automation or digitalization of factories, only a pair of tracking devices need to be used. This makes this solution especially interesting for small and midsized companies without an abundance of either of these resources. The use of RTLS in factories covers a lot of different applications, such as layout planning or adjusting raw material buy orders. [4, 7] Hammerin et al. [8] proposed the use of RTLS for real-time management of production environments. Thiede et al. [9, 10] concentrated on optical and AI-based image recognition RTLS-based approaches capable of identifying humans in factory settings. Wolf et al. [11] derived efficiency data from human-centered ultra wide band (UWB) tracking. While Löcklin et al. [12] focused on the prediction of human movement to avoid accidents. Other technologies, like radio-frequency identification (RFID), were shown by Arkan et al. [13] to work in different scenarios. Küpper et al. [14] investigated the application of 5G for RTLS. Sullivan et al. [18] introduced value stream mapping using UWB as a valid method for the decision making process in manufacturing systems. The visualization, classification of the data, and detection of outliers are some of the contributions that we add to enhance the capabilities of this approach.

While the technical aspects of RTLS are thoroughly explored, the application and data analysis remain ambiguous. In order to derive knowledge and make informed decisions, the raw data has to be processed and displayed in an effective way. Type and quality of visualizations influence our decision-making process. [15] Visual analysis tools are therefore key components of any material flow optimization approach. The scalability of RTLS approaches also needs to be considered. If it is necessary to manually review each data point, automating the data collection phase does not add much efficiency. To bring up important information while minimizing the analyst's workload, an intelligent framework is required.

Typically, a company may want to review its manufacturing processes on a regular basis, but especially after new machines, products, or employees are introduced into the process. The analyst tags certain workpieces with the tracking device and gathers the data automatically. It can then be processed, displayed, and determined what changes are required to maximize productivity (e.g., an additional machine or employee is required, milling must be done differently to produce fewer workpieces that must be reworked, etc.).

To detect these anomalies, simple statistics are often not sufficient. Demonstrated in Fig. 1 is a plot of our workpiece trajectory data. Each column corresponds to a workpiece trajectory. Workpiece trajectories may differ in duration because of traffic building up, the reworking of a product, or any other reason. Based on that, one would expect to spot all outliers this way. However, the total duration correctly indicates an anomaly for trajectory index 3, but misses other anomalies for indexes 0, 5, and 6. Further analysis methods have to be provided to identify them.

To improve on all that and enable a proper analysis of RTLS tracking data, we propose a visual analysis framework and processing pipeline. Our core contributions are

- the generation of a UWB material flow dataset, its filtering, and its preprocessing

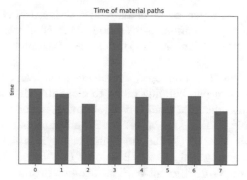

Fig. 1. Total time of material paths of the unfiltered dataset. Anomalies in path 0, 5, and 6 cannot be identified.

- the specification and automated detection pipeline of anomalies for bottleneck identification
- the workpiece specific graph-based trajectory visualization for comparability
- the concept of data type similarity between streamlines in flow fields and RTLS trajectory data
- and the embedding and clustering of workpiece trajectories for automated anomaly detection and pattern identification

In the following, we will chronologically describe the data acquisition and experimental setup before the preprocessing explains how the data needs to be filtered and transformed. The visualization chapter then goes into detail about the analysis and anomaly detection. At last, the limitations and future work are contained in the discussion section before the final conclusion.

2 Method

2.1 Dataset

Since the research on RTLS in the context of material flow is relatively new, there have not been any state-of-the art datasets made available for benchmarks or comparisons. In other domains of research (e.g. computer vision, machine learning, etc.) there is a consensus among researchers to compare their work on the same datasets to enable an objective assessment of the quality of their models. [1–3] This development has not yet taken place in the manufacturing research community. However, it will become increasingly important in the future to establish these types of datasets in order to enable a standardized common ground and build more complex models for automated and reliable systems. The generation of these come with their own set of unique challenges.

Often times, companies will decide to withhold the publication of their data so as not to give competitors insights or other advantages. This isolates the R&D departments of different companies, drives up costs, and hinders innovation. Some companies adopted a middle course, cooperating with universities and sharing insights into non critical processes.

Another challenge is the generation of material flow data itself. Data collection under real-world manufacturing conditions takes time if there is no steady production (as in universities and laboratories). This leads to the current situation with little to no publicly available datasets.

To still enable research on this topic, we recorded our own dataset to demonstrate our approach and provide one building block to close that data gap. It can easily be expanded or modified in the future and serves to provide reproducibility.

There are multiple RTLSs known, each with advantages and disadvantages. [4] Radio-Frequency-Identification (RFID), for example, provides high spatial accuracy but has a very limited operating range (1 m). Bluetooth Low Energy (BLE) suffers from a similar range limitation. With different WiFi localization methods, the range is between 150–200 m, with an accuracy of 1–5 m. Similar accuracy is achieved by 5G, whose range is virtually unlimited for factory scale RTLS. This also applies for GPS, whose accuracy is the worst with 2–10 m and its inherent limitation to outdoor use.

We selected UWB tracking since its precision is the most accurate (~0.5 m) among other technologies and its range can cover a significant part of the factory floor (150–200 m). The tracking was done by a station and a client device. Since UWB tracking is not yet a standard feature for consumer-grade smartphones, a modified Raspberry Pi was used as a handheld device to track the position. It is small enough to be moved together with any other workpiece through a factory.

For our dataset, we used an existing factory hall and set up virtual stations that can represent any kind of material processing, like milling, drilling, or quality control. These stations do not exist in reality, since the actual manufacturing of workpieces would massively exceed the scope of this research and do not contribute directly to the quality of the data. Instead, the stations are simulated using cardboard boxes and tables. Similar to a factory setting, the material is introduced into the factory at some position (the start point). A sketch in Fig. 2 illustrates the qualitative layout of our setup. After the material is present in the manufacturing environment, the tracking device is associated with a single workpiece. It then moves to the first station. In our case, the raspberry pi was taken and moved to one of our artificial stations.

It is then processed at that station. For our setup, this means that we let the tracking device lie near the station just like it would in a real setting while the workpiece is processed. The benefit of our approach is that we can shorten the time span compared to real processing of workpieces. A fixed amount of time, usually in the magnitude of a couple seconds, is enough to simulate the processing of a workpiece. This lets us record more data without any loss in quality. The tracking is then continued from station to station until our fictional product is completed. To enhance the dataset, pre-planned anomalies were introduced at certain steps. This allows algorithms to identify these anomalies and compare their findings with the ground truth. In Table 1 the recorded trajectories and the incorporated anomalies are listed for the first dataset. This clearly identifies what an anomaly is and what an algorithm is supposed to identify. Depending on the manufacturing context, the notion of what constitutes an anomaly might change. For our setup, a significant amount of extra waiting time at a particular station or a different route (rework at the previous station) are considered different from the normal production flow.

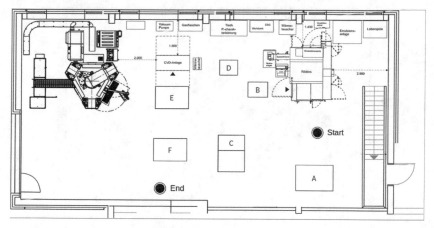

Fig. 2. Sketch of the factory layout for artificial material flow data generation.

Table 1. List of pre-planned anomalies introduced in first dataset. The paths 2, 3, 5, and 8 do not include any anomaly and are thus not shown here.

Material flow path	Anomaly
1	Wait Double Time at Station D
4	After Step 4 go back to repeat step 3 and then 4 again
6	Wait Double Time at Station D
7	Wait Double Time at Station D

We generated three additional datasets, each corresponding to a fictional product with its own path and anomalies, in the same manner as described above. The additional datasets enable us to split it into training and validation data. The robustness and generalizability of any data processing and exploration approach are critical properties. The effects of shifts in scale, noise, time, and other factors can be examined by cross-validation. In the following, we will be using these datasets to present our visual analysis framework and anomaly detection pipeline.

2.2 Preprocessing

Although the precision of UWB tracking is among the most precise technologies available and theoretically resilient to multi-path interferences. [4, 6] It is prone to noise and interferences with metal objects in its path. [5] Disconnects result in duplicates or extreme outliers in the recorded positions for our setup. For the data to be used in any analysis tool, it first has to be cleaned and filtered so that structures can become visible. We did this by removing duplicates and outliers from the trajectories, which allowed them to be identified. All trajectories from the first dataset are shown in Fig. 3, which lets us clearly identify the working stations and rough layout of the factory.

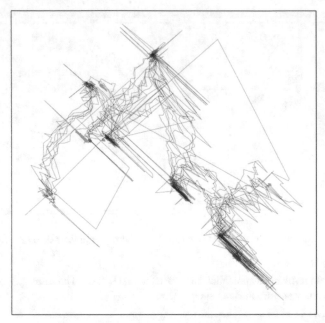

Fig. 3. Unfiltered material flow trajectory dataset with outliers already removed. Paths and working stations can be identified but signal is still very noisy.

However, the signal is still very noisy. To facilitate further processing, filtering has to be applied. While there are many filtering techniques available, we used our knowledge of the system and basic filtering methods to avoid any distortions and preserve the underlying structure.

Since the measurements are taken in the real world, physical and virtual units are correlated. The movement of the workpiece is therefore also bound by the laws of physics. If the acceleration or velocity is outside of expected ranges (e.g. <10 km/h), the recorded point is likely to be an outlier and should be removed. After that, we apply a moving average filter to exclude any high frequency noise.

An important component of the data is centered around the working stations. The length of time a workpiece remains there and the order in which it is visited are useful pieces of information for any type of analysis. For this, we extract stations out of the trajectories by detecting clusters of signals. A cluster is a place where the workpiece stayed for an extended period of time. This way, the working stations and all the points associated with them can be identified. To create the finished material flow path, they are combined, corrected for timestamps, and collapsed if part of a small loop. In Fig. 4 a single filtered trajectory with identified working stations is displayed. It is the basis for further analysis.

This approach may also lead to stations being falsely identified if a workpiece stays in between stations for a longer time. In comparison to other routes, an additional station will pop out immediately. This allows for easy identification of material flow traffic jams in factories.

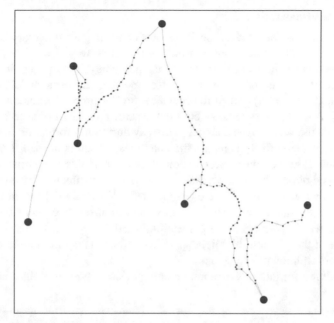

Fig. 4. Single filtered trajectory. Path is can be clearly seen. Stations have been identified.

2.3 Visualization

To quickly allow the identification of the cause of an anomaly, it is important to display the data in a meaningful way. With a visual understanding of why there has been an outlier in the data, it becomes easier to determine what actions need to be taken in order to mitigate the problem. Numerical information about the duration at each station is only useful in the second step.

For this, we chose a graph-based approach. This has several benefits. The general positions of the stations and the trajectory of workpieces remain the same. This lets the user grasp the spatial domain better and understand the scope of the problem. Secondly, the graph-based layout is intuitive because it is used in other applications. The user is therefore already used to it and does not need extensive training to understand what is displayed.

The primary requirement for a material flow visualization in this application is the ability to display and distinguish its properties. For that, we choose a directed graph since the order of workpiece processing is generally relevant. Another property is the timing, which is cumulatively encoded as the thickness of the arrow or circle. It represents the time it took for the workpiece to traverse it. If the trajectory has one of a set of known anomalies that can be detected in the filtered data, it can be marked with a different color to quickly draw attention to it.

2.3.1   Graph Visualization

The benefits of this visualization lie in its ability to display trajectory data without excessive visual clutter. Positional information is still conveyed without details about the exact position of the workpiece. For efficiency analysis purposes, it is irrelevant if a workpiece moves a couple of centimeters more to the right or left of the path. Time information is tightly linked to efficiency. So the time information is displayed cumulatively. A workpiece should be built in the same amount of time regardless of the time of day. To use scale as the channel to display duration information has the benefit that it lets users qualitatively compare the quantities, which is needed if outliers need to be identified. However, the representation of duration as size also comes with known human-centered biases. It is known that an area is underestimated (by an exponent of ~0.7) while other visual stimuli are overestimated. [17] One might want to adjust for this factor to aid visual perception, but this sacrifices absolute comparability. For our visualization, we decided against a perception-based correction. Lastly, the choice to use the color of the glyph to highlight specific elements is natural due to the effect of warning colors on human perception.

All this allows for intuitive exploration and quick assessment of different outliers.

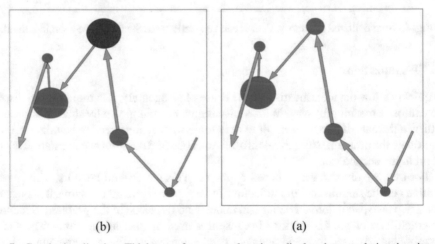

        (b)   (a)

Fig. 5. Graph visualization. Thickness of arrows and stations display the cumulative time it took the material to traverse. Anomaly (double time at top most station) (marked in red) can be identified for left dataset (a) in comparison to right dataset (b).

An example can be seen in Fig. 5, where two trajectories of workpieces from the same product group are displayed. Their position and direction all match up. The layout of the factory can be conceptualized easily. Through careful examination, we are able to identify that (a), the left workpiece trajectory, remains twice as long at the topmost (3rd) station than (b), the right piece. Because excess time on a station is one of the a priori known anomaly types, we can also color the affected node. Not only the presence but also the accuracy of the detected anomaly are sufficient. Neither the recording nor the filtering changed the qualitative scale of the trajectory. The node is twice the size, meaning the workpiece remained there twice as long, which is correct in comparison to the ground truth.

Other anomalies, like additional stations caused by traffic jams or extremely slow transportation times, can be detected in much the same way.

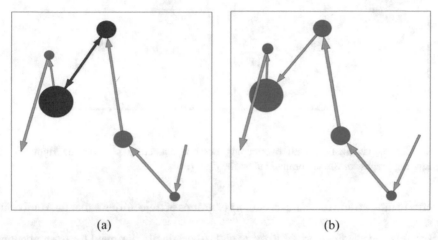

Fig. 6. Graph visualization. Anomaly (revisit station 3 then go back to station 4) (marked in red) can be identified for left dataset (a) in comparison to right dataset (b).

Another example is shown in Fig. 6, where on the left side (a) the workpiece travels back to station 3 before it then continues back to station 4 until finished. In comparison, on the left side (b) the regular path does not involve loops. In our product specification, this is considered an anomaly. The workpiece has some kind of defect, which needs to be fixed at the previous station. But depending on the manufacturing procedure, this may be part of the normal production cycle.

2.3.2 Overlapping Graph Visualization

There are also more advanced visualization techniques that suit the needs of particular applications or improve workflow. Examples are shown in Fig. 7 (a), where additional information in the form of overlaying trajectories may aid in the investigation of certain events.

Increased practicability may be achieved by the overlapping graph visualization of Fig. 7 (b), where less visual memory is required for the comparison between two trajectories.

2.3.3 Embedding Visualization

The graph visualization is useful for single workpiece trajectories. The properties are visualized intuitively. However, for larger amounts of data, it becomes tedious to compare and analyze individual graphs with each other. This requires automatically detecting

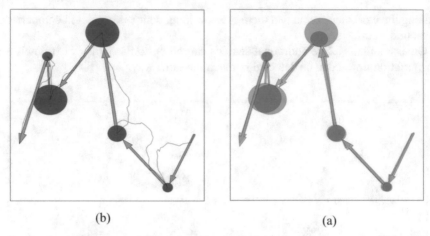

(b)   (a)

Fig. 7. Graph visualization with overlapping detailed trajectory plot on the right (a) and overlapping graphs for direct comparison on the right (b).

patterns in the data and focusing on outliers or groups of trajectories rather than individual ones.

For this, we use the work of Rossl et al. [16] originally designed for the embedding of streamlines. They optimized their embedding using MDS by using the hausdorff distance between two streamlines.

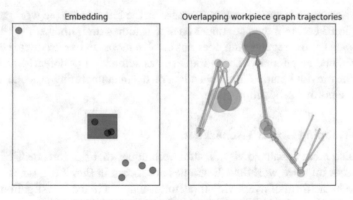

Draw a selection in the embedding to display corresponding graph trajectories.

Fig. 8. Embedding using hausdorff distance and MDS. Cluster in embedding (left) represent similar workpiece trajectories (right). Can be used for pattern recognition and identification of outliers.

Similarly, we can apply this to our data and generate an embedding, as seen in Fig. 8. Workpiece trajectories can be thought of as streamlines of material flow inside a factory. Using this insight, we can apply methods designed for fluid flow and streamlines to our data. Using their approach, the product trajectories get embedded into a lower-dimensional space (here 2D), where Euclidean distance corresponds to similarity. With this, similar workpiece trajectories naturally form clusters in the embedding and unique, dissimilar ones form outliers or anomalies. In Fig. 8, the embedding on the right contains a single point in the top left-hand corner. It corresponds to the trajectory of the workpiece that had to revisit the previous station. But also trends or structures that are more common in the dataset can be identified that way. If the dataset is shifted or roughly equally divided, it may not show any statistical abnormalities, but half the time, small delays are introduced that eventually propagate further. For example, the three points selected in the middle have all spent more time at a station than the rest. Even though they form a considerable portion of the dataset, it is still possible to identify them as different. With this tool, it is possible to detect larger trends and patterns in the data, which can then be individually analyzed using the graph visualization.

3 Discussion

Explorative visual data analysis is important to examine workpiece trajectories of RTLS systems. We showed how automatic processing and visualization can be constructed to aid in the identification of anomalies and production bottlenecks. With the novelty from this paper of pairing technical advancements in RTLS with established visual analysis tools comes the discussion on how such system should be created.

It can be argued that the simplicity of the presented visualizations could be exchanged in favor of more sophisticated target-specific visualizations and proper training of personnel. The filtering and preprocessing of the data also allow for a variety of techniques. Spatio-temporal data might benefit from dedicated trajectory filtering. Advanced approaches were not necessary for our data but may be needed for other factory settings with more metal surfaces causing interferences for UWB receivers. The scalability of our graph visualization is also limited by the number of nodes and arrows that can intersect each other before the result becomes too cluttered. A dynamic alpha value for an interactive exploration of very long paths (high alpha values for nodes close to the selected time) could be one solution. This was not necessary for our data, though. Future work might also inevitably produce more specialized solutions for factory-level feature analysis. Lastly, we expect advancement to be dependent on the availability of public datasets. For this, different kinds of anomalies and other problems in manufacturing can be introduced as an extension to our dataset. Also, our approach is limited to a number of experimental workpiece trajectories. Real manufacturing environments may include additional obstacles like obscuring objects or more volatile movements. Systems for real world applications may have to deal with these additional technical challenges.

Another promising research topic would be the utilization of more advanced methods for analysis, such as machine learning. The application of traditional streamline, flow, or other domain-specific algorithms on workpiece trajectory data has already proven useful and surely holds more opportunities for further optimization.

4 Conclusion

In this paper, a visual analysis and automated processing pipeline for UWB trajectory data was introduced. We used UWB tracking to generate datasets for material flow in manufacturing environments (RTLS). Through filtering and preprocessing, we enabled the automated detection of anomalies and were able to accurately identify bottlenecks. A workpiece trajectory specific graph based visualization allowed the intuitive and quick comparison of individual paths, while bigger datasets could be examined by approaches developed for streamlines in fluid flow visualizations because of its datatype similarity. We showed that cluster selection of embeddings greatly increases the scalability of anomaly detection and enables the systematic examination of factory material flow efficiency. In the future, we expect more work in the automated detection and analysis of this data, together with the rise of industry 4.0 to utilize the computational advancements in other fields and leverage the efficiency of manufacturing factories.

References

1. Grgic, M., Delac, K., Grgic, S.: SCface – surveillance cameras face database. Multimed Tools Appl. **51**, 863–879 (2011). https://doi.org/10.1007/s11042-009-0417-2
2. Deng, J., Dong, W., Socher, R., Li, L.-J., Li, K., Li, F.-F.: ImageNet: a large-scale hierarchical image database. In: IEEE Conference on Computer Vision and Pattern Recognition, pp. 248–255 (2009). https://doi.org/10.1109/CVPR.2009.5206848
3. Cordts, M., et al.: The cityscapes dataset for semantic urban scene understanding. In: Proceedings of the IEEE Conference on Computer Vision and Pattern Recognition (CVPR) (2016)
4. Thiede, S., Sullivan, B., Damgrave, R., Lutters, E.: Real-time locating systems (RTLS) in future factories: technology review, morphology and application potentials. Procedia CIRP **104**, 671–676 (2021). ISSN 2212-8271. https://www.sciencedirect.com/science/article/pii/S2212827121010118. https://doi.org/10.1016/j.procir.2021.11.113
5. Mikoda, M., Kalinowski, K., Ćwikła, G., Grabowik, C., Foit, K.: Accuracy of real-time location system (RTLS) for manufacturing systems. Int. J. Mod. Manuf. Technol. **12**(1) (2020). ISSN 2067-3604
6. Patwari, N., Ash, J.N., Kyperountas, S., Hero, A.O., Moses, R.L., Correal, N.S.: Locating the nodes: cooperative localization in wireless sensor networks. IEEE Signal Process. Mag. **22**(4), 54–69 (2005). https://doi.org/10.1109/MSP.2005.1458287
7. Mütze, A., Hingst, L., Rochow, N., Miebach, T., Nyhuis, P.: Use cases of real-time locating systems for factory planning and production monitoring. In: Proceedings of the Conference on Learning Factories (CLF) 2021 (2021). Available at SSRN: https://ssrn.com/abstract=3857878 or https://doi.org/10.2139/ssrn.3857878
8. Hammerin, K., Streitenberger, R.: RTLS – the missing link to optimizing Logistics Management? (2019). URN: urn:nbn:se:hj:diva-45264. ISRN: JU-JTH-PRS-2–20190057OAI: oai:DiVA.org:hj-45264DiVA, id: diva2:1334693
9. Thiede, S., Ghafoorpoor, P., Sullivan, B.P., Bienia, S., Demes, M., Dröder, K.: Potentials and technical implications of tag based and AI enabled optical real-time location systems (RTLS) for manufacturing use cases. CIRP Ann. **71**(1), 401–404 (2022). ISSN 0007-8506. https://doi.org/10.1016/j.cirp.2022.04.023. https://www.sciencedirect.com/science/article/pii/S0007850622000695
10. Bienia, S., Demes, M., Dreger, J., Dröder, K., Thiede, S.: Functional analysis of an optical real time locating system in production environments. Procedia CIRP **107**, 1107–1111 (2022). ISSN 2212-8271. https://doi.org/10.1016/j.procir.2022.05.116. https://www.sciencedirect.com/science/article/pii/S2212827122004000
11. Wolf, M., et al.: Real time locating systems for human centered production planning and monitoring. IFAC-PapersOnLine **55**(2), 366–371 (2022). ISSN 2405-8963. https://doi.org/10.

1016/j.ifacol.2022.04.221. https://www.sciencedirect.com/science/article/pii/S24058963220
02221

12. Löcklin, A., Ruppert, T., Jakab, L., Libert, R., Jazdi, N., Weyrich, M.: Trajectory prediction of humans in factories and warehouses with real-time locating systems. In: 2020 25th
IEEE International Conference on Emerging Technologies and Factory Automation (ETFA),
pp. 1317–1320 (2020). https://doi.org/10.1109/ETFA46521.2020.9211913

13. Arkan, I., Van Landeghem, H.: Evaluating the performance of a discrete manufacturing process using RFID: a case study. Robot. Comput.-Integr. Manuf. **29**(6), 502–512
(2013). ISSN 0736-5845. https://doi.org/10.1016/j.rcim.2013.06.003. https://www.sciencedi
rect.com/science/article/pii/S0736584513000471

14. Küpper, C., Rösch, J., Winkler, H.: Empirical findings for the usage of 5G as a basis for
real time locating systems (RTLS) in the automotive industry. Procedia CIRP **107**, 1287–
1292 (2022). ISSN 2212-8271. https://doi.org/10.1016/j.procir.2022.05.146. https://www.sci
encedirect.com/science/article/pii/S2212827122004309

15. Eberhard, K.: The effects of visualization on judgment and decision-making: a systematic
literature review. Manag. Rev. Q. **73**(1), 167–214 (2023). https://doi.org/10.1007/s11301-
021-00235-8

16. Rossl, C., Theisel, H.: Streamline embedding for 3D vector field exploration. IEEE Trans.
Visual Comput. Graphics **18**(3), 407–420 (2012). https://doi.org/10.1109/TVCG.2011.78

17. Munzner, T.: Visualization Analysis and Design. CRC Press (2015). ISBN 9781498759717

18. Sullivan, B.P., Yazdi, P.G., Suresh, A., Thiede, S.: Digital value stream mapping: application of
UWB real time location systems. Procedia CIRP **107**, 1186–1191 (2022). ISSN 2212-8271.
https://doi.org/10.1016/j.procir.2022.05.129. https://www.sciencedirect.com/science/article/
pii/S2212827122004139

Author Index

© The Editor(s) (if applicable) and The Author(s) 2023
J. C. Aurich et al. (Eds.): IRTG 2023, *Proceedings of the 3rd Conference on Physical Modeling for Virtual Manufacturing Systems and Processes*, p. 295, 2023.
https://doi.org/10.1007/978-3-031-35779-4

Printed in the United States
by Baker & Taylor Publisher Services